Air Cond ing

Other titles available from E & FN Spon

Facilities Management
K. Alexander

Ventilation of Buildings
H. B. Awbi

Spreadsheets for Architects
L. R. Bachman and D. J. Thaddeus

Building Services Engineering
2nd edition
D. V. Chadderton

Building Services Engineering Spreadsheets
D. V. Chadderton

Naturally Ventilated Buildings
D. T. J. Clements-Croome

Spon's Mechanical and Electrical Services Price Book 1997
Davis Langdon and Everest

Illustrated Encyclopedia of Building Services
D. Kut

Building Energy Management Systems
G. J. Levermore

Energy Management and Operating Costs in Buildings
K. Moss

Heating and Water Services Design in Buildings
K. Moss

International Dictionary of Heating, Ventilation and Air Conditioning
2nd edition
REHVA

Sick Building Syndrome
J. Rostron

Site Management of Building Services Contractors
L. J. Wild

For more information about these and other titles please contact:
The Marketing Department, E & FN Spon, 2–6 Boundary Row, London, SE1 8HN.
Tel: 0171 865 0066.

Air Conditioning

A practical introduction

Second edition

David V. Chadderton

E & FN SPON
An Imprint of Thomson Professional

London · Weinheim · New York · Tokyo · Melbourne · Madras

**Published by E & FN Spon, an imprint of Thomson Professional,
2–6 Boundary Row, London SE1 8HN, UK**

Thomson Science and Professional, 2–6 Boundary Row, London SE1 8HN, UK

Thomson Science and Professional, Pappelallee 3, 69469 Weinheim, Germany

Thomson Science and Professional, 115 Fifth Avenue, New York,
NY 10003, USA

Thomson Science and Professional, ITP-Japan, Kyowa Building, 3F,
2-1-1 Hirakawacho, Chiyoda-ku, Tokyo 102, Japan

Thomson Science and Professional, 102 Dodds
Street, South Melbourne, Victoria 3205, Australia

Thomson Science and Professional, R. Seshadri, 32 Second Main Road, CIT East,
Madras 600 035, India

First edition 1993
Reprinted 1995, 1996

Second edition 1997
© 1993, 1997 David V. Chadderton
Typeset in 10/12pt Times by Pure Tech India Ltd, Pondicherry
Printed in Great Britain by The Alden Press, Oxford

ISBN 0 419 22610 9

A catalogue record for this book is available from the British Library

Library of Congress Catalog Card Number:97-66255

∞ Printed on permanent acid-free text paper, manufactured in
  accordance with ANSI/NISO Z39.48–1992 and ANSI/NISO Z39.48–1984
  (Permanence of Paper).

Contents

Preface

Air Conditioning: A practical introduction (second edition) is a textbook for undergraduate courses in building services engineering, energy engineering, mechanical engineering, BTEC Continuing Education Diploma, Higher National Diploma and Certificate courses in building services engineering and will be of considerable help on National Certificate and Diploma programmes. Heating, ventilating and air conditioning is studied on undergraduate, CED, HND and HNC courses in architecture, building, building management and building surveying and is part of all courses relevant to the design, construction and use of buildings.

The design of air-conditioning systems involves considerable calculation work which is now mainly carried out with dedicated computer software. The engineering principles need to be fully comprehended in the first instance, as are the basic formulae and calculation techniques utilized. The reader is actively involved in the use of such data by the use of worked examples, copious exercises and design assignments.

A spreadsheet has been used for fan, pipe and duct sizing calculations. The formulae used are listed. The reader is encouraged to make full use of spreadsheets where dedicated software is not available. The spreadsheets provided can be edited and easily enlarged.

Each chapter is introduced with the air of learning objectives and a list of the key terms and concepts employed.

Equations and data are taken from the Chartered Institution of Building Services Engineers *Guide*. Only samples of data are reproduced in order to demonstrate how the reader can utilize the reference. Sample data alone will be insufficient for anything other than the set exercises and only the most recent *CIBSE Guide* edition is to be used more widely.

The stages of an air-conditioning design process are clearly stated and are often given in the form of numbered lists. This approach may assist the testing of competences.

Acknowledgements

Those acknowledged below provided valuable advice for the first edition of this book. The overall effort has encouraged this second edition, and I hope that they, and all users, will make good use of the additional material.

Grateful thanks are due to David Ferris BA, senior lecturer and department manager, department of construction, Southampton Technical College. David has been teaching building services engineering for ten years. He refereed each chapter and gave many useful comments from the point of view of the user. David commented that the book is well written, factual and successfully bridges the gap between air-conditioning theory and design practice. It should be well received by students and professional engineers. It will be used as a course reader at technician, higher technician and undergraduate levels. It is an excellent distance-learning package, intensive, informative, compulsive reading, concise, the best he has read on the subject.

Andrew Pettifer, BSc (Hons) Building Services Engineering, University of Bath, senior design engineer with Tilney Simmons and Partners, Southampton, reviewed Chapters 1, 2, 5 and 8. Andrew gained a first-class honours degree, has three years' professional experience and has provided many relevant points from current practice. His contribution has been most beneficial and he is thanked for his efforts.

David Hill, LCIBSE, Eng Tech, senior engineer, Directorate of Technical Services, Southampton City Council, reviewed Chapter 1. David's view is that the systems description and examples will not only be of value as a reference work for students but also as a refresher for practising engineers. David is thanked for his helpful comments.

Bob Collier, senior lecturer in electrical and electronic engineering, Information Systems division, Southampton Institute of Higher Education, is thanked for his review of Chapter 8 and useful comments.

Crowcon Detection Instruments Limited, Abingdon, Oxfordshire, are thanked for the use of photographs of their detectors and for checking the technical content of parts of Chapters 5 and 8.

Victor Bettridge, managing director of Southern Commissioning and Testing Limited is thanked for reviewing Chapter 8. Victor commented that he wished this book had been available when he first started commissioning. He found it comprehensive of the total concept of commissioning. The chapter contains a great deal of information to assist the commissioning engineer to understand the work and carry it out in a professional manner.

Trane (UK) Limited are thanked for the provision of the Trane psychrometric chart This is particularly useful in that cooling coil dehumidification curves are shown.

The reviewers were not asked to check the arithmetic or solutions to examples and questions. The author holds responsibility for these. Many were calculated on spreadsheets. Others have been validated from published data tables and charts. All have been checked at least twice. Higher National Certificate students on the Building Services Engineering (HVAC) course at Southampton Institute of Higher Education have been exposed to prepublication material. They are thanked for the observations offered.

Introduction

This second edition of *Air Conditioning: A practical introduction* continues with the principles established in *Building Services Engineering* (second edition, 1995, E & FN Spon) by the same author. This new edition has an additional chapter on the application of fans to air duct systems. Readers are encouraged to make use of *Building Services Engineering Spreadsheets* by David V. Chadderton (E & FN Spon, 1997), as that book is designed to develop the subjects in greater depth and to make extensive use of commonly available spreadsheet software. *Building Services Engineering* can be referred to for fundamentals such as the meaning of dry- and wet-bulb air temperatures.

Solution procedures are given for many design tasks. These are programmed approaches to complex problems and students find them very helpful. Their provision removes much of the student's need for original thought. However, such capacity must not be discarded as programmes do not automatically encompass every eventuality. They illustrate an approach to the problem that may be easily followed.

Some spreadsheets are provided for entry into the user's computer at college, workplace or home. These should be suitable for any spreadsheet program with the minimum of customization. The user needs to become familiar with the necessary commands for the spreadsheet to be used before trying to utilize these parts of the book. Once such activity has been accomplished, the student will have been converted to a calculation tool of great power and diversity that is only limited by the user's own imagination in solving complex or repetitive calculation tasks.

The reader is challenged to propose suitable designs for air-conditioning problems. The principles are explained within the text. The reader's answers to questions are for discussion with colleagues and tutor. The aim of questions not having model solutions is to build the reader's competence in preparation for tests and practical design work.

It is vital that the reader remembers the normal rules of mathematics when solving numerical examples and questions. Equations are solved in the following order of precedence.

1. Evaluate numbers within brackets.
2. Commence with the innermost brackets.
3. Evaluate numbers raised to powers.
4. Evaluate logarithms.
5. Evaluate numerator and denominator of fractions.
6. Multiplication and division calculations.
7. Addition and subtraction last.

Further details on many topics can be found in *Ventilation of Buildings* by H.B. Awbi (1991), E & FN Spon, *Air Conditioning Engineering* by W. P. Jones (1985), Edward Arnold, and *The Chartered Institution of Building Services Engineers Guide A. B* and *C* (1986).

Units and constants

Système International units are used throughout, and Table 1 shows the basic and derived units with their symbols and common equalities.

Table 1 SI units

Quantity	Unit	Symbol	Equality
mass	kilogramme	kg	
	tonne	t	$1.0\ \text{tonne} = 10^3\ \text{kg}$
length	metre	m	
area	square metre	m^2	
volume	litre	l	
	cubic metre	m^3	$1.0\ \text{m}^3 = 10^3\ \text{l}$
time	second	s	
	hour	h	$1.0\ \text{h} = 3600\ \text{s}$
energy	joule	J	$1.0\ \text{J} = 1.0\ \text{Nm}$
force	newton	N	$1.0\ \text{kg} = 9.807\ \text{N}$
power	watt	W	$1.0\ \text{W} = 1.0\ \text{J/s}$
			$1.0\ \text{W} = 1.0\ \text{Nm/s}$
			$1.0\ \text{W} = 1.0\ \text{VA}$
pressure	pascal	Pa	$1.0\ \text{Pa} = 1.0\ \text{N/m}^2$
	newton/m^2	N/m^2	$1.0\ \text{b} = 100\,00\ \text{N/m}^2$
	bar	b	$1.0\ \text{b} = 10^3\ \text{mb}$
			$1.0\ \text{b} = 100\ \text{kPa}$
frequency	hertz	Hz	$1.0\ \text{Hz} = 1.0\ \text{cycle/s}$
temperature	Celsius	°C	
	kelvin	K	$\text{K} = \text{°C} + 273$
luminous flux	lumen	lm	
illuminance	lux	lx	$1.0\ \text{lux} = 1.0\ \text{lm/m}^2$
Electrical units			
resistance, R	ohm	Ω	
potential, V	volt	V	
current, I	ampere	I	$I = V/R$

Table 2 Multiples and submultiples

Quantity	Name	Symbol
10^{12}	tera	T
10^{9}	giga	G
10^{6}	mega	M
10^{3}	kilo	k
10^{-3}	milli	m
10^{-6}	micro	μ

Table 3 Physical constants

gravitational acceleration	g	9.807 m/s^2
specific heat capacity of air	SHC	1.012 kJ/kg K
specific heat capacity of water	SHC	4.186 kJ/kg K
density of air at 20°C, 1013.25 mb	ρ	1.205 kg/m^3
density of water at 4°C	ρ	1000 kg/m^3
exponential	e	2.718

Symbols

Symbol	Description	Unit
A	area	m^2
A	solar altitude angle	degree
A	absorptivity of glass	
B	solar azimuth angle	degree
b	barometric pressure	bar
$\circ$	angle	degree
C	wall azimuth angle	degree
C	flow coefficient	
D	wall-solar azimuth angle	degree
D	direct irradiance	W/m^2
d	diameter	m or mm
d	suffix for design and diffuse irradiance	
d.b.	dry-bulb temperature	°C
Δp	pressure drop	Pa
Δp_{12}	pressure change from node 1 to node 2	Pa
E	emissivity	
E	slope angle	degree
EL	equivalent length	m
F	Force	N
F	incidence angle	degree
F	surface factor	
FSP	fan static pressure	Pa
FTP	fan total pressure	Pa
FVP	fan velocity pressure	Pa
F_1, F_2	heat loss factors	
F_u	thermal transmittance factor	
F_v	ventilation factor	
F_y	admittance factor	
f	decrement factor	
f	friction factor	
g	gravitational acceleration	m/s^2
g	air moisture content	kg H_2O/kg dry air
g_s	saturation air moisture content	kg H_2O/kg dry air
H	height	m
H	horizontal surface	
H	total pressure drop overcome by pump or fan	Pa

Symbol	Description	Unit
h	time	hours
h	specific enthalpy	kJ/kg
h_{fg}	latent heat of vaporization	kJ/kg
h_o	outdoor-air specific enthalpy	kJ/kg
h_r	room-air specific enthalpy	kJ/kg
h_{so}	outside surface heat-transfer coefficient	W/m^2 K
h_{si}	inside surface heat-transfer coefficient	W/m^2 K
I	electrical current	ampere
I	solar irradiance	W/m^2
I_{DH}	direct solar irradiance on horizontal surface	W/m^2
I_{DV}	direct solar irradiance on a vertical surface	W/m^2
I_{DVd}	design direct solar irradiance on a vertical surface	W/m^2
I_{DHd}	design direct solar irradiance on a horizontal surface	W/m^2
I_{TVd}	design total solar irradiance on a vertical surface	W/m^2
I_{THd}	design total solar irradiance on a horizontal surface	W/m^2
I_{dHd}	design diffuse solar irradiance on a horizontal surface	W/m^2
I_{DSd}	design direct solar irradiance on a sloping surface	W/m^2
I_{TSd}	design total solar irradiance on a sloping surface	W/m^2
I_{NH}	solar irradiance normal to a horizontal surface	W/m^2
I_{NV}	solar irradiance normal to a vertical surface	W/m^2
I_l	long-wave radiation from a surface	W/m^2
J	quantity of energy	joule
k	constant	
k	thermal conductivity	W/m K
kg	mass	kilogramme
kJ	quantity of energy	kilojoule
kVA	kilovolt ampere	
kW	power	kilowatt
kWh	energy consumed	kilowatt-hour
L	length	m
LH	latent heat of evaporation	kW
l	length	m
l	volume	litre
LEL	lower explosive limit	ppm or %
log	logarithm to base 10	
LTHW	low-temperature hot water	
M	mass flow rate	kg/s
MJ	quantity of energy	megajoule
MTHW	medium-temperature hot water	
MW	power	megawatt
m	length	metre
m	mass	kg
mm	length	millimetre
N	air change rate	per hour
N	force	newton
N	number of occupants	

Symbol	Description	Unit
N	rotational speed	Hz
OEL	occupational exposure limit	ppm or %
P	power	W
p	pressure	Pa
p_a	atmospheric pressure	Pa
π	ratio of circumference to diameter of a circle	
ppm	parts per million	
PS	percentage saturation	%
p_s	saturated vapour pressure	Pa
p_s	static pressure	Pa
p_t	total pressure	Pa
p_v	vapour pressure	Pa
p_v	velocity pressure	Pa
p_{sl}	saturated vapour pressure at sling wet-bulb air temperature t_{sl}	Pa
Q	volume flow rate	m^3/s or 1/s
Q_D	direct-transmitted solar irradiance	W
Q_d	diffuse transmitted solar irradiance	W
$Q_{\tilde{t}}$	swing in total heat exchange	W
Q_u	steady-state fabric heat loss	W
$Q_{\tilde{u}}$	cyclic variation in fabric heat loss	W
Q_v	ventilation heat exchange	W
R	overall drive power ratio	
R	resistance, electrical	Ω
R	thermal resistance	m^2 K/W
R	specific gas constant	kJ/kg K
R	reflectivity of glass	
RH	relative humidity	%
ρ	density	kg/m^3
R_a	air space thermal resistance	m^2 K/W
R_{si}	internal surface thermal resistance	m^2 K/W
R_{so}	external surface thermal resistance	m^2 K/W
SH	sensible heat transfer	kW
SHC	specific heat capacity	kJ/kg K
SMR	square of the mean of the square roots	
SQRT(C7)	square root of contents of cell C7	
SR	regain of static pressure	Pa
S/T	sensible-to-total-heat ratio	
s	time	seconds
T	absolute temperature	kelvin
T	torque	Nm
T	total irradiance	W/m^2
T	transmissivity of glass	
t	temperature	°C
t	dry-bulb air temperature	°C
t_a	air temperature	°C

Symbol	Description	Unit
t_{ai}	inside air temperature	°C
t_{ao}	outside air temperature	°C
t_c	resultant temperature at centre of room	°C
t_{dp}	dew point temperature	°C
t_e	environmental temperature	°C
t_{ei}	inside environmental temperature	°C
$t_{\tilde{e}i}$	swing in environmental temperature	°C
t_f	flow temperature	°C
t_g	glass temperature	°C
t_m	mean temperature	°C
t_m	mixed air temperature	°C
t_r	mean radiant temperature	°C
t_r	return temperature	°C
t_r	room temperature	°C
t_s	supply air temperature	°C
t_s	surface temperature	°C
t_{sl}	sling wet-bulb temperature	°C
U	thermal transmittance	W/m^2 K
UEL	upper explosive limit	ppm or %
V	volume	m^3 or 1
V	electrical potential	volt
V	vertical surface	
v	velocity	m/s
v	specific volume	m^3/kg
W	width	m
w.b.	wet-bulb temperature	°C
Y	admittance	W/m^2 K
10^7	10 raised to the power of 7	10^7

1 Air-conditioning systems

INTRODUCTION

Air conditioning means the full mechanical control of the internal environment to maintain specified conditions for a certain purpose. The objective may be to provide a thermally comfortable temperature, humidity, air cleanliness and freshness for the users of the building or it may be to satisfy operational conditions for machinery or processes. The term air conditioning may be used to describe an air-cooling system that reduces excessive temperatures but does not guarantee precise conditions, to minimize capital and operational costs. This is better described as comfort cooling.

The decision to adopt air conditioning is analysed and a list of the possible systems given. So-called free cooling and how to achieve it, is explained. The choice of zones is important to air conditioning and the reasons for their selection are discussed.

The main systems of air conditioning are described but the list is not exhaustive. The large number of possible combinations of components makes each system at least partly unique, commensurate with the diversity of building services engineering.

Drawings of a project building are provided for assignment work, discussion of possible solutions, examination and testing purposes plus the assessment of elements of competence for NVQs, when these are defined. The drawings are not printed to scale and need to be drawn to an appropriate size and scale. They are referred to in subsequent chapters and are sufficiently detailed for most purposes. Detailed dimensions will need to be decided by inference during draughting.

A case study of a recently completed airport demonstrates that multiple systems are used for complex buildings. *Building Services*, the CIBSE journal, is a regular source of practical cases and is recommended reading.

LEARNING OBJECTIVES

Study of this chapter will enable the user to:

1. define the term 'air conditioning';
2. understand the limitations applied to the term 'air conditioning';
3. state the reasons for air conditioning;
4. be able to analyse the apparent need for air conditioning;
5. explain the decision on whether air conditioning is necessary;

6. categorize the available systems;

7. understand how fresh air can be used to cool the building;

8. be aware of low-cost cooling methods;

9. explain to clients and building designers why ambient air ventilation can reduce the need for refrigeration;

10. identify where low-cost cooling is applicable;

11. understand the scheduling of plant air dampers;

12. know why building zones are necessary;

13. know that zone peak heat loads are not simultaneous;

14. state the zones needed for practical designs;

15. understand why some zones are air-pressurized;

16. understand the working principles, applications and limitations of 16 categories of air-conditioning system;

17. produce schematic diagrams of practical systems;

18. explain why a particular system is being proposed for an application by means of description and sketches;

19. apply air conditioning to the project building provided;

20. know the systems used in an airport terminal through a case study.

Key terms and concepts

air conditioning	3	mechanical ventilation	3
air-handling zones	7	mixing box	17
changeover system	15	motorized damper operation	6
chilled ceiling	29	multizone plant	10
Coanda effect	13	orientation	7
containment	8	peak solar gains	7
dilution	14	perimeter heating	14
district cooling	29 ·	process	3
drain pipe	16	recirculation	9
dual duct	17	refrigeration compressor	5
dumping	13	reversible heat pump	17
evaporative cooling tower	5	single duct	9
fan	12	split system	26
fan coil unit	16	stagnation	13
fresh air	5	static pressure	8
fresh-air intake	5	summertime air temperature	3
heat gains	3	terminal unit	11
heating and cooling coils	15	throttling valve	12
independent unit	24	variable air volume	11
induction	14	variable frequency control	11
low-cost cooling	5		

THE DECISION TO AIR CONDITION

Mechanical ventilation includes the moving of air by means of fans, air filtration, heating, humidification, heat reclaim for economy of operation and free cooling that can be obtained from the external atmosphere. The *CIBSE Guide B3* (CIBSE, 1986b, Section B3) states that air conditioning differs from mechanical ventilation by the incorporation of refrigeration. Thus adding mechanical refrigeration equipment and a cooling coil to a mechanical ventilation system turns it into air conditioning. Low-cost air conditioning may only cool inside the building by a specific number of degree Celsius below the outside air temperature or limit the internal percentage saturation to 70%. This is the upper limit for comfort and for the risk of microbiological growth. The term 'air conditioning' does not mean that specified comfort conditions can or will be maintained. Achieving comfort requires the provision of correctly designed, installed and maintained systems.

Air conditioning is included in the total building design for a variety of motives that range from industrial necessity to the prestige to be gained from marketing the building or its use. The overall technical and aesthetic solution includes the use of passive design features to minimize the use of mechanical plant. Good design is visually acceptable as well as being technically competent (Smith, 1991). The environmental and energy cost implications are analysed. Suitable reasons include the following.

1. Unacceptable high internal summer air temperature may result from solar heat gains if the building is not provided with refrigeration.
2. Heat gains occurring within the building, from people, lights, electrical, catering and mechanical equipment, produce uncomfortable air temperatures for the occupants.
3. Occupied areas cannot be satisfactorily supplied with enough fresh air by natural ventilation.
4. In high-rise buildings, prevailing wind pressure may preclude the opening of windows to provide the necessary ventilation.
5. Road traffic, aircraft or train noise close to the building would cause too much disturbance if windows were opened. The building has to be air-sealed from the external environment to limit noise penetration and consequently requires mechanical ventilation and possibly refrigeration.
6. External air pollution requires the building to be sealed.
7. Close control of the internal atmosphere is required for manufacturing processes such as pharmaceuticals, electronics, nuclear components, paper and cotton production.
8. Secure containment of radioactive processes and materials requires that all possible leakages of contaminated air and dust are eliminated. Full mechanical control of the internal environment becomes necessary both for the process and for the personnel.
9. The reliable operation of most microprocessors and electric motors depends upon maintaining a maximum surrounding ambient air temperature of up to 40 °C and the plant room may need to be air-conditioned.

10. Shops, hotels and commercial buildings where customers are admitted may be air-conditioned for their comfort and as a marketing advantage over competitors.
11. Countries within the tropics have air-conditioned buildings, homes, vehicles and cars through necessity.
12. Close control of the internal atmosphere is required for storage and display of works of art, antiques, furniture, fabrics, paintings and paper archives.
13. Sterile conditions are needed for health care.

METHODS OF SYSTEM OPERATION

Air-conditioning systems are categorized by their mode of operation:

1. single-duct variable air temperature with 100% fresh air (SDVATF);
2. single-duct variable air temperature with recirculation of room air (SDVATR). (Figure 1.1 is a schematic layout);
3. single-duct variable air temperature for multiple zones (SDVATM);
4. single-duct variable air volume (SDVAV);
5. single-duct variable air volume and temperature (SDVAVT);
6. single-duct variable air volume with separate perimeter heating system (SDVAVPH);
7. single duct with induction units (SDI);
8. single duct with fan coil units (SDFC);
9. single duct with reversible heat pumps (SDRHP);
10. dual duct with variable air temperature (DDVT);
11. dual duct with variable air volume (DDVAV);
12. independent unit through the wall (IU);

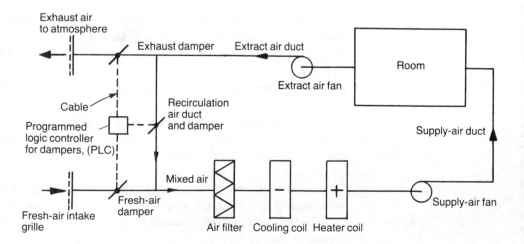

Fig. 1.1 Schematic layout of a single-duct variable air temperature air-conditioning system with recirculation of room air (SDVATR).

13. split system (SS);
14. reversible heat pump (HP);
15. chilled ceiling (CC);
16. district cooling (DC).

LOW-COST COOLING

Cooling of the building can be obtained by passing fresh air through it or by circulating water that has been cooled in an evaporative cooling tower on the roof, through the chilled water coil in the air-handling unit. Direct evaporative cooling from sprayed water within the air-handling plant is used in countries between the tropics. Neither method is free of cost as they both consume electrical energy in the operation of fans and pumps. Maintenance costs are incurred in keeping the system in correct, clean and economical order. Additional capital cost might be incurred for the provision of plant that would not otherwise be needed. The automatic control system has energy consumption with attendant cost and maintenance. The same plant is used as when the refrigeration system is in operation. The cash saving available is from the power consumption of the refrigeration compressor or the heat supplied to the absorption chiller.

Cooling with fresh air use rather than refrigeration means that up to 100% of the room air supply can be by cool outdoor air. The availability of this air depends on the following factors.

1. *Time of year*. Winter, spring and autumn seasons have low external air temperatures. Winter air below 10 °C will have minimum usage.
2. *Time of day*. Sufficiently cool outdoor air will not be available during the warm months in the daytime occupancy hours of offices and shops. Outdoor air above 20 °C will have minimum usage.
3. *Fresh air intake location*. The fresh-air intake grille is positioned in a shaded location above street level where cool and uncontaminated air can be found. The availability of shade depends upon the orientation of the side of the building and the time of day. Underground tunnels or air ducts can be used to lower the temperature of the fresh air intake.
4. *Space availability*. The dimensions of the building's facade that are available for the fresh air intake and exhaust louvres must be sufficient for the passage of 100% of the supply air volume flow rate. Restricted plantroom locations near ground level limit the amount of free cooling that is achievable. Extensively louvred plantroom walls at roof or intermediate floor levels can be utilized.

An evaporative cooling tower lowers the water temperature circulated through it towards the outdoor wet-bulb air temperature. Water evaporation increases the cooling effect of the outdoor air. When the outdoor wet-bulb air temperature is below the supply air temperature needed to cool the room, usable free cooling is potentially available.

Internal rooms within the building that have no direct view or connection with the external walls or roof require continuous cooling through the year. Outdoor air will be suitably cool for much of the year in the UK.

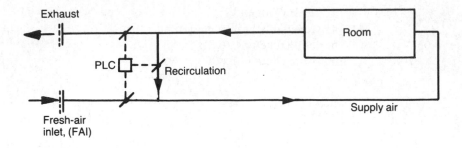

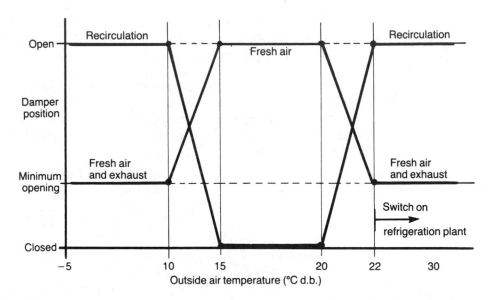

Fig. 1.2   Example of a schedule for the operation of dampers in a SDVATR system.

The amount of free cooling provided is controlled by varying the proportion of fresh air in the room air-supply system. The proportion moves from the minimum up to 100% by the operation of motorized dampers on the fresh air, exhaust and recirculation ducts. The minimum fresh-air intake corresponds to the occupancy or process fresh-air requirement at the peak summer or winter external design conditions. Variable amounts of fresh air are admitted between these extremes. Figure 1.2 shows how the three dampers are scheduled for a single-duct system with recirculation according to outdoor air temperature. The dampers have a linear characteristic that relates flow to opening. The objective is to minimize the use of heating and cooling energy while maintaining the specified internal air conditions. The external air temperatures indicated are sample values that do not apply to all cases and are sensed at the fresh-air intake grille.

The fresh- and exhaust-air dampers are interlinked to have identical openings and they remain at the minimum settings below 10 °C and above 22 °C. These dampers begin opening at 10 °C and are fully open at 15 °C. They remain fully open from 15 °C to 20 °C while cooling is available to remove solar and internal heat gains from the building. At 22 °C the dampers begin to close to reduce the admission of heat gains into the building from the warm ventilation air. They reach their minimum setting at 22 °C and remain there during hot weather.

The recirculation air damper is fully open until the winter air reaches 10 °C. Its opening is ramped down to be fully closed at 15 °C. The fresh-air and exhaust dampers are fully open. The recirculation damper remains closed to allow the building to be flushed with fresh air until 20 °C is reached. The recirculation damper is ramped up to be fully open at 22 °C where it remains static during the summer.

A programmable logic controller (PLC) microprocessor retains these instructions and operates the components. This program can be reset manually or through the wiring from the supervising computer of a building energy-management system (BEMS). The refrigeration system is switched on at 22 °C and is separately control-led. The shape of the damper movement graph shown is for illustration purposes only; each control system is designed according to its unique needs.

AIR-HANDLING ZONES

Areas of the building that have a similar heating, cooling and humidity control plant load are grouped into zones. The south-facing rooms of a modular office block are all exposed to an identical pattern of solar heat gains during the working day in that their peak gains occur simultaneously. Each office has the same occupancy usage. When the south-facing glazing of rooms on the lower storeys is shaded from the direct solar radiation by other buildings, they have lesser heat gains than higher-level rooms on the same side that remain exposed all day. The south facade of the block needs to be separated into two or more zones, each being provided with a different supply air temperature throughout the year. If not, the lower floors would be excessively cooled by supply air that is suitable for the exposed rooms.

Peak solar heat gains through unprotected clear vertical glazing occur at various times and dates (see Table 1.1)

Table 1.1

Orientation	Peak gain (W/m^2)	Sun time (h)	Date
S	510	1300	22 September
E	477	0900	21 June
W	477	1700	21 June
N	161	1300	23 July

Source: CIBSE (1986) Table A9.14 for 51.7 °N latitude

In this table, sun time is Greenwich Mean Time (GMT). British Summer Time (BST) is GMT plus 1 h and is used from the end of March to the end of October.

Interior rooms that have no external glazing experience a constant cooling load that is dependent upon internal heat gains rather than the weather. If each building orientation is grouped into separate zones, the time and date of their peak cooling loads will be different.

Zones can be chosen on the basis of the following.

1. *Time of occurrence of the peak loads*. This enables the air-handling plant to provide a suitable air supply temperature to the whole zone.
2. *Orientation*. Each facade of the building has a unique programme of shading and exposure to solar radiation. Rooms with large areas of glazing are highly vulnerable to the external climate. A building with an irregularly shaped perimeter might have more than one orientation within a zone. A windowless structure is relatively insensitive to cyclic solar radiation heat gains due to its thermal storage capacity and up to 12 h of timelag for heat transfer from outside to inside.
3. *Interior space*. Rooms that have no surfaces which are exposed to the external atmosphere have a daily and year-round constant heat load, usually for cooling.
4. *Height above ground*. Tall buildings are zoned into groups of floor levels owing to their exposure or to reduce the size of the air-distribution ductwork. Below-ground rooms are zoned together.
5. *Containment*. Areas of a building that must be independent of each other – to restrict possible cross-contamination by airborne micro-organisms or radioactive particles – are put into separate zones. Examples are kitchens, hospital operating theatres, X-ray rooms, nuclear, chemical and biological production facilities, research laboratories and clean rooms for pharmaceutical and electronic manufacture.

Containment zones are maintained at a static air pressure that is above or below that of the adjacent rooms or the external atmospheric pressure to ensure an air flow in a specific direction that is into or out of the containment zone. Pressurized airlocks may be used to separate zones. Table 1.2 gives examples of some applications where a specific flow direction must be maintained by suitable positive or negative air pressure when compared to the surrounding areas.

Table 1.2

Zone use	Airflow direction	Static pressure (Pa)
Conditioned office	Out	+ 10
Kitchen, toilet	In	− 10
Operating theatre	Out	+ 10
Airlock between areas	Out	+ 10
Nuclear, biological, chemical	In	− 50
Clean room	Out	+ 10

Air buoyancy, or stack effect, and wind pressure will influence internal air pressure. When less air is extracted than supplied, the room is slightly pressurized and conditioned air will leak outwards into corridors, staircases and eventually outdoors. These airflows are calculated to find the air velocity through gaps around doors and transfer grilles. Conditioned air leaks out of the building. This ensures that the internal conditions remain under control. When more air is extracted than supplied, the room is under a negative pressure and air from surrounding areas or from outdoors will leak into the room. This dilutes the conditioned air and adds to the room heating or cooling load on the plant. This will be undesirable when close

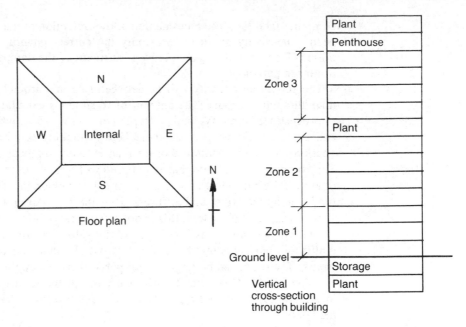

Fig. 1.3 Air-conditioning zones.

environmental control is essential. The ability of a building to maintain a specified static pressure is dependant upon the quality of construction. Air leakages can be calculated from data used in the design of fire escape routes (BSI, 1986).

In conclusion, a zone can be a group of areas, rooms, floors, a whole building, a single room or a single module of a building. A single-duct air-handling system can provide one supply-air condition to a whole building. Terminal temperature or volume regulation ensures that each room can be maintained at its desired state.

Figure 1.3 shows how a simple building design can be separated into zones.

SINGLE-DUCT VARIABLE AIR TEMPERATURE 100% FRESH AIR (SDVATF)

This system is used where contamination produced within the conditioned room is not to be recirculated and a full flushing with outdoor air is to be maintained. A commercial kitchen has a heated supply-air duct and hoods over cooking ranges to exhaust the hot, steamy and grease-laden air to the outdoor atmosphere as quickly as possible. Two-speed fans can be used so that low ventilation rates will meet comfort needs during food preparation but a high extraction can be operated during maximum cooking periods.

SINGLE-DUCT VARIABLE AIR TEMPERATURE WITH RECIRCULATION (SDVATR)

This is a common system adopted for occupied large single-volume rooms, offices, atria, theatres, sports halls, swimming pools, factories, clean rooms and mainframe computer facilities. The recirculation allows retention of the conditioned air that has been expensively produced with only the correct quantity of fresh air being admitted. Free cooling is possible by adjustment of the dampers following schedules of air temperatures.

The supply air quantity will be between four and around twenty air changes per hour through the room. The amount of fresh air is calculated from the occupancy or process needs; it is typically 8 l/s per person for offices where there is no smoking and up to 25 l/s per person where heavy smoking is permitted (CIBSE, 1986a, Section A1). Such admittance of fresh air is likely to create one air change per hour. The room air-change rate that is required to flush out potentially stagnant pockets of air will be between 4 and 20 per hour depending on the design of the air currents produced by the grilles and diffusers. Thus the fresh-air proportion of the ducted air supply will be 25% or less; this is not a fixed ratio, however.

The system only satisfies the conditions for one room and is unsuitable for multiroom buildings that have different but simultaneous air temperature requirements. The room can be large but all parts of it are supplied with identical supply air. The volume flow rates to different parts of the room are in proportion to the localized heating or cooling loads but the same average condition exists throughout the space.

SINGLE-DUCT VARIABLE AIR TEMPERATURE MULTIPLE ZONES (SDVATM)

This is an adaptation of the individual zone single-duct system with recirculation but is designed for multiroom applications such as offices. Additional heater coils, called terminal reheaters, are installed at the room end of the ducts. Each room has

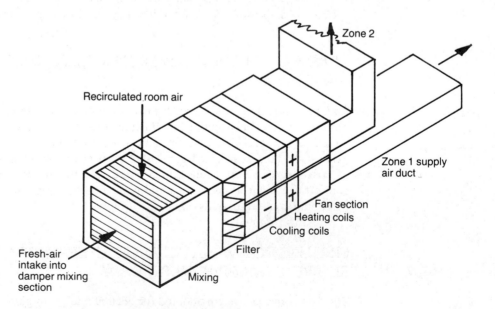

Fig. 1.4   Two-deck multizone air-handling unit.

an air-temperature sensor that modulates the reheat coil water-flow control valve to maintain the desired room condition. The air temperature that leaves the air-handling unit in the plantroom is that for the room requiring the lowest supply-air temperature. The other rooms have a few degrees of reheat.

When rooms can be formed into two or four zones that are to be supplied from the same air handling unit, a multizone plant can be designed to incorporate two or four outlet ducts. Each duct has a reheat coil and zone air-temperature controller. Figure 1.4 shows a multizone air-handling unit. Zone reheat coils can be in the supply air duct as it enters the zone. Several zones can be connected to a single air-handling unit. Each zone has a number of rooms that will be satisfied with one supply air temperature: for example, a row of one-person offices along one facade of the building or a similar row of modules in an open-plan office. The disadvantage of having distributed hot water reheat coils is that additional flow and return piping, controls and access to ductwork are needed.

SINGLE-DUCT VARIABLE AIR VOLUME (SDVAV)

The single-duct variable air volume system has become the standard employed for office buildings because of its economy and controllability when applied to rooms of similar use for comfort air conditioning. It is a cooling system with the air-handling unit providing the lowest air temperature required. Reductions in room cooling load, due to variation in the weather or by the presence of additional heat gains from occupants or electrical equipment, are met by changing the supply-air quantity with terminal variable air volume controllers until the room air temperature is stabilized. The minimum quantity of supply air permitted corresponds to at least the fresh air need of the occupants; thus the turn-down may be from 100% down to around 20%.

It is sensible to reduce the airflow rate through the building during most of the year when the refrigeration and heating loads are less than their design peak values. When the room air temperature falls, the cool supply air volume is reduced until the desired temperature is reached again. This process is repeated in other VAV controllers at a similar time. The characteristic performance of the centrifugal supply air fan in the main air-handling unit is to increase its output static pressure as the volume flow rate is reduced. This is contrary to the room requirements as a higher duct pressure would tend to overcome the control effect of the VAV unit. The main supply air duct static pressure is sensed and used to reduce the speed of the fan by means of a variable-frequency inverter. The fan speed is lowered slowly until the set point of the static pressure controller is established. This is also done with the recirculation air fan. Operating the two fans at less than their maximum speeds greatly reduces the electrical power consumed and the noise generated by them.

The air supply flow rate is varied by a throttling valve and a fan. The conditioned supply air enters the room through a diffuser. This may be a ceiling linear diffuser along the external perimeter. A plenum above the diffuser acts as a header box to distribute the air uniformly along its length. A throttling valve between the plenum and diffuser controls the airflow rate. Figure 1.5 shows the simplified arrangement of three types.

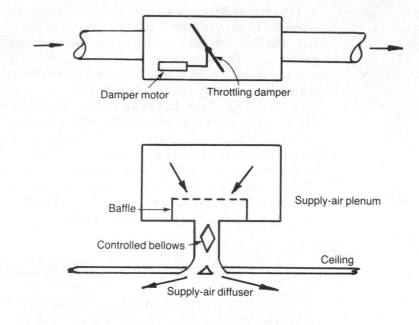

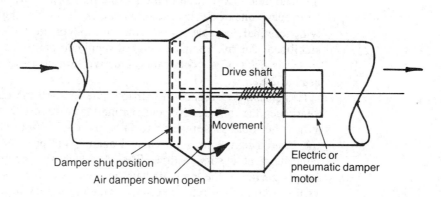

Fig. 1.5 Variable air volume terminal unit controllers.

A fan-powered VAV terminal box is located in each building module. The fan may be used to deliver a constant volume flow rate into the room and have a variable proportion of cool air from the central plant, modulated by the VAV damper. This ensures the maintenance of the correct air distribution through the room. A terminal fan is an additional capital cost, noise and maintenance consideration. Figure 1.6 is a simplified arrangement.

Varying the supply air flow has a strong influence upon the distribution of air into the room and the movement of air within it. The supply air is diffused into the room along the length of the external wall and is blown across the ceiling and down the glazing. Airflow along a surface adheres to that surface and forms a thick boundary

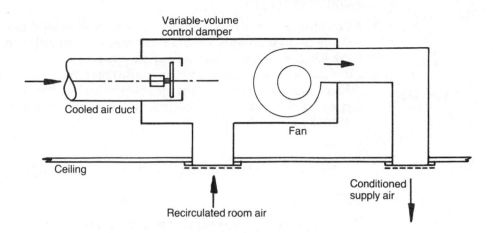

Fig. 1.6 Fan-powered variable air volume terminal unit installed in a false ceiling.

layer. This is known as the Coanda effect after Henri-Marie Coanda (1885–1972), the Romanian-born aeronautical engineer who made a major contribution to fluidic technology in the 1930s. He observed that a fluid stream creates a lowered static pressure adjacent to the surface, thus anchoring the layer to the surface (*Encyclopaedia Britannica*, 1980).

Low air flows created during the normal use of a VAV system can lead to discomfort for a variety of reasons, as follows.

Dumping

Supply air that is cooler than the room air temperature is being throttled to, perhaps, 25% of its design flow rate during mild weather. The combination of low jet speed issuing from the diffuser and the greater density of the cool supply air leads to the air's not staying in contact with the ceiling long enough to mix with the warmer entrained room air, and it falls off the ceiling. The occupants feel cool downdraughts and have cause for complaint. It is necessary to maintain a sufficiently high air velocity jet across the ceiling at the minimum air flow setting to keep the contact. A narrow slot diffuser or a fan-powered unit can achieve this.

Stagnation

Reduced air flows produce less air changes of the room air per hour and less stimulation of the recirculation eddy currents.

Poor distribution

Parts of the room may be starved of sufficient air circulation and become too warm owing to local heat sources such as people, computers, lights and photocopiers.

Dilution

Complaints of stuffiness or smoke-laden air may result if tobacco use is permitted. Odour from the occupants and the materials, solvents and chemicals within the room can become noticeable and can cause a nuisance.

Cases of sick building syndrome (SBS) may be traced to the quantity of air in circulation and its quality. The air volume flow control box of the VAV system is often installed in the false ceiling. It can be situated in a floor void with the supply duct rising to a sill-level grille.

SINGLE-DUCT VARIABLE AIR VOLUME AND TEMPERATURE (SDVAVT)

This is the same VAV system but with the addition of terminal reheat coils to adjust the delivered supply air temperature. The air temperature that leaves the air-handling plant is determined by the zone requiring the most cooling. The use of reheaters adds to the energy consumption of the system as previously cooled air is raised in temperature. The air pressure loss through the heater batteries, and the extra pipes and controls, also increase the capital and operating costs.

SINGLE-DUCT VARIABLE AIR VOLUME PERIMETER HEATING (SDVAVPH)

The separate perimeter heating system will be radiators, radiant panels, convectors or sill-line convectors on a low-pressure hot-water distribution with thermostatically controlled modulating valves. This is to allow each room to be maintained at the correct air temperature while being ventilated with a fixed supply air temperature from the air-handling plant.

The perimeter heaters can be obtrusive in the occupied space. They require careful integration with the furniture, interior design and decoration; they need maintenance; and they result in some loss of lettable floor area. Siting computer workstations near to the heaters has to be avoided owing to possible overheating of the electrical equipment.

SINGLE DUCT WITH INDUCTION UNITS (SDI)

A single duct supplies fresh air only from an air-handling unit in the plantroom into each room. This fresh air is termed the primary air supply. An extract air duct removes 90% to 100% of the fresh air supply quantity and discharges it to the atmosphere. Each room or module has an induction unit beneath the window sill, in the floor or in the ceiling. The primary air is introduced into the induction unit through nozzles that increase the inlet air velocity sufficiently to lower the static air pressure within the induction unit casing. Noise generation is a problem. Room air is induced to flow through the casing owing to the pressure difference. A low-pressure hot-water heating or chilled-water cooling coil conditions the recirculated

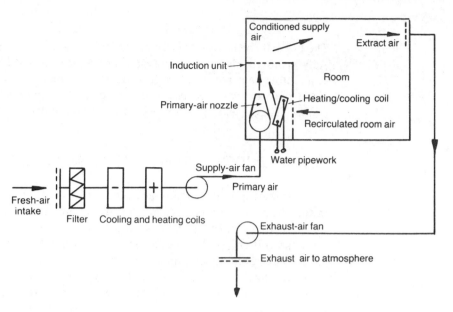

Fig. 1.7 Single-duct induction unit system (SDI).

room air prior to its mixing with the primary air. The mixed air flows into the room through a diffuser. Only the primary air is filtered. The duct system only passes fresh and exhaust air and so is as small as possible for the building when compared with other designs that require ducted recirculation.

The bulk of the sensible heating and refrigeration room load is met from the induction unit coil. The room moisture content can only be controlled from the moisture content of the fresh air supplied and this restricts the humidity control ability of the system. It would generally not be practical to dehumidify the room air with the cooling coil because condensate drain pipework would be needed. A manually operated damper is fitted to sill-line induction units to enable the occupant to vary the heating or cooling effect by adjusting the recirculated room air flow.

The system is used for office accommodation and a typical installation is shown in Fig. 1.7. The water circulation to the heat exchange coil can be of the following types.

Two-pipe changeover

The two-pipe system requires the changeover of the water circulation from the boilers to the refrigeration evaporator by means of a three-way valve in the plant room at a clearly defined season change from winter to spring. The pipework needs to be zoned so that north and south aspects of the building can have warm and cool water simultaneously.

Two-pipe non-changeover

Only chilled water is circulated to the induction unit room coils. All the heating needed is provided by the heater battery of the plantroom air-handling unit if a sufficiently high air temperature is possible. The maximum available is 45 °C.

Four-pipe

The hot and chilled water circuits are maintained at the correct temperatures separately to provide the maximum availability of heating and cooling at all the induction units throughout the year. Each induction unit has two coils, four water pipes and either two two-way valves or one three- or four-port valve with air temperature control. The system is the more costly option but offers the greatest versatility.

SINGLE DUCT WITH FAN-COIL UNITS (SDFC)

A single duct supplies conditioned primary fresh air only from an air-handling unit in the plantroom into each room. An extract air duct removes 90% to 100% of the fresh air supply quantity and discharges it to the atmosphere. Each room or module has a fan coil unit within the false ceiling, within the floor or below the sill, which recirculates room air and mixes it with the primary fresh air. Heating and cooling coils are fitted into the fan coil unit casing. These are supplied from a four-pipe or a two-pipe changeover, hot and chilled water pipe system. Large heating and cooling loads that occur within the occupied space can be met as there are fewer limitations upon coil duties than with other systems owing to the use of a dedicated fan.

The room air can be dehumidified. A drip tray and condensate drain are included in the design. Control of room humidity is poor. Fan coil units are independent of each other. Each fan coil unit creates a zone. The hot and chilled water circulation is supplied at constant temperature from a central plant whereas the primary air ductwork may be zoned.

Drain pipework connects to the foul drain stack through an air gap and a 75 mm deep water trap. An air filter is fitted within the unit. Each fan coil box has to have a removable ceiling panel for regular maintenance of the filter and for access to the fan and controls. The primary air can be supplied into the room through grilles or diffusers rather than into the fan coil unit. The fan noise produced is related to the acoustic design criteria of the air conditioned room. Figure 1.8 shows a typical configuration.

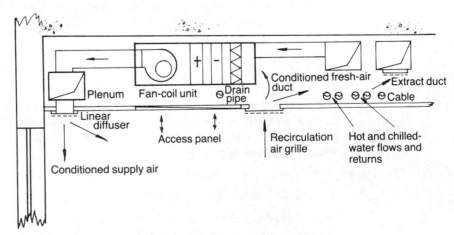

Fig. 1.8 Fan-coil unit air-conditioning system (SDFC).

SINGLE DUCT WITH REVERSIBLE HEAT PUMP (SDRHP)

A single duct supplies conditioned primary fresh air only from an air-handling unit in the plantroom into each room. An extract air duct removes 90% to 100% of the fresh air supply quantity and discharges it to the atmosphere. Each room or module has a water-source reversible heat pump system within the unit casing that recirculates room air and mixes it with the primary fresh air. The terminal unit can be wall-, floor-, ceiling- or underfloor-mounted. It is connected to a two-pipe reverse return water circulation system that is maintained at a constant temperature of, typically, 27 °C throughout the year. The water circuit acts as the heat source in winter and the heat sink in summer. The system is ideally suited for buildings that have simultaneous heating and cooling loads: for example, those with large glazing areas that are oriented both south and north. During cool sunny weather, the excess heat gains through the south glazing require the rooms to be cooled and the heat pumps add heat into the water circuit. The north-facing rooms are suffering heat losses and the heat pumps extract heat from the water to raise the supply air temperature and heat the rooms.

Surplus heat from the building is rejected into the outside atmosphere through the cooling tower or dry cooler. The water is kept at the desired 27 °C. Additional heat is met by the boiler plant to keep the water circulation up to 27 °C. A dead band of 1 °C or more is maintained between the switching points of the cooling tower and boiler to stop overlap of their operation.

Each room or building module has a self-contained refrigeration system comprising a reciprocating or rotary compressor, an air heat-exchange coil, a water heat-exchange coil, a refrigerant reversing valve, controls, filter and recirculation fan. Figure 1.9 shows the arrangement. High noise and maintenance levels are typical. The recirculated room air passing through the coil is heated or cooled by the refrigerant depending upon the position of the reversing valve. During the summer, this coil acts as the refrigerant evaporator. When heating is called for, the room air coil acts as the refrigerant condenser because the reversing valve has exchanged the roles of the two coils.

Humidity control is from the ducted fresh air moisture content as with the other systems described. Close humidity control is unlikely. Condensate drain pipework is needed. The starting and running of refrigeration compressors in each room can be noticeable. They are interlocked electrically to avoid the simultaneous starting of too many and consequent overloading of the wiring and current limiters.

DUAL DUCT WITH VARIABLE AIR TEMPERATURE (DDVT)

The dual duct system supplies air to each room, module or zone through a hot duct and a cold duct. A mixing box regulates the relative amount of each source while keeping the supply air volume flow into the room constant. It is a constant-volume variable-temperature all-air system having no water services outside the air-handling plantroom. Fresh air is mixed into the recirculated room air in the plant room. Figure 1.10 shows the air-handling system and Figure 1.11 the operation of the terminal mixing box.

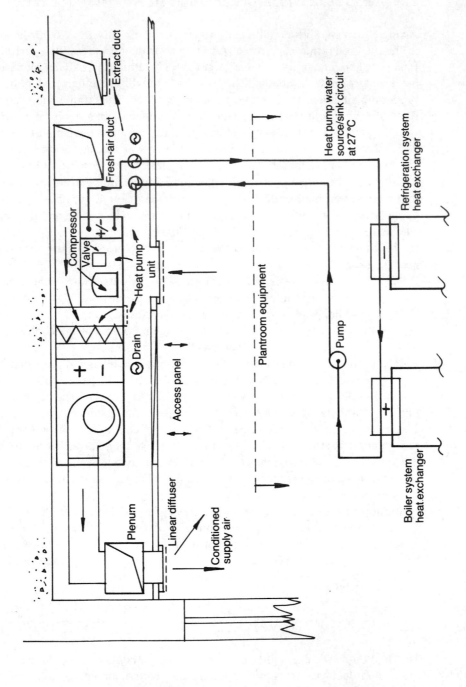

Fig. 1.9 Single-duct system with reversible heat pump (SDRHP).

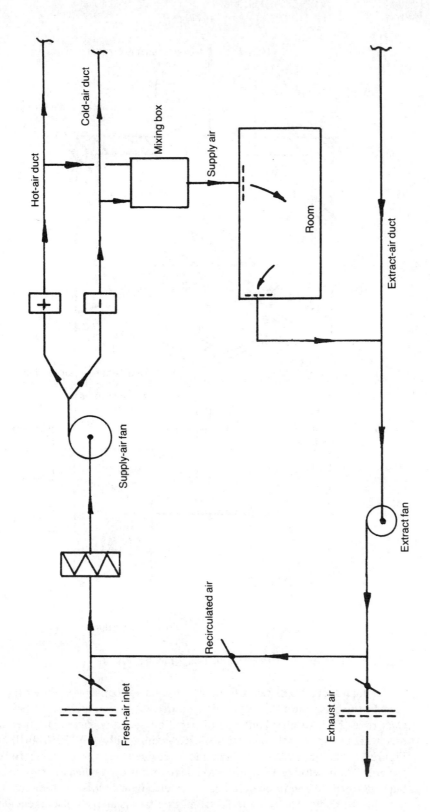

Fig. 1.10 Dual-duct variable air-temperature system (DDVT).

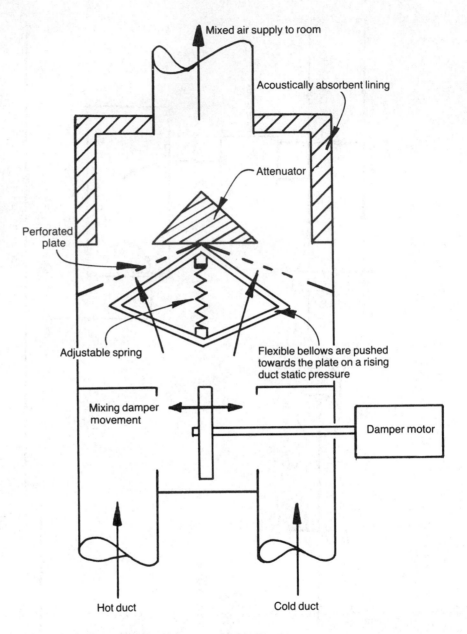

Fig. 1.11   Operating principle of a dual-duct mixing box.

The dual-duct system has two supply-air ducts of equal size. Either the hot or cold duct has to pass the full supply air quantity. A recirculation duct passes air from each room back to the plant. The distribution space occupied by three air ducts is much greater than that for water-based systems with only fresh air being ducted. Figure 1.12 demonstrates the comparable service shaft space needed for the various systems. High-velocity air-distribution ductwork is used when duct spaces are limited but this leads to additional fan and air turbulence noise generation transmitted through the ducts. The room terminal mixing box is also a noise attenuator/silencer.

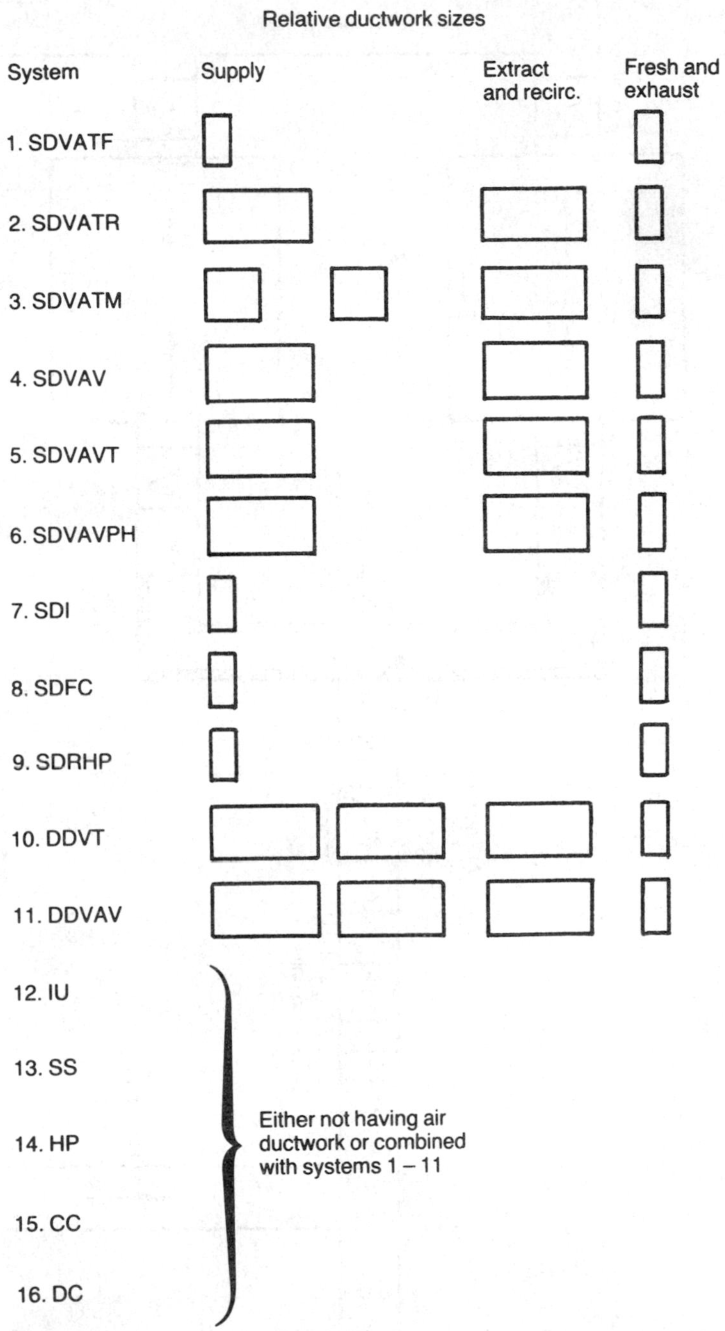

Relative ductwork sizes

System	Supply		Extract and recirc.	Fresh and exhaust
1. SDVATF				
2. SDVATR				
3. SDVATM				
4. SDVAV				
5. SDVAVT				
6. SDVAVPH				
7. SDI				
8. SDFC				
9. SDRHP				
10. DDVT				
11. DDVAV				
12. IU				
13. SS				
14. HP		Either not having air ductwork or combined with systems 1 – 11		
15. CC				
16. DC				

Fig. 1.12 Comparative air-duct spaces.

Similar treatment may be necessary for the air ducts and plant room. Figure 1.13 indicates where acoustic attenuation may be fitted to an air-duct system. Terminal attenuators have the important function of minimizing crosstalk between rooms, which can otherwise be both annoying and a loss of security.

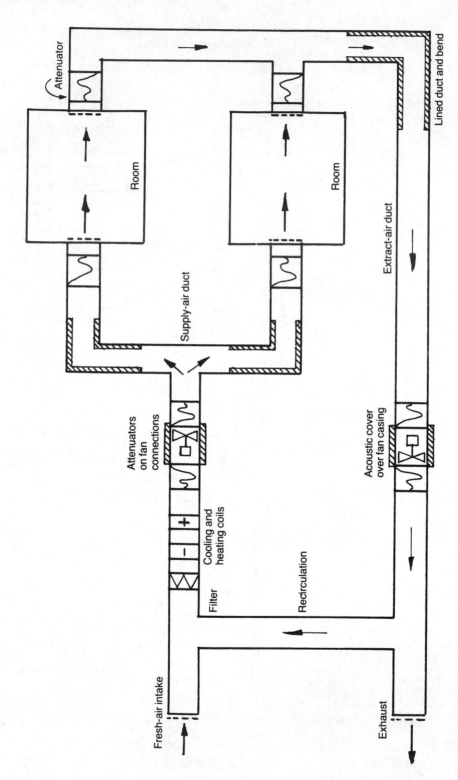

Fig. 1.13 Likely places for acoustic attenuation.

The dual duct system provides for simultaneous heating and cooling of adjacent zones and provides the maximum system flexibility. A mixing box can supply a small-perimeter room or a large space such as a conference room where the supply air from the box is distributed to several diffusers or grilles. All the ductwork is contained within the ceiling void as it is impractical anywhere else because of its size. Only the supply and extract grilles and diffusers are visible in the room. An open-plan office, airport lounge, banking hall, conference or exhibition area is suitable for the dual duct system as the occupancy varies during the day and temporary concentrations of the occupants are accommodated easily.

The hot- and cold-duct air temperatures are scheduled in relation to the external weather conditions for economy, and Figure 1.14 shows a simplified illustrative programme. In winter, the hot air duct is maintained at 30 °C to be able to offset room heat losses. When the outdoor air reaches 10 °C, the reduced heat loss and solar heat gains allow a lower supply air temperature and reduced heating energy consumption. The hot-duct temperature is reduced until it is at 23 °C and the heater

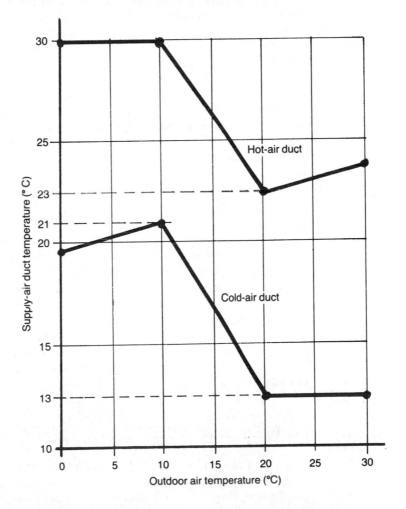

Fig. 1.14 Possible schedule for dual duct supply-air temperatures.

coil is fully off. This is the minimum hot-duct air temperature that can be achieved by the mixture of recirculated room air and outdoor air at 20 °C. Further increases in outdoor air temperature elevate the mixed temperature in the hot duct. During winter, the cold-duct air temperature is the product of cold outdoor air mixing with warm recirculated room air and may be 19 °C at 0 °C outside. Increasing outdoor air temperatures raises the cold duct to 21 °C at 10 °C externally. At this time of year, solar and internal heat gains produce a cooling load on the plant and the fresh-air intake volume can be increased to provide low-cost cooling. At around 15 °C outside, the refrigeration plant is activated, the fresh-air dampers are closed to their minimum setting and the cold-duct air temperature is gradually reduced to its summer value of 13 °C.

The supply fan in the air-handling plant provides a constant total-volume flow rate to the hot and cold ducts. As the mixing boxes vary the quantities taken from each duct, the static pressures at the inlets to the box will vary. This is because a reduced airflow in one duct yields a lower pressure drop due to friction; while the fan static pressure stays the same, the static air pressure in the duct must rise. The excess static pressure is absorbed by a regulator in each box to maintain system balance and avoid crossflow between the hot and cold ducts. The regulator is a flexible diaphragm that is exposed to the difference between the static pressures in the room and in the supply duct. It is preset by the manufacturer to allow the design airflow at a specific inlet static air pressure of between 200 Pa and 2000 Pa and can be adjusted on site. When in use, an increasing inlet-duct air pressure pushes the diaphragm forwards to produce a smaller aperture for the airflow into the room. The increased frictional resistance of the aperture absorbs the excess pressure and maintains a constant supply airflow volume. A reduction in supply-duct air pressure allows the diaphragm to move back, open the aperture and provide less frictional resistance.

DUAL DUCT WITH VARIABLE AIR VOLUME (DDVAV)

The VAV terminal unit consists of a split plenum that receives air from both hot- and cold-supply ducts. The plenum is fitted within a false ceiling and conditioned air is admitted into the zone through a perimeter slot or other diffuser. The hot air is blown towards the external glazing and may be maintained at a constant-volume flow rate throughout the year to counteract the heat gains and losses there. The cold air slot is directed towards the inner part of the room and has its volume flow rate varied according to the cooling requirement. In other respects, the system is the same as previously described for the VAV and DD systems. Figure 1.15 shows the terminal unit.

INDEPENDENT UNIT (IU)

These are free-standing air-conditioning units containing refrigeration, electrical or water heating, air filtration, controls and a fresh air supply ducted from the local outside air. They are often suitable for installation in existing buildings when a multizone ducted system is impractical. They are characterized by their obtrusive

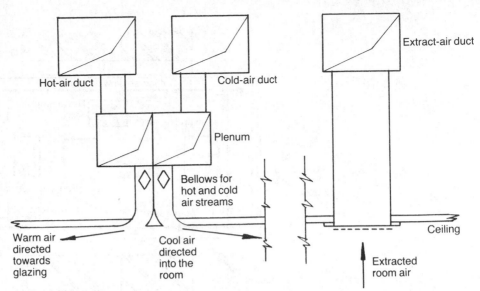

Fig. 1.15 Terminal unit for DDVAV system.

position within the conditioned room, compressor and fan noise, low air-filtration capacity, prominent appearance on the external building facade and by their being influenced by outside traffic noise and wind pressure.

Independent units can be installed through the wall or window to facilitate heat-pump operation and provide summer refrigeration and winter heating. They are used in houses, hotels, offices, computer suites, retail, commercial or industrial buildings. Figure 1.16 shows a through-the-wall air conditioner during summer operation. The heat extracted from the recirculated room air is raised in temperature by the refrige-

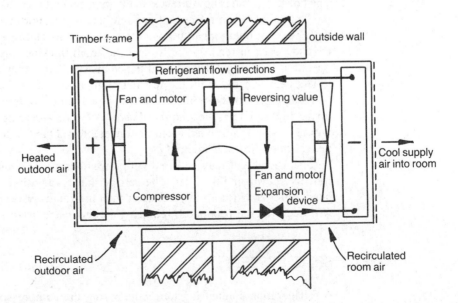

Fig. 1.16 Through-the-wall packaged air-conditioning unit operating in room-cooling mode.

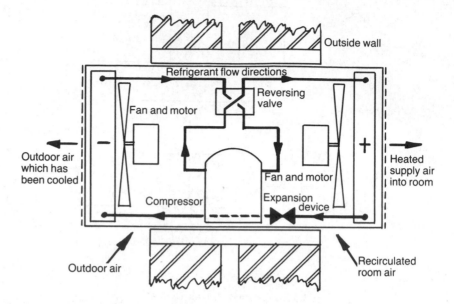

Fig. 1.17 Through-the-wall packaged air-conditioning unit operating in the room-heating mode; may be described as heat-pump mode.

ration compressor and is removed from the building by heating the external air. Figure 1.17 shows the same system running with the flow of refrigerant being reversed at the changeover valve and heating the inside air of the building. The refrigerant can only flow one way through the compressor. It can flow in either direction through the expansion device. This is a coil of capillary tube to create a large pressure drop. Flow through the heat-exchange coils can be in either direction. The hot refrigerant is now directed to the room air coil to be condensed and this becomes the heating coil. The refrigerant pressure and temperature drop through the expansion device and the outdoor part becomes the cooling coil. Heat is extracted from the outdoor air and raised in temperature to heat the room, making this the heat-pump mode of operation. Additional ducted conditioned air ventilation may be used.

The heat rejected from the refrigeration condenser of larger (over 10 kW) independent units can be removed with a piped water system connected to a cooling tower. Figure 1.18 is the simplified layout of a free-standing unit that may be located within the air-conditioned room or concealed and have a ducted air circulation with or without a fresh air inlet.

A packaged unit may be fitted with a weatherproof cabinet and installed on a flat roof as shown in Fig. 1.19. The refrigerant evaporating temperature in all independent units will normally be lower than the room-air dew point and condensation will occur on the room coil, requiring a condensate drain.

SPLIT SYSTEM (SS)

A refrigeration condensing unit, comprising the compressor, air-cooled condenser, condenser fan, controls and weather-proof casing, is located outdoors, usually on

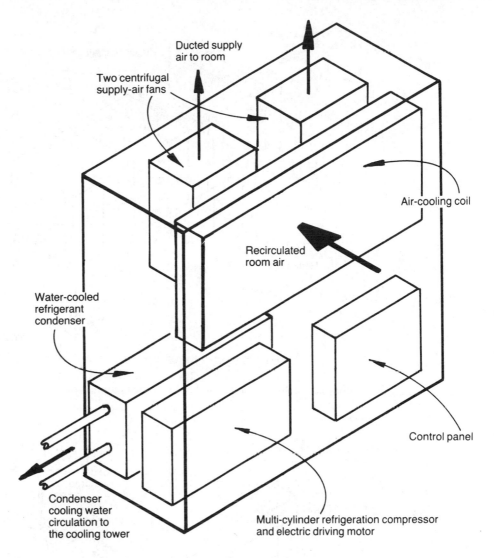

Fig. 1.18 Free-standing air-conditioning unit components.

the roof of an office, shop, restaurant or computer room. Liquid refrigerant leaves the outdoor unit and is piped indoors to the room cooler that has the expansion valve, refrigerant evaporator cooling coil and controls. Evaporated refrigerant gas is piped back to the condensing unit to be compressed and recycled.

The room cooler has a fan to circulate room air through the heat exchanger and a temperature control that is manually adjustable. A condensate drain is needed from the room cooling and dehumidifying coil. No humidification is possible and room air temperature control is by step control of the fan speed, on-off fan operation, on-off compressor switching, varying the refrigerant volume flow rate (VRV) by compressor motor speed control with a variable frequency controller (VFC) or multistage unloading of the larger compressor systems.

Figure 1.20 shows a typical split-system installation. Its advantage is that the noise-producing compressor and condenser fan are located away from the

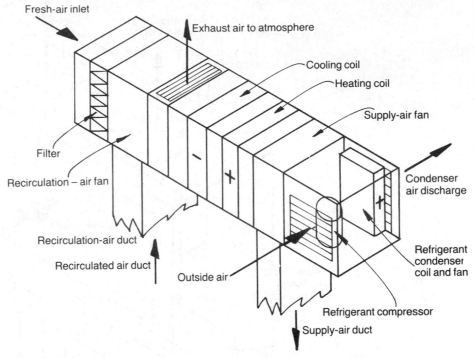

Fig. 1.19   Roof-mounted packaged air-conditioning unit.

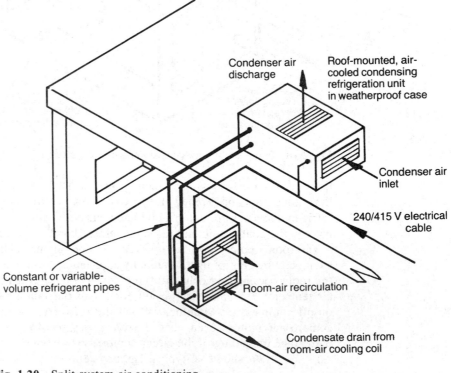

Fig. 1.20   Split-system air-conditioning.

conditioned room. The two units can be separated by several floors of a building. The indoor unit often has high fan and airflow noise. The principle advantage is low cost. These systems are most suitable for adding into existing buildings.

REVERSIBLE HEAT PUMP (HP)

All refrigeration systems are heat pumps. When the useful heat energy is taken from the condenser to warm the interior of the building, such a system is termed a heat pump. There are independent units (IU) and split-system (SS) types used for heating the building from one energy source, electricity. The packaged heat pump can be roof-mounted above a retail premises to provide year-round cooling and heating with a single-duct air distribution with recirculation.

CHILLED CEILING (CC)

Metal ceiling panels with chilled water pipes attached to them are either suspended below the ceiling or incorporated into the false ceiling design. They provide up to $150 \, W/m^2$ of convective and radiant cooling in offices and may provide a degree of comfort where solar gains are small (CIBSE, 1992). Low-cost cooling can be achieved by circulating the water only through an evaporative cooling tower and not using refrigeration. Ventilation, heating and air-conditioning systems are dealt with separately. Surface condensation, dust collection, appearance and decoration are attendant problems.

DISTRICT COOLING (DC)

The centralized supply of cooling services is available in a similar manner to district heating. A shopping, residential and office building development can have a central refrigeration plant that circulates chilled water to each rented space. A heat meter integrates the total water flow with the inlet and outlet water temperatures to summate the kWh of cooling energy purchased. Each user is billed separately for the energy consumed. The plant operator provides and maintains the service for the energy charge. The consumer has fan-coil air conditioners or ducted air handling as appropriate.

PROJECT BUILDING

Figures 1.21 to 1.28 show the design for an office building for Si joule plc who require the whole building to be air-conditioned. A set of scale drawings should be created for practical work. The site location is to be chosen by the reader to be locally convenient so that appropriate climate data can be selected. Questions are posed as to the suitability of different systems and, in later chapters, about other aspects of air conditioning. Model solutions are not provided as the designs should

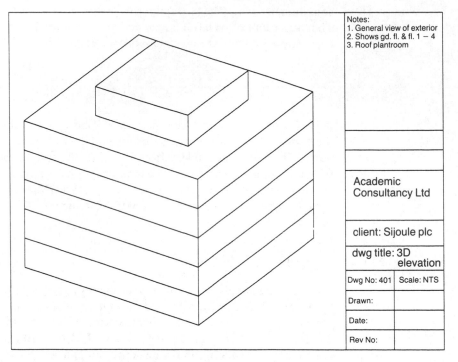

Fig. 1.21 General view of project building.

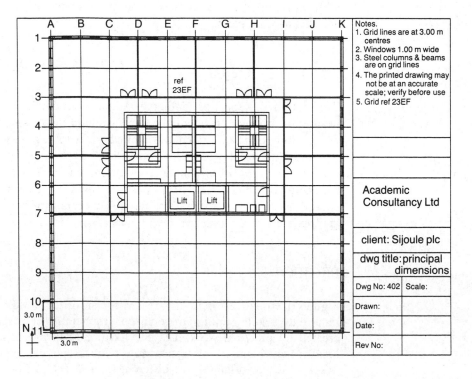

Fig. 1.22 Grid layout and dimensions.

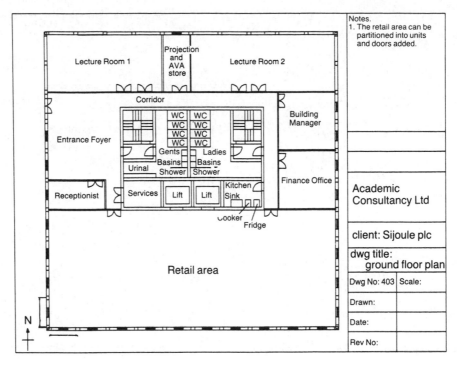

Fig. 1.23   Ground-floor plan.

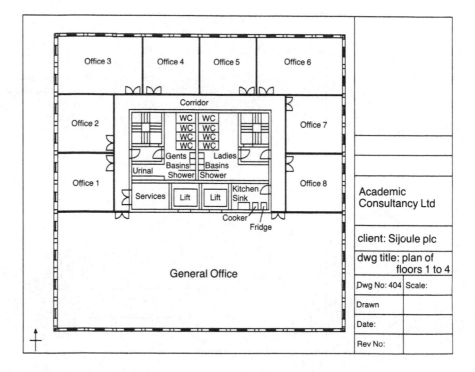

Fig. 1.24   Intermediate floors.

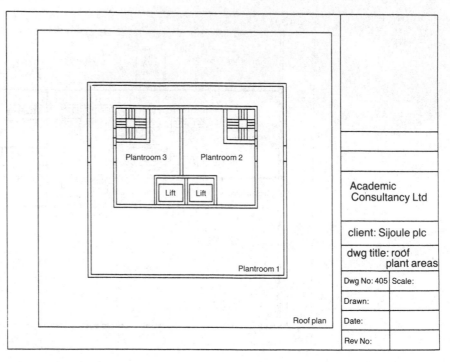

Fig. 1.25 Roof plantroom areas.

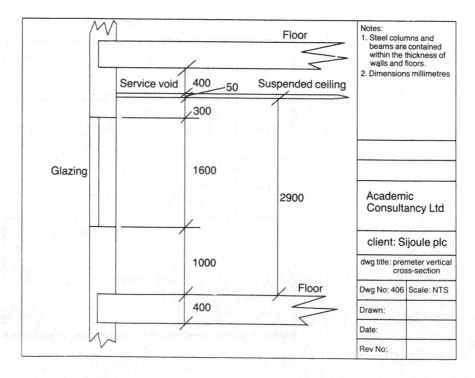

Fig. 1.26 Perimeter vertical cross-section.

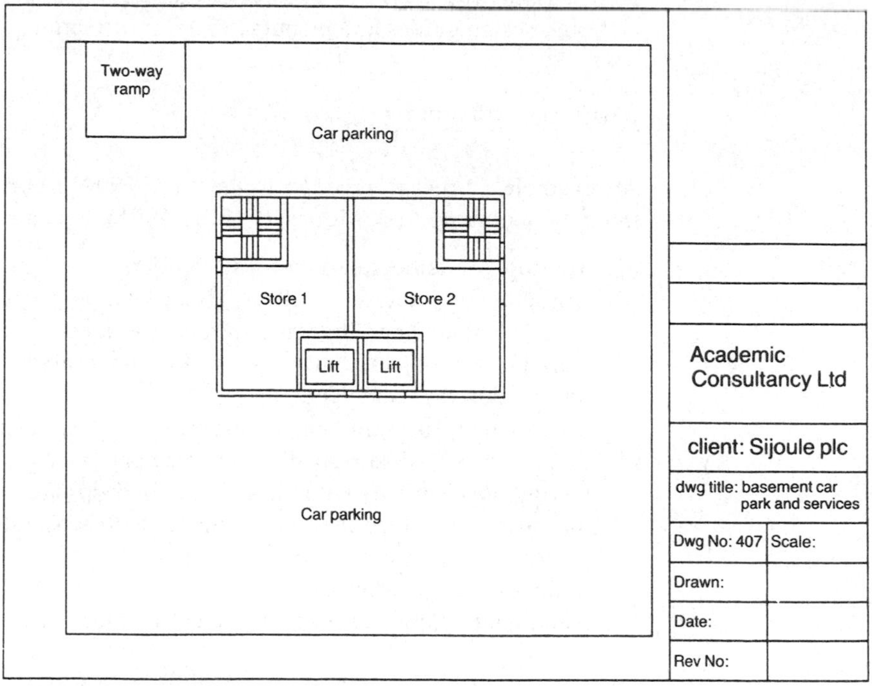

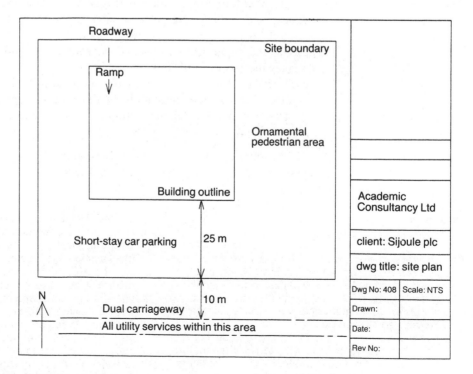

Fig. 1.27 Basement plan.

Fig. 1.28 Site plan.

be discussed with the tutor and colleagues, and use should be made of reference material, design guides and manufacturers' information.

AIRPORT SYSTEM

An example of recent practice is the Euro-Hub airport terminal at Birmingham International Airport (B. Morris) (CIBSE, 1991), which has the following features:

1. rooftop low-temperature hot-water boilers;
2. constant-temperature, variable-volume heating water circulation to VAV reheat coils, plant air-handling units and space heaters;
3. variable water temperature weather-compensated heating circulation to perimeter skirting convectors;
4. rooftop reciprocating compressor water-chilling refrigeration plant;
5. fan-assisted VAV air-conditioning terminal boxes in public areas;
6. ground-level offices with fan-coil air conditioning;
7. the main concourse area has a single-duct constant-volume recirculation air-conditioning system;
8. toilet extract ventilation;
9. ventilation plant for kitchens, electric motor rooms and baggage-handling areas;
10. the ducted fresh air enters the plant through activated carbon filters to remove aircraft engine fumes;
11. the terminal building is maintained at positive static pressure above the outside atmosphere to stop the possible ingress of aircraft engine exhaust fumes through boarding gates and external doors;
12. arrival and departure passenger levels each have four-zone air-handling units serving the VAV controllers;
13. each air-handling unit has variable-speed supply and extract fans that are controlled to maintain constant main air- duct static pressures;
14. the fresh air supply is regulated from air-quality sensors to adjust for occupancy levels.

The total building cost was £2609/m^2 of floor area. Heating, cooling, ventilation and air conditioning cost £121/m^2. Other mechanical and electrical services cost £144.25/m^2.

Questions

The answers to descriptive questions are within the text of the chapter. The reader is advised to extend the range of study material to the CIBSE *Guides*, manufacturers' literature, *OPUS* and to discussions with colleagues. Answers to numerical questions are given in the **Answers** section on pp.304–307.

1. Discuss the statement 'this building is air conditioned' with reference to what this will mean, the systems that must be installed and the possible satisfaction of user thermal comfort.

2. Explain the difference between mechanical ventilation and air conditioning.
3. Define the term 'low-cost air conditioning'. State the ways in which it can be achieved.
4. List the reasons for the air conditioning of the different categories of building, such as residence, office, retail, containment and manufacturing, and write notes to explain each reason.
5. Study the drawings of the Si joule plc building and write lists and notes on the heating, ventilating and air-conditioning systems that are likely to be needed. List the assumptions that you make on the location and use of the building. Discuss

your results with colleagues. This exercise is suitable for groups of students to formulate different proposals and then present them to the class.

6. Summarize in your own words the considerations involved in making use of outdoor air to cool a building without the use of refrigeration. State the limitations inherent in such designs and the ways in which outdoor air can be cooled without refrigeration.

7. A single-duct air-conditioning system with recirculation is to have variable quantities of fresh air admitted into an office building when it is between 12 °C and 23 °C. Outside these air temperatures, the minimum outdoor air of 25% of the supply quantity is to be used. Outdoor air at 14 °C can be 100% of the supply air quantity to the rooms. When the fresh air reaches 21 °C outside, its proportion must commence reduction towards the minimum at 23 °C. Draw a graph of the damper operation to scale and explain fully the sequence of operation of the system during both increasing and reducing outdoor air temperatures.

8. Sketch the air-handling arrangement for a single-duct variable air temperature recirculation system, state the necessary components and explain how the design is used to satisfy the room cooling and heating loads.

9. List the reasons for creating air-conditioning zones and the zones that may be formed for a variety of comfort, industrial process and containment applications.

10. Create suitable air-conditioning zones for the Si joule plc building, stating the reasons, and draw them on the plans and elevations.

11. Calculate the peak solar heat gains through the unprotected windows for each side of the Si joule plc building using only the glazing heat gain data in the *Air-handling zones* section of this chapter. State the dates and times of these occurrences and the influence that they will have on the building peak refrigeration plant capacity.

12. State the reasons for creating different air static pressures in rooms or zones, explain how such differences can be maintained and state pertinent applications.

13. List the applications of the 16 types of air-conditioning system and discuss the suitability of each with colleagues.

14. Form syndicates of four students to propose air-conditioning designs for the Si joule plc building. Each member of the group is to assume a different role.
    (a) The building owner is to decide the usage of each area and the standard of internal environmental control to be achieved. Make limitations upon the intrusiveness, likely cost, maintenance or appearance of the systems to be proposed.
    (b) The consulting engineer, who is engaged for a fee calculated from a fixed percentage of the total cost of the installation, is to use the independence of this position to propose designs that will fully satisfy the client's stated requirements.
    (c) The design and installation contractor is engaged in a competitive situation against other, similar companies and is to propose designs that maximize the likelihood of gaining the contract.
    (d) The manufacturer of a wide range of air-conditioning products who offers a design service free of charge to the client.

    The three engineers are to formulate proposals for different air-conditioning designs and present them to the client with arguments in favour of their company's suitability. Presentations should be made on acetate sheets of the building drawings with coloured sketches of the systems proposed on the plans, sections and elevations for the benefit of the whole student group. It may be possible to integrate such group activity with architectural, building and surveying student groups.

15. 'The single-duct air-conditioning system provides the basic design for the other configurations that overcome its limitations.' Discuss this statement and state the ways in which it can be adapted for multizone applications.

16. The variable air volume system has become very popular for office accommodation. Explain its principles of operation and limitations. Include the topics of room air circulation, zone volume control, economy control of the fans, duct air static pressure modulation and the satisfaction of user thermal and aural comfort.

17. State the advantages that can be gained by using a water- source heat-pump air-conditioning system capable of simultaneous cooling and heating of adjacent zones. List suitable applications. Comment upon the maintenance required for such systems.

18. Explain with the aid of sketches and sample graphs how the supply-duct air temperatures are controlled in a dual-duct air- conditioning system to provide the maximum operational economy while satisfying the user's comfort needs.

19. List the applications for independent air-conditioning units, split systems, reversible heat pumps, chilled ceilings and district cooling, commenting upon their design characteristics and maintenance requirements. Acquire manufacturers' literature of a variety of equipment and apply them to the Si joule plc building.

20   Make sketches of the building your are familiar
     with and propose suitable designs for air-condi-
     tioning systems, drawing the components on
     plans and elevations.

2 Psychrometric design

INTRODUCTION

The properties of humid air are introduced and the use of the CIBSE psychrometric chart is thoroughly described in logical and manageable steps. Worked examples are used to demonstrate the calculations needed to find the physical properties of humid air. Calculated properties are compared with tabulated values and those read from the psychrometric chart. A supply of CIBSE psychrometric charts and access to the *CIBSE Guide* are needed for completion of the chapter. The use of sketch charts is introduced for explanatory purposes and for use during preliminary design.

Sufficient accuracy can be achieved for most learning purposes by means of the calculation techniques used, but for precise data and professional use, students should use either published CIBSE data or commercial computer programs.

The psychrometric processes of heating, mixing, cooling, dehumidification and humidification are described. Example calculations of conditions and complete processes are detailed; these processes are fundamental to air-conditioning systems.

LEARNING OBJECTIVES

Study of this chapter will enable the user to:

1. state the constituents of humid air;
2. understand the composition of atmospheric pressure;
3. use a sketch of the CIBSE psychrometric chart;
4. read a CIBSE psychrometric chart;
5. understand, use, calculate and read (from tables and charts) the properties of humid air – dry- and wet-bulb air temperatures, moisture content, percentage saturation, specific enthalpy, specific volume and density;
6. calculate and find the dew-point temperature of air;
7. calculate the properties of humid air from formulae;
8. understand the meaning of 'saturation state';
9. calculate and use air vapour pressure;
10. find all the physical properties of humid air with readings taken from a sling psychrometer;
11. know what affects the density of humid air;

12. understand the psychrometric processes of heating, mixing, cooling, humidification and dehumidification;

13. calculate the air-conditioning plant loads of heating, cooling and humidity-control processes;

14. understand the meaning of 'cooling coil contact factor';

15. compare steam humidification with that by water sprays.

Key terms and concepts

atmospheric pressure	38	relative humidity	41
dew point	45	specific enthalpy	44
dry-bulb temperature	38	specific volume	45
moisture content	38	vapour pressure	41
percentage saturation	39	wet-bulb temperature	41

PROPERTIES OF HUMID AIR

Understanding the use of the thermal properties of humid air is fundamental to a study of air conditioning. Formulae, data and psychrometric charts from the *CIBSE Guide, Volume A* (CIBSE, 1986a) will be referred to for the normal atmospheric pressure of 101 325 Pa (1013.25 mb) at sea level. All calculations, examples and questions in this book will use normal atmospheric pressure and the sling psychrometer temperatures unless stated otherwise. The standard of measurement is the kg of dry air, so specific volume and enthalpy are both per kg dry air. The reader requires a supply of CIBSE psychrometric charts to use during the worked examples, questions and for practical design work. A sketch of the psychrometric chart will suffice for much of the work and Fig. 2.1 can be reproduced for plotting approximate states and lines representing air-conditioning processes. In the early stages of design calculations, it is easier to use a freehand sketch than to be limited by the rigour of plotting exact locations. Precise data can be plotted subsequently on a full-size chart. Figure 2.2 is a reduced psychrometric chart that may be used to find data. It is reproduced by permission of the Chartered Institution of Building Services Engineers. There are computer programs to calculate and plot psychrometric states and lines. They have not replaced the need to handle the data and charts manually.

The uppermost curve of the psychrometric chart demonstrates the increasing capacity of air to hold moisture in suspension as the air temperature increases. This is the 100% saturation curve. Thermodynamic states to the left of the 100% curve represent fully saturated air (wet fog plus a pool of water on the floor), while those to the right are conditions that can be plotted from dry- and wet-bulb air temperatures. The chart is plotted from two linear axes. The horizontal scale is that of dry-bulb air temperature (t °C) with equally spaced increments of 1 °C with 0.5 °C subdivisions. The right-hand-side vertical scale is that of air moisture content in

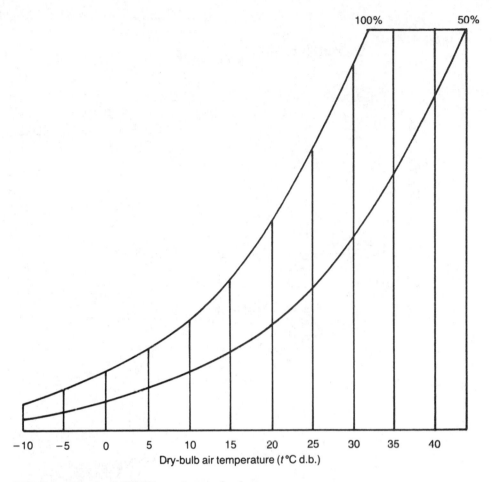

Fig. 2.1 Sketch of CIBSE psychrometric chart.

kg H_2O per kg of dry air (kg/kg). At each value of dry-bulb temperature, air can sustain a maximum moisture content, in the form of water vapour, of g_s kg/kg. If the air is less than saturated, then its moisture content will be g kg/kg, and it will have a percentage saturation PS of

$$PS = 100 \times g/g_s \%$$

EXAMPLE 2.1

Air in an occupied room is at a temperature of 20 °C d.b. and has a moisture content of 0.007 376 kg/kg. When air at 20 °C d.b. is fully saturated, it can hold 0.014 75 kg/kg. Calculate the percentage saturation of the room air. Check the answer with a psychrometric chart and the data tables of the properties of humid air (CIBSE, 1986c).

$$g = 0.007\,376 \text{ kg/kg} \quad \text{and} \quad g_s = 0.014\,75 \text{ kg/kg}$$

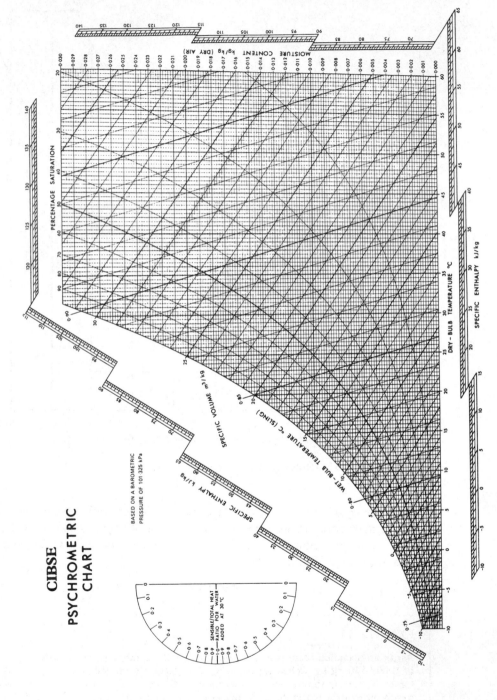

Fig. 2.2 CIBSE psychrometric chart (© CIBSE, London, 1989).

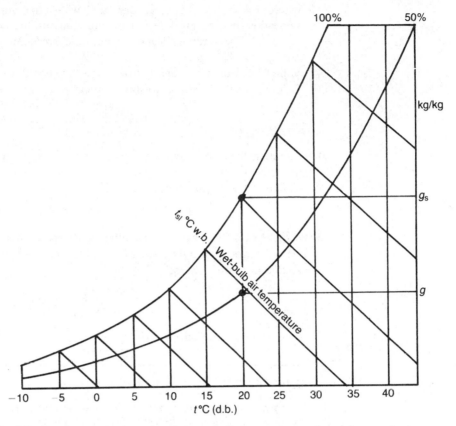

Fig. 2.3 Psychrometric chart showing wet-bulb temperature and moisture content.

$$PS = 100 \times \frac{0.007\,376}{0.014\,75}\,\%$$

$$= 50\,\%$$

Relative humidity, RH%, is used extensively and is found from the ratio of vapour pressure to saturation vapour pressure at the dry-bulb temperature:

$$RH = 100 \times \frac{p_v}{p_s}\,\%$$

The wet-bulb air-temperature scale, t_{sl} °C, is linear but sloping at 35 ° downwards to the right and has 1 °C increments. Wet- and dry-bulb temperatures coincide at 100% percentage saturation as shown in Fig. 2.3. Plotting an air condition with these two values allows all other data to be calculated or read from the chart or data tables.

Moisture content is calculated from the air vapour pressure, p_v Pa, and the air vapour pressure at saturation, p_s Pa. The procedure is as follows.

1. Find the values of t °C and t_{sl} °C.
2. Saturation vapour pressure p_s kPa is found from

$$\log p_s = 30.590\,51 - 8.2 \times \log T + 2.4804 \times 10^{-3}\,T - \frac{3142.31}{T}$$

where $T = (°C + 273.15)$ K. Dry- and wet-bulb temperatures, t, are equal at saturation, but the air is less than saturated and the percentage saturation is to be found. The saturated vapour pressure corresponds to the dry-bulb temperature, so $t = t$ °C d.b.

Vapour pressures will nearly always be between 0 mb and 70 mb.

3. Calculate the saturation moisture content g_s from

$$g_s = \frac{0.621\,97\,p_s}{(1013.25 - p_s)} \quad kg/kg$$

4. Calculate the saturation vapour pressure p_{sl} mb at the wet-bulb temperature from the formula above.

5. Calculate the actual vapour pressure p_v using the dry-bulb t and wet-bulb t_{sl} from

$$p_v = p_{sl} - 1013.25 \times 6.66 \times 10^{-4} \times (t - t_{sl}) \quad mb$$

If t_{sl} is below 0 °C, calculations relate to ice rather than water and other formulae are used. We shall only concern ourselves with liquid states.

6. Actual moisture content:

$$g = \frac{0.621\,97\,p_v}{1013.25 - p_v} \quad kg/kg$$

7. Percentage saturation $PS = 100\dfrac{g}{g_s}$ %

EXAMPLE 2.2

A sling psychrometer shows that the air condition in an occupied room is 22 °C d.b. and 17 °C w.b. Calculate the percentage saturation and relative humidity. Check the values with published data.

At the dry-bulb temperature

$$\log p_s = 30.590\,51 - 8.2 \times \log(22 + 273.15) + 2.4804 \times 10^{-3} \times (22 + 273.15) - \frac{3142.31}{22 + 273.15}$$

$$\log p_s = 0.4214$$

$$\text{and } p_s = 2.638 \text{ kPa}$$

$$g_s = \frac{0.621\,97\,p_s}{1013.25 - p_s}$$

$$= \frac{0.621\,97 \times 2.638}{101.325 - 2.638}$$

$$= 0.016\,63 \text{ kg/kg}$$

Using the wet-bulb temperature $t_{sl} = 17$ °C

$$\log p_s = 30.590\,51 - 8.2 \times \log(17 + 273.15) + 2.4804 \times 10^{-3}$$

$$\times (17 + 273.15) - \frac{3142.31}{17 + 273.15}$$

$$\log p_s = 0.2867$$

$$p_s = 1.935$$

$$= 1.935 \, \text{kPa}$$

$$p_v = p_s - 101.325 \times 6.66 \times 10^{-4} \times (t - t_{sl}) \, \text{kPa}$$

$$= 1.935 - 101.325 \times 6.66 \times 10^{-4} \times (22 - 17) \, \text{kPa}$$

$$= 1.6 \, \text{kPa}$$

$$g = \frac{0.621\,97 \, p_v}{101.325 - p_v} \, \text{kg/kg}$$

$$= \frac{0.621\,97 \times 1.6}{101.325 - 1.6}$$

$$= 0.01 \, \text{kg/kg}$$

$$PS = 100 \times g/g_s$$

$$= 100 \times \frac{0.010}{0.016\,63} \, \%$$

$$= 60\%$$

$$RH = 100 \frac{p_v}{p_s}$$

$$= 100 \times \frac{1.6}{2.638}$$

$$= 60.7\%$$

Calculated data agree with those published.

The density ρ kg/m^3 of humid air is the reciprocal of specific volume v m^3/kg and it depends upon the following factors.

Dry-bulb temperature

This changes significantly throughout most air-conditioning systems because of mixing, heating and cooling coil processes, the turbulent heating effect of the fan, heat generation by the fan motor and heat exchanges with the ambient environment through the duct wall.

Atmospheric pressure

Density varies linearly with total atmospheric pressure, increasing as pressure increases and reducing as pressure reduces. Height above sea level of a building determines the local atmospheric pressure. The weather can change the normal value by up to 10% owing to high or low atmospheric pressure variations. Increases in wind velocity produce reductions in atmospheric (static) pressure according to Bernoulli's theorem. Prevailing wind force on the side of a building temporarily creates positive or negative pressure changes. These possible changes are usually ignored. Standard psychrometric data are usually employed.

Moisture content

Water vapour in air exists at its own partial pressure. Water vapour has a lighter molecular mass (18.015 kg/kmol) than dry air (29 kg/kmol) (Rogers and Mayhew, 1987). Thus the greater the amount of water vapour in dry air, the less the mixture weighs. As g kg H_2O/kg dry air increases, so does the specific volume v m^3/kg, and density ρ kg/m^3 reduces.

The standard condition for stating air density is 0 °C d.b. and 101.325 kPa atmospheric pressure, but 1.013 25 bar, 1013.25 mb, 101 325 Pa, 101 325 N/m^2, 10.33 m H_2O and 760 mmHg are also used. The standard density of dry air can be found from the general gas law

$$Pv = mRT$$

$$\rho = \frac{m}{v} = \frac{P}{RT}$$

$$= 101\,325\,\frac{N}{m^2} \times \frac{kg\,K}{287.1\,kJ} \times \frac{1}{273\,K}$$

$$= 1.293\,kg/m^3$$

EXAMPLE 2.3

Find the densities of humid air at 25 °C d.b. when it is at 20% saturation and then when it is at 23 °C w.b.

From CIBSE (1986c), Table C1, at 25 °C d.b. and 20% saturation:

$$v = 0.8498\,m^3/kg$$

$$\rho = 1/0.8498\,kg/m^3$$

$$= 1.177\,kg/m^3$$

From the same table, at 25 °C d.b., 23 °C w.b.:

$$v = 0.8672\,m^3/kg$$

$$\rho = 1.153\,kg/m^3$$

Reading the psychrometric chart shows specific volume lines at 73 ° sloping downwards to the right, spaced at 0.01 m^3/kg, of 0.85 and 0.87 m^3/kg for these air conditions.

The specific enthalpy of humid air can be approximated from

$$h = 1.0048t + g\,(2500.8 + 1.863t)\ kJ/kg$$

where 1.0048 is the specific heat of dry air at constant pressure at 293 K and 2500.8 is the specific enthalpy of dry saturated steam at 0 °C (Rogers and Mayhew, 1987). For the air condition in Example 2.2

$$h = 1.0048 \times 22 + 0.01 \times (2500.8 + 1.863 \times 22)\ kJ/kg$$

$$= 47.52\,kJ/kg$$

The data table shows 47.64 kJ/kg. The 0.3% error is due to the simplification and rounding of decimal places made. CIBSE data are to be used where accuracy has to be assured.

The specific volume of humid air can be found from the general gas law

$$Pv = mRT$$

where P = absolute pressure of dry air (Pa), = atmospheric pressure − vapour pressure = $101\,325 - p_v$ (Pa); v = specific volume of 1 kg of dry air (m³/kg); m = 1 kg dry air; R = specific gas constant, 0.2871 kJ/kg K, 287.1 J/kg K (Rogers and Mayhew, 1987); T = absolute gas temperature (K). So

$$v = mR\frac{T}{P}\ \text{m}^3/\text{kg}$$

EXAMPLE 2.4

Calculate the specific volume of humid air at 22 °C d.b., 17 °C w.b. and vapour pressure 16 mb.

---

1 bar = 100 000 Pa, 1 mb = 100 Pa, 1 Pa = 1 N/m², 1 J = 1 Nm.

$$p_s = 16\,\text{mb} \times 100\,\text{Pa/mb}$$

$$= 1600\,\text{Pa}$$

dry air partial pressure $P = 101\,325 - 1600\,\text{Pa}$

$$= 99\,725\,\text{Pa}$$

$$= 99\,725\,\text{N/m}^2$$

$$v = 1\,\text{kg} \times 287.1\frac{\text{J}}{\text{kg K}} \times 295\,\text{K} \times \frac{\text{m}^2}{99\,725\,\text{N}} \times \frac{1\,\text{Nm}}{1\,\text{J}}$$

$$v = 0.8493\,\text{m}^3\ \text{per kg dry air}$$

This agrees with tabulated data.

Specific volume can be read from the psychrometric chart as shown in Fig. 2.4. The dew-point temperature t_{dp} of humid air can be found:

1. by inspecting the tabulated data;
2. by moving horizontally leftwards from the air condition on the psychrometric chart until the 100% curve is reached (this means the temperature at which the moisture held in the air will begin to condense);
3. by using an equation having sufficiently close agreement with the 100% saturation curve that enables direct calculation of the dew point from the saturated air vapour pressure p_s Pa:

$$t_{dp} = 14.62\ln(p_s/600.245)\ °\text{C}$$

EXAMPLE 2.5

Find the dew-point temperature of humid air at 22 °C d.b., 17 °C w.b. and vapour pressure 16 mb using the three methods described.

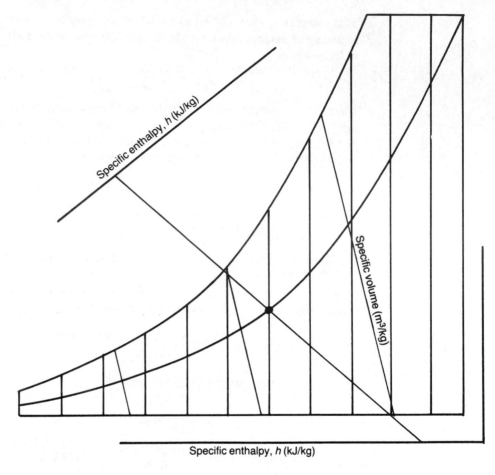

Fig. 2.4 Psychrometric chart showing specific enthalpy and specific volume.

Method 1
22 °C d.b. and 17 °C w.b.: reference to the table shows a dew point of 14 °C.

Method 2
Figure 2.5 demonstrates that the dew point is 14 °C.

Method 3

$$t_{dp} = 14.62 \ln(p_s/600.245) \text{ °C}$$

The p_s of 1600 Pa becomes the saturated vapour pressure as the local air temperature has been lowered to the dew point that lies on the 100% curve; so

$$p_s = p_v$$

$$= 1600 \text{ Pa}$$

$$t_{dp} = 14.62 \ln(1600/600.245) \text{ °C}$$

$$t_{dp} = 14.3 \text{ °C}$$

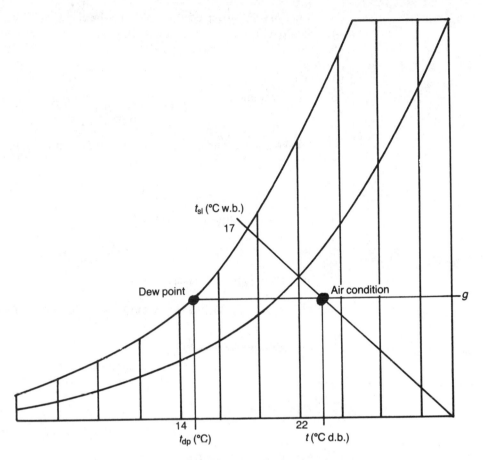

Fig. 2.5 Psychrometric chart showing the location of dew-point temperature.

The three methods agree.

When the dew point is known, the air vapour pressure can be found from a transposition of the curve-fit equation

$$p_s = 600.245 \exp(0.0684 \, t_{dp}) \, \text{Pa}$$

For $t_{dp} = 14.3 \, °C$

$$p_s = 600.245 \times \exp(0.0684 \times 14.3) \, \text{Pa}$$

$$p_s = 1596 \, \text{Pa}$$

If the further decimal places in the dew point value are used, the original 1600 Pa is calculated.

Summary of formulae

The quantities and formulae used are as follows.

1. $t \, °C$ dry-bulb temperature from a mercury-in-glass thermometer.
2. $t_{sl} \, °C$ wet-bulb temperature from a mercury-in-glass thermometer covered by a saturated wick.

3. Sling psychrometer used for t °C dry-bulb and t_{sl} °C wet-bulb.
4. Sea-level atmospheric pressure 101 325 Pa, 1013.25 mb.
5. Absolute temperature:

$$T = (°C + 273.15) \text{ K.}$$

6. Gas constant:

$$R = 287.1 \text{ J/kg K}$$

7. Air moisture content g kg H_2O per kg of dry air, kg/kg.
8. Percentage saturation:

$$PS = 100 \frac{g}{g_s} \%$$

9. Relative humidity:

$$RH = 100 \frac{p_v}{p_s} \%$$

10. Saturation vapour pressure p_s kPa:

$$\log p_s = 30.59051 - 8.2 \log T + 2.4804 \times 10^{-3} T - \frac{3142.31}{T}$$

11. Saturation moisture content:

$$g_s = \frac{0.62197 p_s}{1013.25 - p_s} \text{ kg/kg}$$

12. Vapour pressure:

$$p_v = p_{sl} - 1013.25 \times 6.66 \times 10^{-4} \times (t - t_{sl}) \text{ mb}$$

13. Actual moisture content:

$$g = 0.62197 \frac{p_v}{1013.25 - p_v} \text{ kg/kg}$$

14. Density:

$$\rho = \frac{p}{RT} \text{ kg/m}^3$$

15. Specific volume $v = \frac{1}{\rho}$ m³/kg
16. Specific enthalpy:

$$h = 1.0048\, t + g\, (2500.8 + 1.863\, t) \text{ kJ/kg}$$

PSYCHROMETRIC PROCESSES

Air conditions and processes of change are represented by coordinates and lines on the psychrometric chart shown in simplified form in Fig. 2.6. A typical room condition of 21 °C d.b. and 50% saturation is shown along with the other properties. When calculating the heating or cooling loads on coils, the inlet air specific volume v_1 m³/kg is used to convert the air volume flow rate into mass flow rate (kg/s). Changes in specific volume through the coil are dealt with by the coil manufacturer.

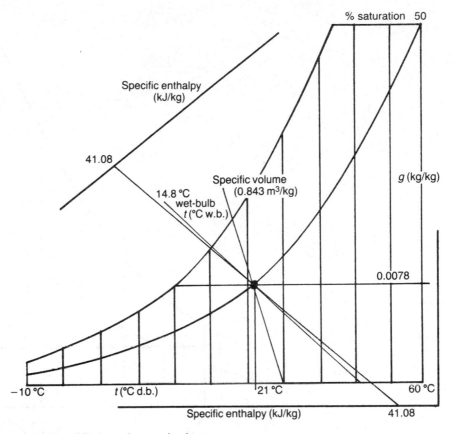

Fig. 2.6 Simplified psychrometric chart.

HEATING

Heating of air up to 40 °C d.b. is performed by low- or high-pressure hot-water finned pipe heater, battery, electric resistance elements or fuel-fired heat exchangers. Figure 2.7 shows that air at condition 1 is being raised in dry-bulb temperature and in specific enthalpy to condition 2, while the moisture content remains constant. The percentage saturation reduces because the saturation moisture content increases exponentially while the actual moisture content remains at the original level. The specific volume increases as the air expands; conversely, the air density reduces.

EXAMPLE 2.6.

1 m³/s of outdoor air at −2 °C d.b., −3 °C w.b. and 80% saturation enters a heating coil to be raised to 30 °C d.b. and supplied to an office. Calculate the heating load in kW, sketch the process on a psychrometric chart and identify all the properties of the air before and after heating.

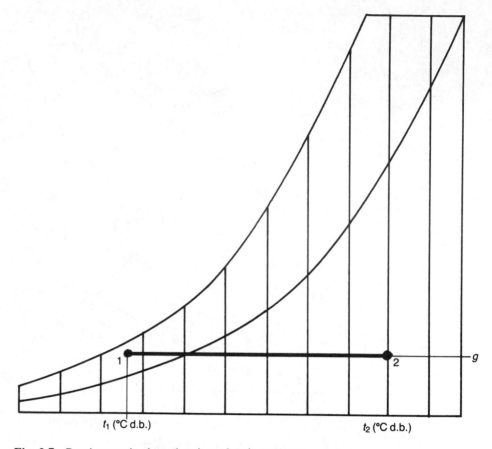

Fig. 2.7 Psychrometric chart showing a heating process.

Tabulated data for entering air at condition 1: − 2 °C d.b.; 80% saturation; − 3 °C w.b.; 0.002 564 kg/kg; 4.392 kJ/kg; 0.7708 m³/kg; 0.4141 kPa vapour pressure; − 4.6 °C dew point.

Tabulated data for heated air at condition 2: 30 °C d.b.; the moisture content is known to be the same as at 1, so is 0.002 564 kg/kg; the table for 30 °C d.b. has air with this moisture content at around 10% saturation; linear interpolation between 8% and 12% produces 13.6 °C w.b;, 37.17 kJ/kg; 0.862 m³/kg; 0.4141 mb vapour pressure; − 4.1 °C dew point that should be the same as at 1, − 4.6 °C. The erroneous dew point demonstrates that linear interpolation of exponential curve data suffers from inaccuracy.

The mass flow of air into the heater battery,

$$M = \frac{1 \text{ m}^3/\text{s}}{0.7708 \text{ m}^3/\text{kg}}$$

$$= 1.297 \text{ kg/s}$$

specific enthalpy rise $= h_2 - h_1$

$$= 37.17 - 4.392 \text{ kJ/kg}$$

$$= 32.778 \text{ kJ/kg}$$

$$\text{heater input power} = 1.297 \frac{\text{kg}}{\text{s}} \times 32.778 \frac{\text{kJ}}{\text{kg}} \times \frac{\text{kW s}}{\text{kJ}}$$

$$= 42.513 \text{ kW}$$

Values can also be calculated for condition 1:

$$\log p_s = 30.59051 - 8.2 \times \log(-2 + 273.15) + 2.4804 \times 10^{-3}$$

$$\times (-2 + 273.15) - \frac{3142.31}{-2 + 273.15}$$

$$\log p_s = -0.2781$$

and

$$p_s = 0.527 \text{ kPa}$$

$$g_s = \frac{0.62197 \, p_s}{101.325 - p_s}$$

$$= \frac{0.62197 \times 0.527}{101.325 - 0.527}$$

$$= 0.00325 \text{ kg/kg}$$

Using the wet bulb temperature, $t_{sl} = -3 \,^\circ\text{C}$:

$$\log p_s = 30.59051 - 8.2 \times \log(-3 + 273.15) + 2.4804 \times 10^{-3}$$

$$\times (-3 + 273.15) - \frac{3142.31}{-3 + 273.15}$$

$$\log p_s = -0.31$$

$$p_s = 0.489 \text{ kPa}$$

$$p_v = p_s - 1013.25 \times 6.66 \times 10^{-4} \times (t - t_{sl}) \text{ mb}$$

$$= 0.489 - 101.325 \times 6.66 \times 10^{-4} \times (-2 - -3) \text{ mb}$$

$$= 0.422 \text{ kPa}$$

$$g = \frac{0.62197 \, p_v}{101.325 - p_s} \text{ kg/kg}$$

$$= \frac{0.62197 \times 0.422}{101.325 - 0.422}$$

$$= 0.0026 \text{ kg/kg}$$

$$\text{PS} = 100 \frac{g}{g_s}$$

$$= 100 \frac{0.0026}{0.00325} \%$$

$$= 80\%$$

$$h = 1.0048t + g\,(2500.8 + 1.863t)\ \text{kJ/kg}$$
$$= 1.0048 \times -2 + 0.0026 \times (2500.8 + 1.863 \times -2)\ \text{kJ/kg}$$
$$= 4.48\ \text{kJ/kg}$$
$$v = 287.1 \times \frac{273 - 2}{101\,325 - 0.4223 \times 1000}\ \text{m}^3/\text{kg}$$
$$= 0.771\ \text{m}^3/\text{kg}$$
$$t_{dp} = 14.62 \times \ln\left(\frac{422.3}{600.245}\right)\ °\text{C}$$
$$= -5.1\ °\text{C}$$

Values that are calculated for condition 2:

$$\log p_s = 30.59051 - 8.2 \times \log(30 + 273.15) + 2.4804 \times 10^{-3}$$
$$\times (30 + 273.15) - \frac{3142.31}{30 + 273.15}$$
$$\log p_s = 0.627$$

and

$$p_s = 4.24\ \text{kPa}$$
$$g_s = \frac{0.621\,97\,p_s}{101.325 - p_s}$$
$$= \frac{0.621\,97 \times 4.24}{101.325 - 4.24}$$
$$= 0.027\,16\ \text{kg/kg}$$

Using the wet bulb temperature, $t_{sl} = 13.6\ °\text{C}$, read from the tables as this cannot be calculated:

$$\log p_s = 30.590\,51 - 8.2 \times \log(13.6 + 273.15) + 2.4804 \times 10^{-3}$$
$$\times (13.6 + 273.15) - \frac{3142.31}{13.6 + 273.15}$$
$$\log p_s = 0.192$$
$$p_s = 1.556\ \text{kPa}$$
$$p_v = p_s - 101.325 \times 6.66 \times 10^{-4} \times (t - t_{sl})\ \text{kPa}$$
$$= 1.556 - 101.325 \times 6.66 \times 10^{-4} \times (30 - 13.6)\ \text{kPa}$$
$$= 0.449\ \text{kPa}$$
$$g = \frac{0.621\,97\,p_v}{101.325 - p_v}\ \text{kg/kg}$$
$$= \frac{0.621\,97 \times 0.449}{101.325 - 0.449}$$
$$= 0.002\,77\ \text{kg/kg}$$

$$PS = 100 \frac{g}{g_s}$$

$$= 100 \frac{0.002\,77}{0.027\,16} \%$$

$$= 10.2\%$$

$$h = 1.0048t + g\,(2500.8 + 1.863\,t)\ \text{kJ/kg}$$

$$h = 1.0048 \times 30 + 0.0027 \times (2500.8 + 1.863 \times 30)\ \text{kJ/kg}$$

$$= 37.05\,\text{kJ/kg}$$

$$v = 287.1 \times \frac{273 + 30}{101325 - 4.49 \times 100}\ \text{m}^3/\text{kg}$$

$$= 0.862\,\text{m}^3/\text{kg}$$

$$t_{dp} = 14.62 \times \ln\left(\frac{449}{600.245}\right) °\text{C}$$

$$= -4.2\,°\text{C}$$

The results show sufficiently good agreement.

COOLING

Chilled water, brine or refrigerant is passed through finned-pipe cooling coils in the air stream. If the coolant is above the air dew-point temperature, no condensation takes place, the process is sensible cooling only and the air remains at a constant moisture content. Figure 2.8 also shows the case when the coolant is below the air dew point and dehumidification takes place. This is the normal case for comfort air conditioning.

The precise path of the air being cooled from 1 to 2 is not easily defined as sensible and latent heat exchanges are taking place that involve the transfer of moisture mass from the humid air to the drip tray. Water is being squeezed out of the air quickly when particles of air contact the cold surfaces, but more slowly by forced convective cooling of the air that passes between the pipes and fins. A dehumidifying coil operates wet and the exposed water surfaces increase the cooling area. There is little to be gained by attempting to plot detailed states between 1 and 2, unless a coil is being designed. The end conditions need only be considered. A slight curve is generated by the reducing performance of the second and subsequent rows of finned tubes in the cooling coil (Jones, 1985).

When designing the cooling and dehumidification process, the air condition that enters the coil, point 1, is known from the room and external air conditions, but the coil exit condition is not. The moisture content required upon leaving the coil, g_2, will be known, but an amount of variation may be allowable owing to a tolerance in the room percentage saturation of up 5% either side of the target 50%. For economy of refrigeration plant and operating costs of the air-conditioning system,

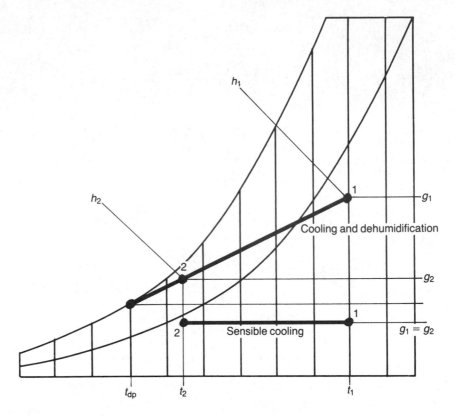

Fig. 2.8 Cooling and dehumidification psychrometric processes.

the refrigerant temperature needs to be as high as possible, consistent with the plant capacity to meet the cooling load.

The air entering the coil will be cooled towards the saturation curve. A straight line will be used to represent this process. The intersection of the line with the 100% curve will be the dew-point temperature of the cooling coil. This dew point will be close to the lowest of the coolant temperatures and this should be found in the last row of finned pipes. There will be a small temperature rise between the coolant and coil dew point due to heat transfer through the pipe wall and along the fin. When refrigerant is within the pipes, it evaporates at constant temperature and pressure within the coil at around 5–10 °C for comfort systems but subzero for food chillers. Chilled water is often supplied to a coil at 6 °C and leaves at 10 °C. The coil dew point will be within 0.5 °C of the lower coolant value. Incomplete mixing of the turbulent air flowing over the coil will mean that not all the air will be subjected to the lowest attainable temperature. The leaving condition, position 2, will be at about 90% of the distance between the entering and dew-point states. This 90% is known as the coil contact factor; its value depends upon coil design in terms of pipe and fin spacing, surface corrugations, number of rows of coils, direction of flow of the cooling medium, staggering of pipes in the rows and the average face velocity of the air entering from the duct, around 2.5 m/s. Contact factor can be scaled along the three linear axes of dry-bulb temperature, moisture content or specific enthalpy. Dry-bulb temperature is normally used.

EXAMPLE 2.7

Summer outdoor air at 32 °C d.b. and 22 °C w.b. is cooled and dehumidified by a chilled-water cooling coil before being passed into an office at a temperature of 16 °C d.b. The cooling coil has a dew point of 5 °C. Sketch the psychrometric cycle. Find the outside air and supply-air properties. Calculate the contact factor for the coil and the amount of moisture removed per kg of dry air.

The procedure is as follows.

1. Sketch Fig. 2.9 to show the cooling process.
2. Plot the data on a real chart and sketch.
3. Read the tabulated data for condition 1: 32 °C d.b., 22 °C w.b., 40% saturation, 0.012 31 kg/kg, 63.7 kJ/kg, 0.8812 m³/kg, 1.957 kPa vapour pressure, 17.2 °C dew point.
4. Read the properties for point 1 from the chart and compare for accuracy with the tabulated figures.
5. Plot the cooling-coil dew point of 5 °C on the 100% curve.

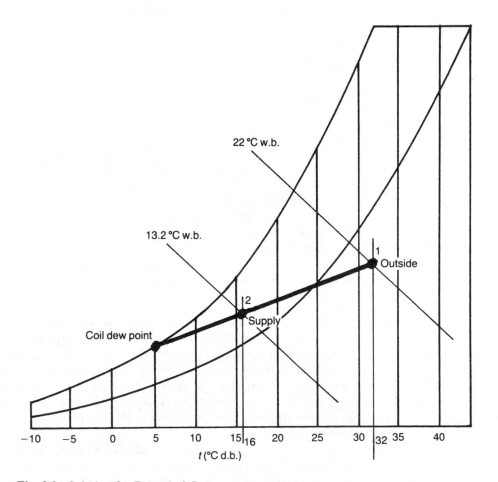

Fig. 2.9 Solution for Example 2.7.

6. Use the cooling-coil manufacturers' performance curve or draw a straight line from point 1 to the coil dew point in the first instance.
7. Plot the intersection of this cooling process with a vertical from 16 °C d.b. This is point 2 and represents the air condition leaving the coil.
8. Read the wet-bulb temperature of point 2 from the chart, 13.2 °C w.b.
9. Read the tabulated properties of point 2, interpolating as necessary: 16 °C d.b., 13.2 °C w.b., 73% saturation, 0.0083 kg/kg, 37.11 kJ/kg, 0.83 m³/kg, 1.329 kPa, 11.2 °C dew point.
10. Check the same values on the chart.
11. The contact factor of the cooling coil is found by ratio along the dry bulb scale:

$$\text{contact factor} = 100 \times \frac{32 - 16}{32 - 5}$$

$$= 59\%$$

12. The amount of moisture removed from the air as it is cooled:

$$\text{condense collected} = g_1 - g_2 \text{ kg/kg}$$

$$= 0.01231 - 0.0083 \text{ kg/kg}$$

$$= 0.00401 \text{ kg/kg}$$

MIXING

The mixing of air streams can take place when:

1. the fresh air drawn into an air-handling unit is mixed with recirculated room air prior to conditioning and supply into the building;
2. supply air is blown through supply grilles or diffusers into the room;
3. exhaust air is discharged from the building into the external atmosphere;
4. the discharge air from an external dry or evaporative cooling tower is released into the atmosphere;
5. steam from cooking equipment is released;
6. opening doors between rooms connects areas held at different temperature, humidity or pressure states;
7. air is extracted from different rooms and is mixed in the recirculation ductwork;
8. air from different parts of the same room is drawn towards the extract grill and into ductwork.

Whenever such a mixing process takes place, changes occur in the air condition. Severe cases may produce condensation if highly humid warm vapour discharges from cooking ranges and is mixed with cool air from outdoors or another room. Deposition of grease, condensation or dust particles can occur. In comfort air conditioning, room air extracted into the return-air duct may or may not accurately represent the room-air condition around the occupants owing to the relative mixing of air from different parts of the room. All examples are treated on their merits and importance.

The blending of two airstreams is often an adiabatic mixing process. Little or no heat is transferred outwards from the process or inwards to it. Either the air-handling unit has an insulated casing or the process is sufficiently rapid and is within a small

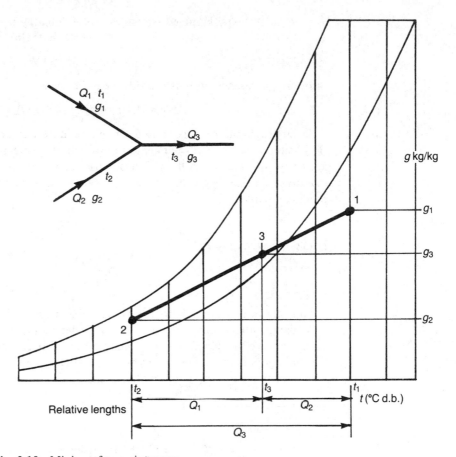

Fig. 2.10 Mixing of two airstreams.

space. Figure 2.10 shows a mixing process where there is continuity of mass of air and it is adiabatic.

Mixed mass flow = mass flow of stream 1 + mass flow of stream 2
$$M_3 = M_1 + M_2 \text{ kg/s}$$

It is important to summate mass flow rates for accurate results although an approximation can be made by adding volume flow rates. The inaccuracy here will be due to the changes in air density as the temperatures change during the mixing process.

$$Q_3 = Q_1 + Q_2 \text{ m}^3\text{/s}$$

Mass and volume flow rates are calculated from

$$M \text{ kg/s} = Q \text{ m}^3\text{/s} \times \rho \text{ kg/m}^3$$

and additive mass flows are

$$Q_3 \, \rho_3 = Q_1 \, \rho_1 + Q_2 \, \rho_2 \text{ kg/s}$$

By conservation of energy at the mixing, an enthalpy balance will be

$$\text{total enthalpy in mixed stream 3} = \text{total enthalpy of stream 1}$$
$$+ \text{total enthalpy of stream 2}$$

$$M_3 h_3 = M_1 h_1 + M_2 h_2 \text{ kW}$$

Similarly, a balance of moisture flows can be made:

$$M_3 g_3 = M_1 g_1 + M_2 g_2 \text{ (kg } H_2O)/s$$

An approximation to the mixed air temperature can be made from the volume flow rates and dry bulb temperatures of the two streams, often with sufficient accuracy for plotting on the psychrometric chart, that is to within 0.5 °C d.b., by making the assumption that the air densities and specific heat capacities remain practically constant.

$$Q_3 t_3 = Q_1 t_1 + Q_2 t_2$$

Divide each side by Q_3:

$$t_3 = \frac{Q_1}{Q_3} t_1 + \frac{Q_2}{Q_3} t_2$$

This shows that the mixing process is a linear ratio of the dry-bulb temperature scale using either volume or mass flow rates depending upon the accuracy required. These relative line lengths are indicated in Fig. 2.10. It is similarly possible to scale the moisture content and specific humidity scales. The mixed air condition lies on a straight line connecting the two end states.

EXAMPLE 2.8

Outdoor air at -1 °C d.b. and 100% saturation enters a mixing box at 1 m³/s when measured at the outdoor air condition. Recirculated room air enters the mixing box at 22 °C d.b. and 50% saturation at 2 m³/s at the room condition. Calculate the mixed air condition using accurate and approximate methods and plot the data on a psychrometric chart.

The solution procedure is as follows.

1. Plot the room air condition, 1, and outdoor air, 2, on a sketch of the psychrometric chart.
2. Plot the conditions on a real chart.
3. Connect the two states with a straight line.
4. Estimate the mixed air volume flow rate Q_3:

$$Q_3 = Q_1 + Q_2 \text{ m}^3/s$$

$$= 2 + 1 \text{ m}^3/s$$

$$= 3 \text{ m}^3/s$$

5. Carry out an approximation to the mixed air temperature:

$$t_3 = \frac{Q_1}{Q_3} t_1 + \frac{Q_2}{Q_3} t_2$$

$$t_3 = \frac{2}{3} \times 22 + \frac{1}{3} \times -1$$

$$t_3 = 14.3\,^{\circ}\text{C d.b.}$$

6. Calculate the mass flow rate of air from the room:

$$M_1 \text{ kg/s} = Q_1 \text{ m}^3/\text{s} \times \rho_1 \text{ kg/m}^3$$

$$\rho_1 = 1/v_1$$

From the tables and chart for room air at 22 °C d.b. and 50% saturation:

$$v_1 = 0.847\,\text{m}^3/\text{kg}$$

$$\rho_1 = 1.1806\,\text{kg/m}^3$$

$$M_1 = 2\,\text{m}^3/\text{s} \times 1.1806\,\text{kg/m}^3$$

$$= 2.361\,\text{kg/s}$$

7. Calculate the mass flow rate of air from outdoors:

$$v_2 = 0.7748\,\text{m}^3/\text{kg}$$

$$M_2 = \frac{1\,\text{m}^3/\text{s}}{0.7748\,\text{m}^3/\text{kg}}$$

$$= 1.291\,\text{kg/s}$$

8. Find the total mass flow rate:

$$M_3 = M_1 + M_2 \text{ kg/s}$$

$$= 2.361 + 1.291\,\text{kg/s}$$

$$= 3.652\,\text{kg/s}$$

9. Read the moisture contents from tables:

$$g_1 = 0.008\,366\,\text{kg/kg}$$

$$g_2 = 0.003\,484\,\text{kg/kg}$$

10. Calculate the moisture content of the mixed air:

$$M_3 g_3 = M_1 g_1 + M_2 g_2 \text{ kg/s}$$

$$3.652\,g_3 = 2.361 \times 0.008\,366 + 1.291 \times 0.003\,484$$

$$g_3 = \frac{0.024\,25}{3.652\,\text{kg/kg}}$$

$$= 0.006\,64\,\text{kg/kg}$$

11. Plot the intersection of the mixing process line previously drawn on the chart and the moisture content of the mixed air.
12. Read the mixed air temperature from the chart, 14 °C d.b.
13. Read the percentage saturation from the chart, 66%.
14. Check that the tabulated value for air at 14 °C has a corresponding moisture content and percentage saturation. It does to four significant decimal places.

15.   Read the specific enthalpy of the two air streams:

$$h_1 = 43.39\,\text{kJ/kg}$$

$$h_2 = 7.702\,\text{kJ/kg}$$

16.   Calculate the specific enthalpy of the mixture:

$$M_3 h_3 = M_1 h_1 + M_2 h_2 \,\text{kW}$$

$$3.652\, h_3 = 2.361 \times 43.39 + 1.291 \times 7.702$$

$$h_3 = 30.77\,\text{kJ/kg}$$

17.   Check h_3 with the table, 30.77 kJ/kg.
18.   Read v_3 from the table, $v_3 = 0.8217\,\text{m}^3/\text{kg}$
19.   Calculate the mixed air mass flow:

$$Q_3 = 3.652\,\text{kg/s} \times 0.8217\,\text{m}^3/\text{kg}$$

$$= 3\,\text{m}^3/\text{s}$$

STEAM HUMIDIFICATION

Adding moisture to air is most hygienically done with steam injection from electric-resistance heaters alongside the air duct. Potable cold water is taken from the mains drinking-water pipework at around 10 °C, below the temperature at which micro-organisms in the water become active. Raising water to 100 °C and then boiling it into steam at atmospheric pressure avoids bacterial contamination of the air if condensation of the injected moisture does not take place within the ducts. The supply of steam is carefully controlled for economy and health reasons. The specific enthalpy of steam is higher than that of the air being humidified. A little sensible heating takes place. The air moisture content increases. The dry-bulb temperature of the air remains close to its lower moisture-content value.

Consider the humidification to be an adiabatic process and there being no condensation of the steam. Figure 2.11 shows the general arrangement. Mass flow rate, enthalpy and moisture flow rate balances can be made:
Mass flow rates

$$M_1 + M_3 = M_2 \,\text{kg/s}$$

Enthalpy equation

$$M_1 h_1 + M_3 h_g = M_2 h_2 \,\text{kW}$$

Moisture mass flows

$$M_1 g_1 + M_3 = M_2 g_2 \,\text{kg/s}$$

The moisture contents g_1 and g_2 are known from the required air conditions. The mass flow rate of steam, M_3, needs to be found. The mixed mass flow rate M_2 is unknown. To find M_3:

$$g_1 M_1 + M_3 = M_2 g_2$$
$$= (M_1 + M_3)\, g_2$$
$$= g_2 M_1 + g_2 M_3$$

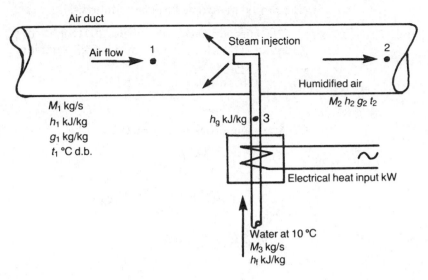

Fig. 2.11 Steam humidification.

By rearrangement:

$$M_3 - g_2 M_3 = g_2 M_1 - g_1 M_1$$

$$M_3 (1 - g_2) = M_1 (g_2 - g_1)$$

$$M_3 = M_1 \frac{g_2 - g_1}{1 - g_2} \text{ kg/s}$$

Calculate the leaving air mass flow rate M_2 and enthalpy h_2:

$$M_2 = M_1 + M_3 \text{ kg/s}$$

$$M_1 h_1 + M_3 h_g = M_2 h_2$$

h_1 is read from the chart or tables and h_g can be taken at 0 °C, 2500.8 kJ/kg (Rogers and Mayhew, 1987) although the water may normally be a few degrees warmer. h_2 can be found:

$$h_2 = \frac{M_1 h_1 + M_3 h_g}{M_2} \text{ kJ/kg}$$

The dry-bulb temperature of the humidified air can be calculated from the specific enthalpy formula:

$$h = 1.0048\, t + g\, (2500.8 + 1.863\, t) \text{ kJ/kg}$$

EXAMPLE 2.9

Outdoor air at a flow rate of 0.5m³/s, 20 °C d.b. and 10 °C w.b. leaves a preheater battery and is humidified to a moisture content of 0.0074 kg/kg by steam injection. Find the steam flow rate, humidified air specific enthalpy, leaving air condition and humidifier heater input power.

The procedure adopted can be as follows.

1. The inlet air data from tables: 20 °C d.b., 10 °C (10.1 used) w.b, 0.003 54 kg/kg, 29.1 kJ/kg, 0.8348 m³/kg, 24% saturation.

2. $M_1 = \dfrac{0.5\,\mathrm{m^3/s}}{0.8348\,\mathrm{m^3/kg}}$

   $= 0.599\,\mathrm{kg/s}$

3. After humidification, $g_2 = 0.0074\,\mathrm{kg/kg}$

4. $M_3 = M_1 \dfrac{g_2 - g_1}{1 - g_2}\,\mathrm{kg/s}$

   $= 0.599 \times \dfrac{0.0074 - 0.003\,54}{1 - 0.0074}\,\mathrm{kg/s}$

   $= 2.329 \times 10^{-3}\,\mathrm{kg/s}$

5. $M_2 = M_1 + M_3\,\mathrm{kg/s}$

   $= 0.599 + 2.329 \times 10^{-3}\,\mathrm{kg/s}$

   $= 0.601\,\mathrm{kg/s}$

6. $h_2 = \dfrac{M_1 h_1 + M_3 h_g}{M_2}\,\mathrm{kJ/kg}$

   $= \dfrac{0.599 \times 29.1 + 2.329 \times 10^{-3} \times 2500.8}{0.601}\,\mathrm{kJ/kg}$

   $= 38.69\,\mathrm{kg/s}$

7. Calculate the mixed-air temperature from

   $$h_2 = 1.0048\,t_2 + g_2\,(2500.8 + 1.863\,t_2)\,\mathrm{kJ/kg}$$

   $$38.69 = 1.0048\,t_2 + 0.0074\,(2500.8 + 1.863\,t_2)$$

   $$38.69 = 1.0048\,t_2 + 18.506 + 0.0138\,t_2$$

   $$38.69 - 18.506 = t_2\,(1.0048 + 0.0138)$$

   $$20.184 = 1.0186\,t_2$$

   $$t_2 = 19.82\,°\mathrm{C\ d.b.}$$

This is indistinguishable from the inlet air dry bulb temperature of 20 °C. The air is humidified at virtually constant dry-bulb temperature.

8. Plot the two air conditions on a chart (Fig. 2.12) and connect them with a straight vertical line.

9. Calculate the heater input power Q:

   $$Q = M_1\,(h_2 - h_1)\,\mathrm{kW}$$

   $$= 0.599 \times (38.69 - 29.1)\,\mathrm{kW}$$

   $$= 5.744\,\mathrm{kW}$$

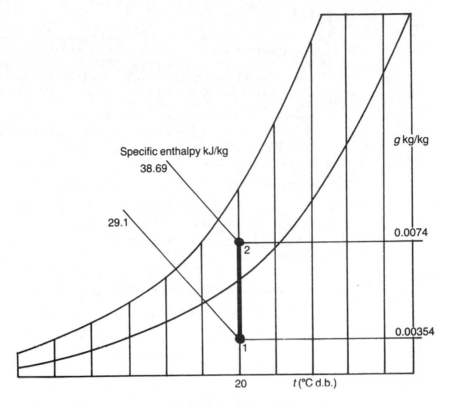

Fig. 2.12 Steam humidification in Example 2.9.

10. The heater power can be estimated from the heat input to the steam when there are no losses:

$$Q = M_3 \, h_g \text{ kW}$$

$$= 2.329 \times 10^{-3} \times 2500.8 \text{ kW}$$

$$= 5.824 \text{ kW}$$

DIRECT-INJECTION HUMIDIFICATION

The direct injection of potable water into the room air through atomizing nozzles connected to overhead pipework can be used in industrial premises such as paper and cotton factories. The distribution system is separated from the drinking-water system through a storage tank so that a pumped and pressurized supply can be assured. This avoids the use of water-spray humidification within the air-handling plant and its attendant hygiene maintenance. There is some evaporative cooling of the room air as the water turns from atomized liquid into vapour and the dry-bulb air temperature will reduce. The calculations are similar to those for steam injection.

EXAMPLE 2.10

Water at 4 b gauge pressure and 12 °C is sprayed into a factory 50 m × 30 m and 5 m high where paper is being produced. The factory air saturation must not drop below 70% at a dry-bulb temperature of 20 °C. The winter outdoor air temperature is expected to fall to − 3 °C d.b. at − 6 °C w.b. on the worst day. A mechanical ventilation system flushes the factory with 1.5 air changes per hour of heated fresh air. Calculate the air preheater battery duty and the expected water supply rate into the humidification pipe and nozzle system. Note the data for each air condition. Compare the results with those from a preheater and steam humidifier that are both in the air-handling plant.

A similar procedure to the previous example is adopted.

1. The factory air data from tables: 20 °C d.b., 70% saturation, 16.5 °C w.b, 0.010 33 kg/kg, 46.32 kJ/kg, 0.8438 m³/kg.
2. Factory volume $v = 50 \times 30 \times 5 \text{ m}^3$

$$= 7500 \text{ m}^3$$

Ventilation rate $Q = 7500 \times 1.5/3600 \text{ m}^3/\text{s}$

$$= 3.125 \text{ m}^3/\text{s at factory condition}$$

$$M_1 = \frac{3.125 \text{ m}^3/\text{s}}{0.8438 \text{ m}^3/\text{kg}}$$

$$= 3.704 \text{ kg/s}$$

3. After humidification, $g_2 = 0.01033 \text{ kg/kg}$.
4. Outdoor air from tables: − 3 °C d.b., − 6 °C w.b., 40% saturation, 0.001 178 kg/kg, − 0.076 kJ/kg, 0.7663 m³/kg. Outdoor air enters the factory at $g_1 = 0.001 178 \text{ kg/kg}$. The water spray mass flow rate M_3 is

$$M_3 = \frac{M_1 (g_2 - g_1)}{1 - g_2} \text{ kg/s}$$

$$= 3.704 \times \left(\frac{0.01033 - 0.001178}{1 - 0.01033} \right) \text{ kg/s}$$

$$= 34.253 \times 10^{-3} \text{ kg/s}$$

5. $$M_2 = M_1 + M_3 \text{ kg/s}$$

$$= 3.704 + 34.253 \times 10^{-3} \text{ kg/s}$$

$$= 3.738 \text{ kg/s}$$

6. The specific enthalpy of the water that is sprayed into the room air is read from tables as saturated water at 12 °C, $h_f = 50.38 \text{ kJ/kg}$ (CIBSE, 1986C). The absolute pressure of water does not affect h_f. In this case, h_f replaces h_g.

$$h_2 = \frac{M_1 h_1 + M_3 h_g}{M_2} \text{ kJ/kg}$$

$$46.32 = \frac{3.704 h_1 + 34.253 \times 10^{-3} \times 50.38}{3.738} \text{ kJ/kg}$$

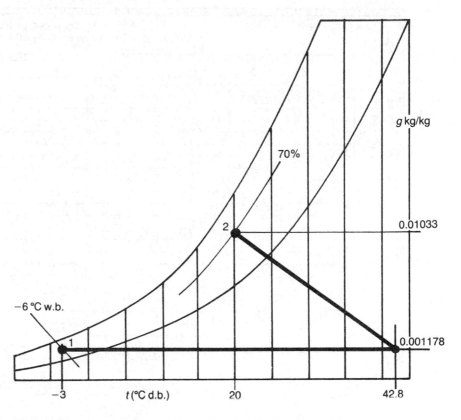

Fig. 2.13 Heating and water-spray humidification for Example 2.10.

h_1 has to be found as this is the preheated air condition prior to humidification.

$$46.32 \times 3.738 = 3.704\, h_1 + 1.726$$

$$h_1 = 46.28\,\text{kJ/kg}$$

This is within calculation accuracy of the final air condition h_2. Thus the factory air is humidified adiabatically, meaning at constant specific enthalpy.

7. Plot the outdoor and factory air conditions on a psychrometric chart (see Fig. 2.13).
8. Draw a line across the chart at the room specific enthalpy, 46.3 kJ/kg.
9. Draw a line horizontally from the outdoor air condition to intersect with the 46.3 kJ/kg specific enthalpy line.
10. The intersection occurs at 42.8 °C d.b. This is the preheated factory air temperature prior to humidification.
11. Calculate the preheater battery power, Q.

$$Q = M_1 \text{ kg/s} \times (h_2 - h_1) \text{ kJ/kg}$$

$$= 3.704 \times (46.32 - - 0.076)\,\text{kW}$$

$$= 171.851\,\text{kW}$$

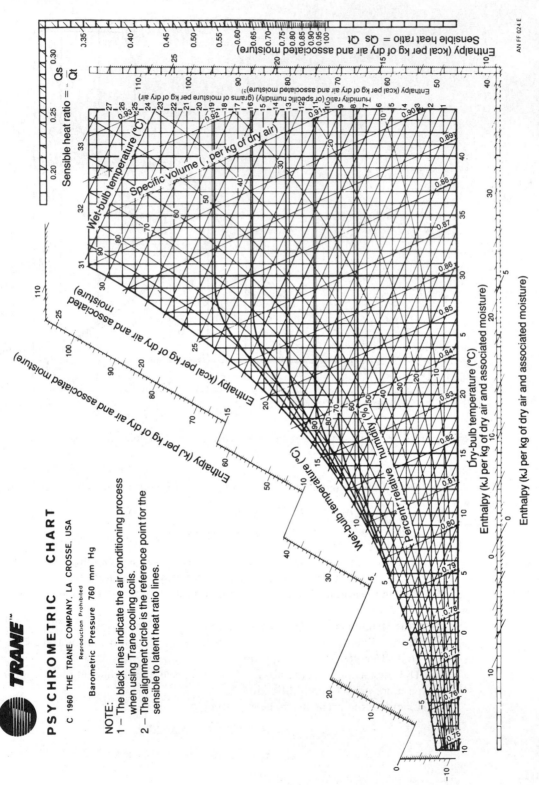

Fig. 2.14 The cooling and dehumidification performance of coils is demonstrated by the thick lines curving down to the left from the horizontal. (Reproduced by courtesy of Trane (UK) Limited.)

12. The alternative humidification system utilizing a steam humidifier in the air-handling plant would require the preheater to raise outdoor air to 20 °C. The heater battery power, Q, is given by

$$Q = 3.704 \times (23.11 - - 0.076)\,\text{kW}$$

$$= 85.881\,\text{kW}$$

13. The steam boiler input power Q is given by

$$Q = M_3\, h_g\,\text{kW}$$

$$= 34.253 \times 10^{-3}\,\text{kg/s} \times 2500.8\,\text{kJ/kg}$$

$$= 85.66\,\text{kW}$$

14. The total heat input from the preheater and steam boiler, Q, is given by

$$Q = 85.881 + 85.66\,\text{kW}$$

$$= 171.541\,\text{kW}$$

This is the same as for the preheater only design. The advantage of the steam injection system is that it would be impractical to preheat the factory air to 42.8 °C to enable it to be adiabatically humidified.

The Trane psychrometric chart is reproduced as Fig. 2.14. The cooling and dehumidifying performance of a cooling coil can be plotted on the thick lines curving down to the left from the horizontal. This is more realistic than drawing a straight line between the expected end states of the air.

Questions

These questions require the use of CIBSE psychrometric charts and/or data tables. Use the simplified sketch psychrometric chart for rough work and for recording the data for each question.

1. Air in a room is at 22 °C d.b. and 15 °C w.b. Find the following conditions from a chart and verify them from the data tables: percentage saturation, specific volume, dew point, specific enthalpy and moisture content.
2. Outdoor air at 0 °C d.b. and 100% saturation is heated with a low-pressure hot-water battery to 30 °C d.b. Sketch the psychrometric cycle on a chart and identify all the condition data for both entry and exit states.
3. Air at 12 °C d.b. and 20% saturation is heated through an increase of 25 K and then adiabatically humidified to 90% saturation. Sketch the cycle and identify all the condition data for the end states.
4. Summer outdoor air at 30 °C d.b., 21 °C w.b. is cooled to 12 °C d.b. and 90% saturation. Sketch the psychrometric cycle and identify all the condition data for the end points.
5. Outdoor air at 28 °C d.b., 22 °C w.b. passes through an HFC134a direct-expansion cooling coil where the refrigerant evaporates at 5 °C. Assume that the air-side dew point of the coil is 5°C and that 100% contact factor is maintained. Sketch the psychrometric cycle and identify all the condition data for the coil air-on and air-off states. Calculate the reductions in specific enthalpy and moisture content per kg of air produced.
6. Recirculated room air at 21 °C d.b., 50% saturation is mixed in equal amounts with summer outdoor air at 31 °C d.b., 22 °C w.b. Sketch the mixing process and state the condition of the mixed air.
7. An air-handling unit receives recirculated room air at 23 °C d.b., 50% saturation and a flow rate of 2 m^3/s, and fresh air at -5 °C d.b., 100% saturation with a flow rate of 0.5 m^3/s. The mixed air is heated with a 1PHW heater battery to 35 °C d.b. and then adiabatically humidified to 24 °C d.b.

   Sketch the psychrometric cycle, identify each condition point and calculate the heater battery load in kW.
8. Calculate the heater battery duty when air is heated from 10 °C d.b., 8 °C w.b. to 40 °C d.b. when the inlet air volume flow rate is 3 m^3/s. Use the inlet specific volume to calculate the air mass flow rate.

9. A cooling coil has an air-inlet condition of 29 °C d.b., 21.8 °C w.b. and an air-outlet state of 13 °C d.b., 90% saturation. Sketch the cycle and mark on the sketch all the condition data. Calculate the cooling duty of the coil when the inlet air flow to the coil is $4 \, \text{m}^3/\text{s}$.

10. A single-duct air-conditioning system takes 1.5 m^3/s of external air at -3 °C d.b., 80% saturation and mixes it with $5 \, \text{m}^3/\text{s}$ of recirculated room air at 20 °C d.b., 50% saturation. The mixed air is heated to 32 °C d.b. prior to being supplied to the rooms. Sketch the cycle and calculate the heater battery duty.

11. An air-handling unit mixes $0.8 \, \text{m}^3/\text{s}$ of fresh air at 32 °C d.b., 23 °C w.b. with $4 \, \text{m}^3/\text{s}$ of recirculated room air at 22 °C d.b., 55% saturation. The mixed air passes through a chilled-water cooling coil whose dew point is 6 °C d.b. Incomplete contact between the air and dew-point surfaces causes 10% of the mixed air to bypass the cooling effect. This 10% air flow mixes with the 90% that contacts the wet surfaces and is cooled to the coil dew point.

    Sketch the mixing and cooling process and identify all the data. Calculate the refrigeration capacity of the cooling coil in kW and tonne refrigeration (given that 1 tonne refrigeration is 3.517 kW) and the rate of moisture removal from the air in kg/h.

12. Outside air at -5 °C d.b., 80% saturation enters a preheater battery and leaves at 24 °C d.b. The fresh-air inlet volume flow rate is $2 \, \text{m}^3/\text{s}$. Find the outdoor-air wet-bulb temperature and specific volume, the heated-air moisture content and percentage saturation. Calculate the heater battery duty.

13. A cooling coil has chilled water passing through it at a mean temperature of 10 °C. An air flow of 1.5 m^3/s at 28 °C d.b., 23 °C w.b. enters the coil and leaves at 15 °C d.b. Find the leaving air wet-bulb temperature and percentage saturation. Calculate the refrigeration capacity of the coil.

14. $2 \, \text{m}^3/\text{s}$ of air that has been recirculated from an air-conditioned room is at 22 °C d.b., 50% saturation. It is mixed with $0.5 \, \text{m}^3/\text{s}$ of fresh air that is at 10 °C d.b., 6 °C w.b. Calculate the dry-bulb air temperature and moisture content of the mixed air. Plot the process on a psychrometric chart and read the specific enthalpy, specific volume and wet- bulb temperature of the mixed air.

15. The cooling coil of a packaged air conditioner in a hotel bedroom has refrigerant in it at a temperature of 16 °C. Room air at 31 °C d.b., 40% saturation enters the coil and leaves at 20 °C d.b. at a flow rate of $0.5 \, \text{m}^3/\text{s}$. Is the air dehumidified by the conditioner? Find the room-air wet-bulb temperature and specific volume. Calculate the total cooling load in the room.

16. Air in an occupied room is measured to be 24 °C d.b. and 16 °C w.b. with a sling psychrometer. Calculate the following physical properties and verify them from CIBSE tables and the psychrometric chart, commenting upon any differences found: saturation vapour pressure, saturation moisture content, vapour pressure, moisture content, percentage saturation, specific volume, density, specific enthalpy, dew point.

17. Outdoor air is at 1 °C d.b. and 0.5 °C w.b. Calculate the physical properties of the air and verify them from CIBSE tables and the psychrometric chart.

18. Outdoor air is at 32 °C d.b. and 22 °C w.b. Calculate the physical properties of the air and verify them from CIBSE tables and the psychrometric chart.

3 Heating and cooling loads

INTRODUCTION

The air-conditioning plant is designed to maintain the specified internal air temperature and humidity when the expected heat gains and losses occur. The CIBSE method of calculation of these plant heat loads is explained with the minimum of background theory. If refrigeration is not to be employed, the summer heat gains will produce an elevated internal air temperature. This summertime indoor temperature is calculated.

The sources of heat gains are identified and the areas of shading on a building are derived from first principles. Trigonometry is used to find the components of the solar irradiance, that is, at a right angle to the irradiated surface as this is used to calculate the heat gain. The formulation of design total irradiance is analysed for vertical, horizontal and sloping surfaces. The heat transmission through glazing and the effect of different types of glass are explained.

Sol-air temperature is introduced for the calculation of heat gains through opaque structures. The ideas of transient heat flow into and out of the thermal storage capacity of the walls, roof and floor lead to the use of a 24 h mean heat flow and a swing, or cyclic variation, to the mean.

The plant cooling load can be found from these methods. Repeated analysis produces the hourly cooling requirements for zone loads and the prediction of the balance temperature for the building. This is a measure of energy efficiency.

LEARNING OBJECTIVES

Study of this chapter will enable the user to:

1. understand the balance between heat gains and losses in an air-conditioned building;

2. identify and calculate all the sources of heat gain;

3. understand the cyclic behaviour of heat gains;

4. decide when a cooling system is needed;

5. calculate heat gains for applications in various countries;

6. relate the sun position to a building;

7. manipulate trigonometric ratios for design calculations;

8.   calculate heat gains with direct solar and diffuse sky irradiances;
9.   understand the terms 'altitude', 'azimuth' and 'incidence';
10. calculate solar gains to horizontal, vertical and sloping surfaces;
11. calculate heights and surface areas of buildings using trigonometry;
12. calculate areas of shade;
13. find the design total irradiance;
14. understand the use of sol-air temperature;
15. know the types of glass used and their thermal properties;
16. calculate glass temperature and heat transfer;
17. understand the use of mean heat flow and cyclic swing in heat transfer through opaque structures;
18. use time lag, decrement factor and thermal admittance in heat-gain calculations;
19. assess the cooling load of a room and a building;
20. understand the use of sensible and latent heat gains;
21. calculate the peak summertime temperature in a building;
22. find the balance outdoor air temperature for a building.

Key terms and concepts

absorptivity	93	plant cooling load	101
air temperature	71	properties of glass	95
altitude	77	reflectance	95
azimuth	77	resultant temperature	71
balance temperature	111	sensible and latent heat	102
decrement factor	99	shaded area	82
design total irradiance	91	shading devices	72
diffuse irradiance	81	slope	73
direct irradiance	83	sol-air temperature	93
electrical equipment	101	solar gains	71
emissivity	93	solar position	77
environmental temperature	71	sun position	85
glass	94	surface factor	99
glass temperature	96	swing in temperature	99
heat gain	71	thermal admittance	72
incidence	77	thermal balance	94
infiltration	71	thermal transmittance	72
intermittent heat gain	71	total heat gain	72
latitude	77	transmittance	72
opaque structure	72	trigonometry	72
peak summertime environmental		wall solar azimuth	77
temperature	71		

SOLAR AND INTERNAL HEAT GAINS

The thermal conditions within a building result from a balance between:

1. heat loss to the cooler external environment;
2. thermal storage within the fabric of the structure;
3. heat gains from solar radiation through glazing;
4. heat gains from solar radiation on to the opaque parts of the structure such as walls and the roof;
5. conduction heat flows through the glazing;
6. conduction heat flows through the opaque walls and roof;
7. the infiltration or direct injection of hot outdoor air into the rooms;
8. conduction or ventilation heat gains from surrounding rooms;
9. heat gains from the internal use of the building (these are from the occupants, lighting, electric motors, electrical appliances, computers, cooking equipment, water heating for washing purposes, animals, furnaces, industrial processes, gas-fired cooking and water heating, electrical and gas-fired incinerators);
10. heat transfers between hot or cold services pipes or air ducts and surrounding rooms or the movement of hot or cold products between rooms.

These heat transfers are intermittent, meaning that they all change in value during the day and will cause the room air temperatures to be continuously varied. Solar heat gains are cyclical with the Earth's rotation and the annual nature of weather patterns. The calculation of heat losses for finding the size of central-heating radiators and boilers is usually based upon steady-state heat flows from the building on the coldest day. This would be unacceptable for air-conditioning cooling loads. The thermal comfort of the occupants, or the satisfaction of the conditions for other reasons, depends upon the net effect of these heat flows. The combination of large heat gains to an enclosure at a time of high outdoor air temperature and intense solar radiation, together with poor ventilation, may lead to overheating of the internal air, insufficient removal of bodily heat production, excessive sweating and discomfort.

Air temperature t_{ai} is insufficient to describe the thermal effects produced by the combination of convection and radiation heat exchanges within the building. The external surfaces are heated by solar irradiance, thus raising the internal room surfaces above the enclosed air temperature. Discomfort can be produced from the low-temperature radiation from the walls and ceiling plus the glare from windows. Environmental temperature t_{ei} combines some of these:

$$t_{ei} = 0.33\, t_{ai} + 0.67\, t_m \ \text{°C}$$

where t_{ai} = dry-bulb air temperature (°C), t_m = mean surface temperature (°C).

Mean surface temperature is not calculated in this context as the heat-gain formula includes appropriate correction factors. Resultant temperature t_c is used to specify the desired comfort condition:

$$t_c = 0.5\, t_{ai} + 0.5\, t_m$$

The upper limit for the room environmental temperature for normally occupied buildings is 27 °C (CIBSE, 1986a, Section A8). The external design air temperatures for comfort in offices in London (CIBSE, 1986a, Table A2.22) may be chosen as 29 °C d.b., 20 °C w.b. Higher outdoor air temperatures are often achieved. The indoor limit of 27 °C will be exceeded frequently in naturally ventilated buildings from the combination of the infiltration of outdoor air, solar and internal heat gains. In parts of the world where high radiation intensity and continuously higher external temperatures are common, for example Sydney with 35 °C d.b, 24 °C w.b., the necessity for refrigeration can be recognized.

Solar heat gains are calculated for the estimate of the peak summertime environmental temperature produced within a naturally ventilated building or when air conditioning is to be used because of insufficient low-cost cooling, for calculation of the refrigeration cooling power needed. The peak solar heat gain through the glazing in summer will often be considerably greater than the sum of all the other heat gains. The heat transfers through the thermal storage effect of the opaque parts of the structure may be small at the time of the peak solar gain through the glazing, or even negative, meaning a net loss from the room. A quick heat-gain assessment may justifiably ignore the opaque gain/loss. The solar heat gains to a building are found from:

1. the angular position of the sun in relation to each face of the building;
2. the intensity of the solar radiation incident upon each external surface;
3. the surface areas exposed to the sun;
4. the date and the sun time.
5. the area of the glazing and walls that are shaded from the direct solar radiation either by nearby buildings or by shading devices or projections from the subject building;
6. type of glazing (clear, heat-absorbing, reflective or tinted);
7. provision of external shading devices;
8. internal blinds or curtains for shading or solar intensity reduction;
9. the absorption of solar radiation by the opaque parts of the structure (this is related to external surface colour);
10. the storage and transmission of solar heat gains through the structure, using transmittance U W/m^2 K and admittance Y W/m^2 K.

Shaded areas receive only blue-sky diffuse solar radiation that is non-directional and equal on all the faces of the building simultaneously. Those areas that are unshaded receive both the directly transmitted line-of-sight and diffuse radiation from the sun. There is a time delay of up to 12 h between the occurrence of solar radiation falling on an opaque surface and the heat gain appearing within the room and influencing the resultant temperature. Remember that resultant temperature is 50% of the dry-bulb air temperature plus 50% of the mean surface temperature. The room surfaces that are exposed to external walls and roofs will be warmed by the solar heat gains that are stored within the bricks, steel and concrete, subsequently heating the room air.

The total heat gain experienced by a room is found from the sum of:

1. instantaneous gains from the solar transmission through the glazing;
2. heat released from the structure due to solar radiation some hours previously;
3. thermal transmittance through the glazing and opaque structure from the higher outdoor air temperature;
4. infiltration of warmer outdoor air directly into the room;
5. sources of heat generated within the room from the people, lights and powered equipment such as computers.

SUN POSITION

It is convenient to consider that the sun moves around the stationary building. The position of the sun relative to the irradiated surface is defined by five angles, as follows.

1. *Solar altitude*. This is the vertical angle from the earth's horizontal plane up to the centre of the sun, $A°$.
2. *Solar azimuth*. This is the compass orientation of a vertical plane through the sun measured clockwise from north on the plane of the earth. In plan view, this is the angle between north and the sun position, $B°$.
3. *Wall azimuth*. This is the compass angle from a line drawn at right angles from a vertical wall to north, $C°$. A right angle from a surface is called normal to the surface. A sloping wall or glazing has an azimuth but a flat roof does not.
4. *Wall solar azimuth*. This is the difference between the solar and wall azimuth angles, $D°$.
5. *Slope*. This is the vertical angle from a normal line from the surface to the horizontal, $E°$. For example, a vertical wall or window has a slope of $0°$ and a flat roof has a slope of $90°$.
6. *Incidence*. This is the combined effect of the other angles and represents the path of the solar radiation striking the surface, $F°$.

It is necessary to use trigonometry in the analysis of the solar position in relation to the building and for the calculation of shaded areas. This is a good time to revise the concepts of trigonometric ratios in mathematics. Only three ratios are needed: sine, cosine and tangent. Figure 3.1 is used to relate these to a right-angled triangle. If design problems produce non-right-angled triangles, it may help to divide the shape into right-angled triangles before attempting any calculations. Figure 3.1(a) shows a right-angled triangle where the points X, Y and Z have angles x, y and $90°$ and the names of the sides in respect to the angle $x°$ are opposite vertically, adjacent horizontally and hypotenuse the slope. The opposite and adjacent labels would reverse if angle $y°$ were being considered. The trigonometric ratios needed are:

$$\text{sine } x° = \frac{\text{opposite}}{\text{hypotenuse}} = \frac{O}{H} \qquad \sin x = O/H$$

$$\text{cosine } x° = \frac{\text{adjacent}}{\text{hypotenuse}} = \frac{A}{H} \qquad \cos x = A/H$$

$$\text{tangent } x° = \frac{\text{opposite}}{\text{adjacent}} = \frac{O}{H} \qquad \tan x = O/A$$

EXAMPLE 3.1

Figure 3.1(b) shows a 12-storey building that has been surveyed to find its height. The surveyor measured an angle of $25°$ from the ground to the top of the building at a distance of 80 m from it. Calculate the building height.

The tangent ratio can be used, $\tan x = O/A$.

O = building height H m

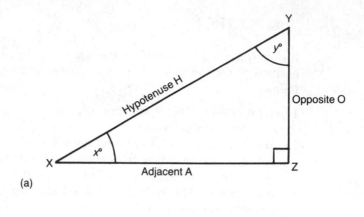

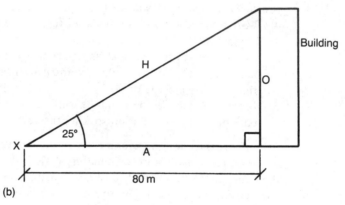

Fig. 3.1 Trigonometry of a right-angled triangle: (a) definition of sides for angle X; (b) calculation of the height of a building.

A = distance from the building, 80 m

$$\tan 25° = \frac{\text{height } H \text{ m}}{80 \text{ m}}$$

Rearrange the formula to find H m.

$$H = 80 \text{ m} \times \tan 25°$$

Enter 25 into your calculator, making sure that it is set to degree format, and press the tangent key, tan or tan x, tan $25° = 0.4663$; multiply this by 80 m and see that the height of the building is 37.305 m. Divide this by 12 storeys and find that each storey is 3.109 m high.

EXAMPLE 3.2

A recent trend has been to design buildings with sloping sides as depicted in Fig. 3.2. The maintenance contractor needs to find the surface area of the sloping face of the building for its cleaning cost. The face slopes at 75 ° to the horizontal and the roof line is at 50 ° from the ground at a distance of 60 m. Calculate the vertical height of the building, the length of the sloping face and the sloping face area.

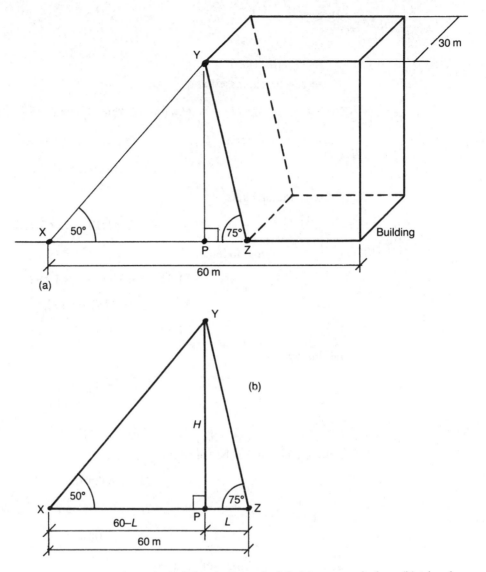

Fig. 3.2 Sloping surface of a building in Example 3.2: (a) surveyor's data; (b) triangles.

Figure 3.2(b) is drawn at a normal to the foot of the building. Divide the triangle XYZ into two right-angled triangles by inserting a vertical line YP. The length of YP is the same for both the triangles XYP and YZP. Divide the base of the triangles into the unknown length L and $(60 - L)$. The common vertical side YP is the height of the building H m. Two tangent formulae can be written and solved as simultaneous equations:

$$\tan 50 = \frac{H}{60 - L}$$

$$H = (60 - L) \tan 50$$

$$\tan 75 = \frac{H}{L}$$

$$H = L \tan 75$$

These can be expressed as

$$H = (60 - L)\tan 50 = L \tan 75$$

Evaluate the tangents:

$$(60 - L) \times 1.192 = L \times 3.732$$

Remove the brackets:

$$60 \times 1.192 - L \times 1.192 = L \times 3.732$$

Bring the L terms together:

$$71.52 = L \times 3.732 + L \times 1.192$$

$$71.52 = L \times (3.732 + 1.192)$$

$$71.52 = L \times 4.924$$

Divide by 4.924:

$$\frac{71.52}{4.924} = L$$

$$L = 14.525\,\text{m}$$

Now the height can be found:

$$H = L \times \tan 75$$

$$H = 14.525 \times \tan 75$$

$$H = 54.208\,\text{m}$$

and the length of the sloping face:

$$\cos 75 = \frac{14.525}{\text{YZ}}$$

Rearranging:

$$\text{YZ} = \frac{14.525}{\cos 75}$$

$$= 56.12\,\text{m}$$

the sloping face area $= 56.12\,\text{m} \times 30\,\text{m}$

$$= 1683.6\,\text{m}^2$$

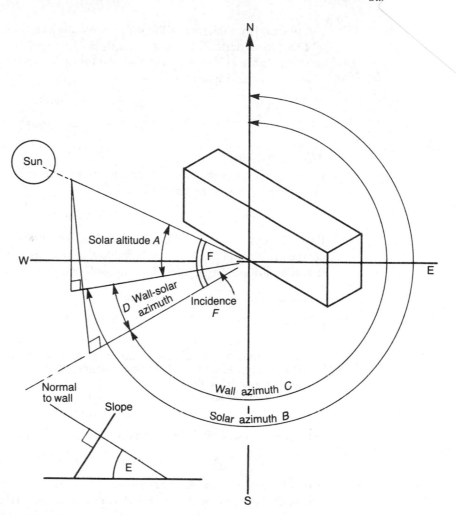

Fig. 3.3 Solar and wall orientation.

The five angles that define the orientation of a surface are shown in Fig. 3.3. Tabulated data on the solar positions is available (CIBSE, 1986a, Table A2.23). Latitude gives the vertical angle above or below the Equator of a location on earth: for example, London is 51 ° 29′ N, meaning 51 ° and 29 minutes north of the Equator, or 51.5 ° N as there are 60 minutes in a degree; Sydney is 33 ° 52′ S; Hong Kong is 22 ° 18′ N; and Singapore is almost on the Equator at 1 ° 18′ N.

The angle of incidence of radiation on a vertical surface is found from

$$\cos F = \cos A \cos D$$

where F = angle of incidence, A = angle of solar altitude, D = wall solar azimuth angle.

EXAMPLE 3.3

A vertical wall in London faces southeast. At 0900 h sun time on 22 May the solar altitude is 44 ° and solar azimuth 113 °. Calculate the wall solar azimuth and incidence angles.

A normal from the face of the wall points southeast and this has an azimuth angle of 135 ° from north. The solar azimuth is 113 ° so the difference is the wall solar azimuth $(135 - 113°) = 12°$. The wall is almost facing the sun in the morning.

To find the incidence, use

$$\cos F = \cos A \cos D$$

$$= \cos 44 \times \cos 12$$

$$= 0.7036$$

Leaving this number in the calculator display, the angle whose cosine is 0.7036 is to be found. This is the $\cos^{-1}$ function:

$$\cos^{-1}(0.7036) = x°$$

Press the $\cos^{-1}$ key and 45.3 ° is found. Check that it is correct by pressing the cosine key to find $\cos 45.3° = 0.7036$. Notice that the incidence is a combination of the solar altitude being altered by the wall solar azimuth.

EXAMPLE 3.4

A wall of an office in Plymouth, latitude 50 ° 21′ N, faces southwest and slopes at 10 ° forwards from the vertical as an architectural feature. The sun time is noon on 21 June, the solar altitude is 64 ° and the solar azimuth is 180 °. The intensity of the direct solar radiation normal to the sun is 900 W/m^2. Prove the formula for solar incidence and calculate the incidence angle upon the wall.

Figure 3.4 shows the plan positions of the sun and wall.

$$\text{Wall solar azimuth angle } D = 225° - 180° = 45°$$

The solar radiation needs to be resolved to a value that is normal to the wall surface. The angle between the lines normal to the sun and normal to the wall is the incidence angle F. View 1 would show a vertical triangle along a north–south line and the sun's altitude of 64 ° as depicted in Fig. 3.5.

$$\cos 64 = \frac{Z}{I}$$

So

$$\text{side } Z = I \cos 64$$

From Fig. 3.4

$$\cos 45 = \frac{Y}{Z}$$

So

$$\text{side } Y = Z \cos 45$$

and as Z is known

$$\text{side } Y = I \cos 64 \times \cos 45$$

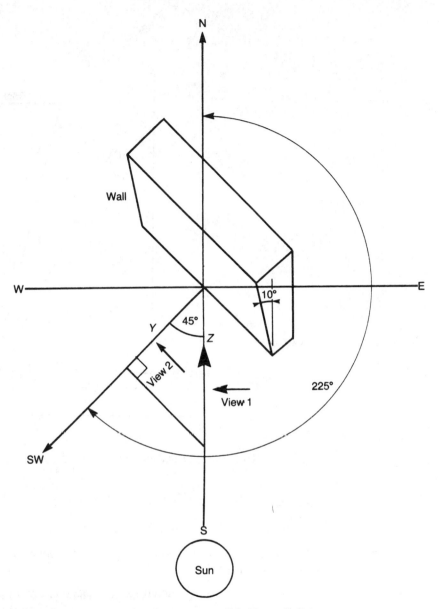

Fig. 3.4 Incidence of solar irradiance on a wall in Example 3.4.

The face of the building is tilted downwards at 10° so the component of the solar radiation that is normal to it is line X. Line Z has an equal length along the ground as shown and this becomes the hypotenuse of the lowest triangle (see Fig. 3.6).

$$\sin 80° = \frac{X}{Y}$$

So

$$\text{side } X = Y \sin 80$$

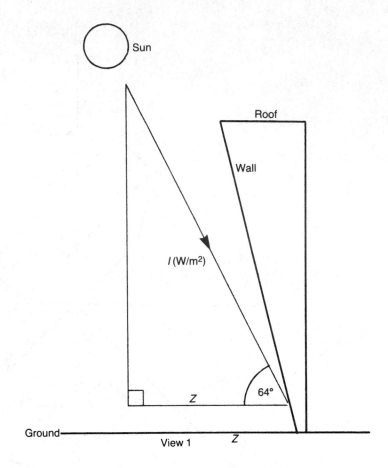

Fig. 3.5   Solar altitude for Example 3.4.

and as Y is known

$$\text{side } X = I \cos 64 \times \cos 45 \times \sin 80$$

where side X represents that component of the solar radiation that is normal to the face of the wall, I_x.

$$I_x = 900 \text{ W/m}^2 \times \cos 64 \times \cos 45 \times \sin 80$$

$$= 900 \times 0.438 \times 0.707 \times 0.985 \text{ W/m}^2$$

$$= 274.5 \text{ W/m}^2$$

Referring to Fig. 3.3, the incidence angle F is that between the normal from the surface, I_x, and the normal from the sun, I. This triangle slopes outward from the surface of the paper, and

$$\cos F = \frac{I_x}{I} = \frac{1 \times \cos 64 \times \cos 45 \times \sin 80}{I}$$

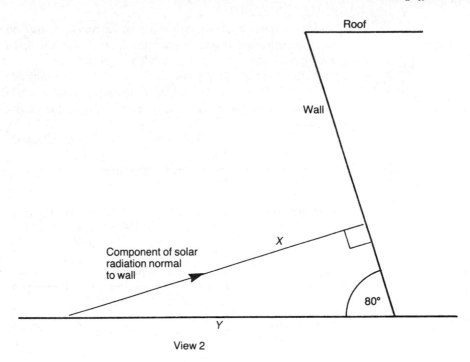

Fig. 3.6 Solar incidence on the sloping wall in Example 3.4.

so

$$\cos F = \cos 64 \times \cos 45 \times \sin 80$$

$$= 0.438 \times 0.707 \times 0.985$$

$$= 0.305$$

and to find the incidence angle

$$\cos^{-1} 0.305 = 72.2°$$

Bearing in mind that

$$64° = \text{solar altitude angle } A$$

$$45° = \text{wall solar azimuth angle } D$$

$$80° = \text{slope angle of the surface } E$$

then

$$\cos F = \cos A \times \cos D \times \sin E$$

and this can be generally used for other problems.

SHADING EFFECTS

Shade upon windows and walls greatly reduces the incident solar irradiance by eliminating the direct line-of-sight heat flow and only leaving the blue-sky diffuse irradiance. Deliberate shading is produced by external louvres, balconies, verandahs and protruding columns while fortuitous shading may be temporarily created by nearby buildings, vegetation or landscape. A window that is recessed from the face of the building will be partly shaded from the side owing to the wall solar azimuth and from the top owing to the combination of solar altitude and wall solar azimuth. Figure 3.7 shows that the width of shade from the side wall, W, can be found from the wall solar azimuth angle D and the depth of the recess, R, such that

$$\tan D = \frac{W}{R} \quad \text{and} \quad W = R \tan D$$

EXAMPLE 3.5

A window is 2.5 m long and 1.5 m high and has a vertical column alongside that projects 250 mm from the face of the building. The wall azimuth is 220 ° and the solar azimuth is 180 °. Calculate the areas of glass exposed to direct and diffuse irradiances from the shade cast across the its side.

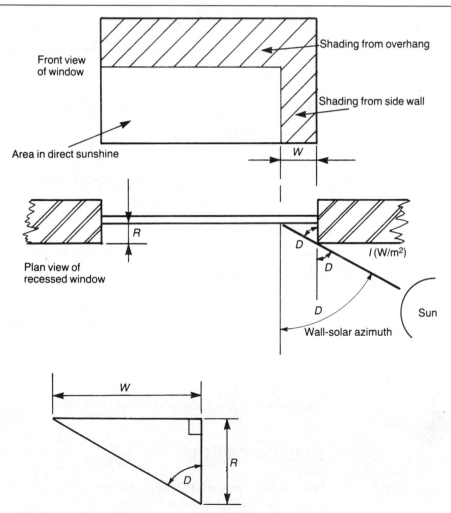

Fig. 3.7   Angles for calculation of vertical shading.

$$\text{wall solar azimuth } D = 220° - 180°$$

$$= 40°$$

$$\text{width of side shade } W = 250\,\text{mm} \times \tan 40$$

$$= 209.8\,\text{mm}$$

$$= 0.21\,\text{m}$$

$$\text{side shaded area} = 0.21\,\text{m} \times 1.5\,\text{m}$$

$$= 0.315\,\text{m}^2$$

$$\text{total area of glazing} = 2.5\,\text{m} \times 1.5\,\text{m}$$

$$= 3.75\,\text{m}^2$$

$$\text{unshaded area of glazing} = 3.75 - 0.315\,\text{m}^2$$

$$= 3.435\,\text{m}^2$$

Thus the glass exposed to direct irradiance is 3.435 m^2 while the glass exposed to blue sky diffuse irradiance is the whole window area of 3.75 m^2.

Vertical shading, caused by the overhanging projection of a balcony, external blind, slats or louvres and recessing of the glazing, is calculated from a combination of the wall solar azimuth and the solar altitude. The greater the wall solar azimuth, the deeper the shadow for a constant value of solar azimuth and this can be tested with a card folded into an L and the sunshine or a lamp by turning the vertical surface at increasingly acute angles from the light source. Figure 3.8 shows the basis for calculating the height of the window's horizontal shaded portion.

From the plan view $(R/Y) = \cos D$ and $Y = R/\cos D$ where Y is the length of the horizontal component of the irradiation. View AA reveals a vertical triangle where

$$\frac{h}{Y} = \tan A$$

and

$$h = Y \tan A$$

Substitute for Y and

$$h = R\,\frac{\tan A}{\cos D}$$

EXAMPLE 3.6

A window 2.5 m long and 1.5 m deep is recessed 250 mm from the face of a building. Calculate the unshaded glass area at 1000 h GMT when the solar altitude is 52 ° and the solar azimuth is 131 ° on 21 May. Wall azimuth is 195 °. Calculate the unshaded area of the glass.

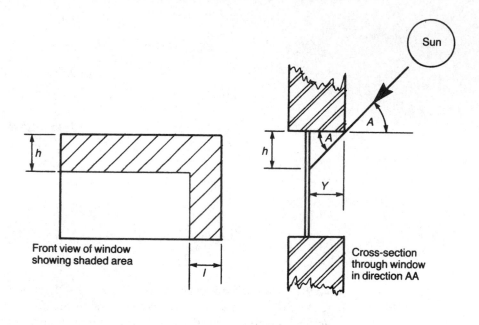

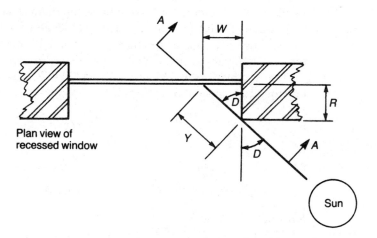

Fig. 3.8   Derivation of horizontal shading depth.

$$\text{Side shade length } W = R \tan D$$

$$\text{Wall solar azimuth } D = 195° - 131°$$

$$= 64°$$

$$W = 250 \times \tan 64$$

$$= 512.6 \, \text{mm}$$

$$= 0.513\,\text{m}$$

$$h = R\,\frac{\tan A}{\cos D}$$

$$= \frac{250 \times \tan 52}{\cos 64}$$

$$= 729.9\,\text{mm}$$

$$= 0.73\,\text{m}$$

$$\text{length of unshaded area} = 2.5 - 0.513\,\text{m}$$

$$= 1.987\,\text{m}$$

$$\text{height of unshaded area} = 1.5 - 0.73\,\text{m}$$

$$= 0.77\,\text{m}$$

$$\text{unshaded area} = 1.987 \times 0.77\,\text{m}^2$$

$$= 1.53\,\text{m}^2$$

Areas of shade on a window or a complete building can be found by calculation or by illuminating a scale model of the subject building and its surroundings from a lamp positioned at the correct altitude and azimuth angles corresponding to the time of day. The refrigeration cooling load and the size of each room terminal cooling unit are calculated from the peak heat gain that normally occurs at the time of the greatest solar irradiation upon the unshaded glazing. The deliberate use of shading and the calculation of adventitious shading from nearby buildings can significantly reduce the refrigeration plant load. Compare the exposure of a building to solar irradiance at two different locations using the data in Table 3.1 (CIBSE, 1986a).

Table 3.1   Solar positions

GMT (h)	Near Brisbane 30° S 21 December		Near London 50° N 21 June	
	Altitude	Azimuth	Altitude	Azimuth
06	12	69	18	74
07	24	76	27	85
08	37	82	37	97
09	50	88	46	110
10	62	96	55	128
11	75	112	61	151
12	84	180	64	180
13	75	248	61	209
14	62	264	55	232
15	50	272	46	250
16	37	278	37	263
17	24	284	27	275
18	12	291	18	286

Notice the reversal of the dates of the seasons between the northern and southern latitudes and the greater solar altitude when closer to the Equator where it is overhead at noon on 21 March and 21 September. Overhanging roofing traditionally

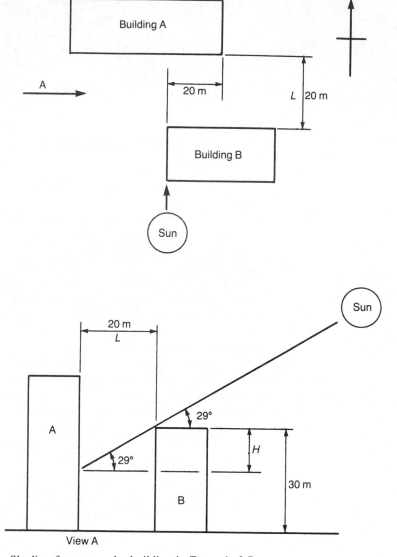

Fig. 3.9   Shading from a nearby building in Example 3.7.

provides completely shaded windows near Brisbane and this also serves to collect rainwater for storage. Variations in shade across the facades of a building occur throughout the day and this has an effect upon the cooling load that has to be matched by the automatic control system.

EXAMPLE 3.7

On 20 February at noon sun time, the solar altitude is 29 °. Figure 3.9 shows the relative positions of two nearby buildings. Calculate the area of shade on building A caused by building B.

At noon the sun is due south so shade width is 20 m. The height of shade on building $A = (30 - H)$ m

$$\tan 29 = \frac{H}{L}$$

$$H = L \tan 29$$

$$= 20 \tan 29$$

$$= 11.09 \, \text{m}$$

$$\text{shaded area} = 20 \, \text{m} \times (30 - 11.09) \, \text{m}$$

$$= 378.2 \, \text{m}^2$$

EXAMPLE 3.8

Figure 3.10 shows the site plan and elevation of building A, that is to be air-conditioned, and building B that is across the street and will cast shadows on to A. The site is at a latitude of 50 ° N. Calculate the solar irradiance normal to each surface of building A at 1500 h GMT on 21 March when the solar altitude is 27 ° and solar azimuth is 232 ° and the areas of roof and walls that are exposed to direct and diffuse irradiance. The basic direct solar irradiance is 700 W/m^2 and the basic diffuse irradiance is 72 W/m^2.

Figure 3.11 shows the geometry needed to find the areas of shade on the roof and front face of building A. The leading edges of building B cast the shadows.
In plan

$$\tan 58 = \frac{Y}{20}$$

$$Y = 20 \tan 58$$

$$= 32 \, \text{m}$$

From view Z the true length of the horizontal component, T, of the irradiance upon the roof can be found.

$$\tan 27 = \frac{15}{T}$$

$$T = \frac{15}{\tan 27}$$

$$= 29.439 \, \text{m}$$

The leading edge of the roof of building B casts a shadow on to the roof of building A. The depth of this shadow is found in plan as

$$\sin 58 = \frac{S}{29.439}$$

$$S = \sin 58 \times 29.439$$

$$= 24.966 \, \text{m}$$

The shadow projects 24.966 − 20 m on to the roof, 4.966 m.

The front face of A is completely shaded. The northwest face and roof of A is partly in direct irradiance.

The southeast and northeast faces of A are completely shaded from direct irradiance.

The roof receives direct irradiance on the horizontal surface that is the vertical component of the basic direct irradiance of 700 W/m^2 as indicated in Fig. 3.12:

$$I_{DH} = 700 \text{ W/m}^2 \times \sin 27$$

$$= 318 \text{ W/m}^2$$

A normal from the left wall of building A faces 20 ° N of W and has a wall azimuth of 290 ° and the direct solar irradiance upon the exposed part of the vertical surface is I_{DH}.

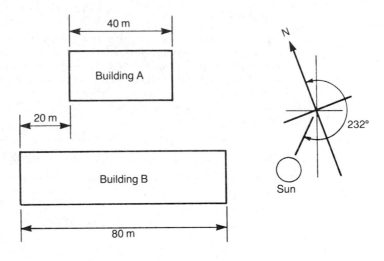

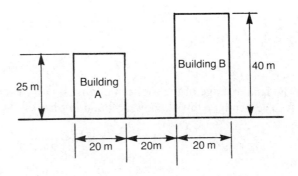

Fig. 3.10   Shading cast on to building A in Example 3.8.

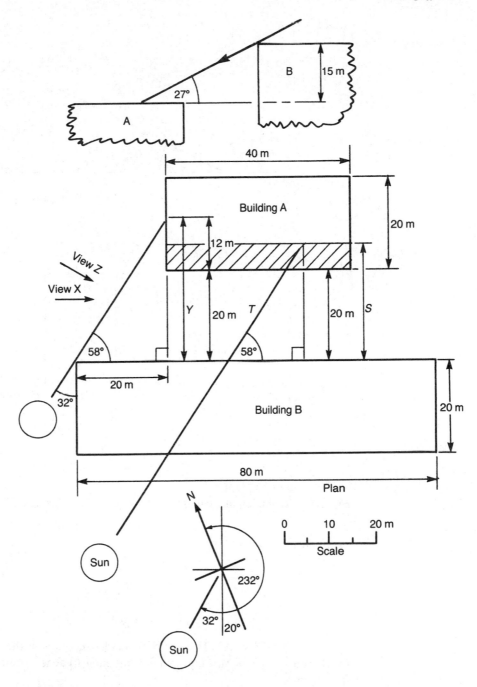

Fig. 3.11   Solution of shading Example 3.8.

The wall solar azimuth angle *D* is given by:

$$D = 290° - 232°$$

$$= 58°$$

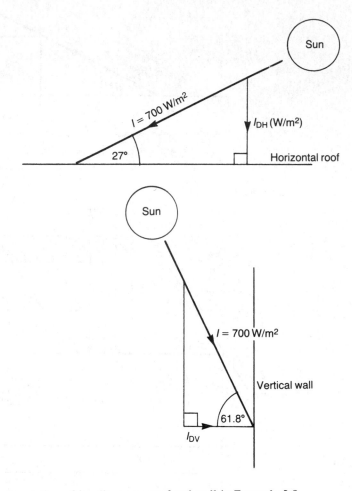

Fig. 3.12   Resolution of irradiance on roof and wall in Example 3.8.

From previous work:

$$\cos F = \cos A \cos D$$

$$= \cos 27 \times \cos 58$$

$$= 0.472$$

and $\cos^{-1} 0.472 = 61.8°$. The angle of incidence between the vertical wall and the irradiance is $61.8°$. The horizontal component that is normal to the wall can be found from

$$I_{DV} = I \cos F$$

$$= 700 \text{ W/m}^2 \times 0.472$$

$$= 330 \text{ W/m}^2$$

The shaded surfaces receive diffuse sky irradiance of 72 W/m^2.

DESIGN TOTAL IRRADIANCE

The design total solar irradiance upon a vertical surface I_{TVd} comprises the direct line-of-sight I_{DVd} value plus half the diffuse sky irradiance I_{dHd} and that reflected from the ground. The first suffix is for either the total T, direct D or diffuse d irradiance; the second suffix denotes a horizontal surface H or a vertical wall V; and the third suffix d shows that the irradiance will be used for cooling load design calculations after all the correction factors have been included. The ground reflected irradiance is 0.2 of the total of the direct and diffuse values in temperate and humid tropical localities but 0.5 in arid regions.

The design total irradiance on a vertical surface, I_{TVd} is:

$$I_{TVd} = I_{DVd} + 0.5\,I_{dHd} + 0.5 \times 0.2\,I_{THd} \text{ W/m}^2$$

The design total irradiance upon a horizontal surface I_{THd} is the sum of the direct and diffuse irradiances:

$$I_{THd} = I_{DHd} + I_{dHd} \text{ W/m}^2$$

These are found in CIBSE (1986a), Table A2.27, and have peak values of 625 W/m^2 for an east wall at 0800 h on 21 June, 850 W/m^2 on a roof at noon on 21 June and I_{dHd} of 265 W/m^2 at noon on 23 July.

A sloping surface has a combination of I_{DHd} and I_{DVd} depending upon the inclination from the horizontal, angle S. When S is between 0 ° and 90 °, the design direct irradiance upon the sloping surface I_{DSd} is found from

$$I_{TSd} = I_{DHd} \cos S + I_{DVd} \sin S \text{ W/m}^2$$

Only the direct irradiance has angular effects. The diffuse I_{dHd} exists equally in all directions and is added to the direct components to form the total irradiance:

$$I_{TSd} = I_{DSd} + I_{dHd} \text{ W/m}^2$$

EXAMPLE 3.9

Prove the formula for the irradiance upon a sloping surface and calculate the I_{TSd} for a roof pitch of 20 ° facing southwest at 1400 h on 23 July in southeast England when I_{THd} is 715 W/m^2, I_{TVd} is 580 W/m^2 and I_{dHd} is 230 W/m^2 from CIBSE (1986a), Table A2.27.

First, find the direct solar irradiances on the horizontal and vertical surfaces from the tabulated totals:

$$I_{THd} = I_{DHd} + I_{dHd} \text{ W/m}^2$$

so

$$I_{DHd} = I_{THd} - I_{dHd}$$

$$= 715 - 230$$

$$= 485 \text{ W/m}^2$$

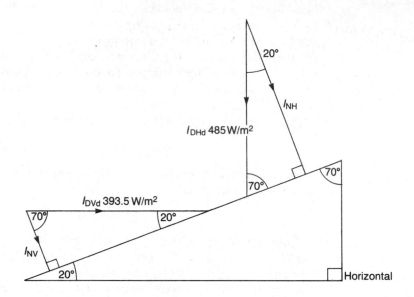

Fig. 3.13   Finding the direct solar irradiance upon a sloping surface in Example 3.9.

Second, find the direct solar irradiance on a vertical surface from

$$I_{TVd} = I_{DVd} + 0.5\,I_{dHd} + (0.5 \times 0.2)\,I_{THd}\ \text{W/m}^2$$

$$I_{DVd} = I_{TVd} - 0.5\,I_{dHd} - (0.5 \times 0.2)\,I_{THd}$$

$$= 580 - 0.5 \times 230 - 0.5 \times 0.2 \times 715$$

$$= 393.5\ \text{W/m}^2$$

The direct irradiance normal to the sloping surface will be a combination of the horizontal and vertical components as shown in Fig. 3.13. Let I_{NV} and I_{NH} be the components that are normal to the surface from the vertical and horizontal irradiances. The three triangles have equal angles.

$$\sin 20 = \frac{I_{NV}}{I_{DVd}}$$

$$I_{NV} = I_{DVd} \sin 20$$

and

$$\cos 20 = \frac{I_{NH}}{I_{DHd}}$$

$$I_{NH} = I_{DHd} \cos 20$$

The sum of I_{NV} and I_{NH} is the direct irradiance normal to the sloping surface, I_{DSd}.

$$I_{\mathrm{DSd}} = I_{\mathrm{NH}} + I_{\mathrm{NV}}$$

$$= I_{\mathrm{DHd}} \cos 20 + I_{\mathrm{DVd}} \sin 20$$

This proves the previously stated formula,

$$I_{\mathrm{DSd}} = I_{\mathrm{DHd}} \cos S + I_{\mathrm{DVd}} \sin S \ \mathrm{W/m^2}$$

$$I_{\mathrm{DSd}} = I_{\mathrm{DHd}} \cos 20 + I_{\mathrm{DVd}} \sin 20$$

$$= 485 \cos 20 + 393.5 \sin 20$$

$$= 590.3 \ \mathrm{W/m^2}$$

The total irradiance on the sloping surface is the sum of the direct and diffuse values:

$$I_{\mathrm{TSd}} = I_{\mathrm{DSd}} + I_{\mathrm{dHd}} \ \mathrm{W/m^2}$$

$$= 590.3 + 230$$

$$= 820.3 \ \mathrm{W/m^2}$$

SOL-AIR TEMPERATURE

Sol-air temperature, t_{eo}, is that outdoor temperature that, in the absence of solar irradiance, would give the same temperature distribution and rate of heat transfer through the wall or roof as exists with the actual outside air temperature and solar irradiance (CIBSE, 1986a). Thus it is an artificial, calculated temperature and it is used in the calculation of steady and transient heat transfers into the building from the continually changing hot external climate. Solar heat gains to opaque, solid building components, due to their radiation absorptivity, raise the temperature of the solid, and this can easily be verified by touch. The heated surface transmits low-temperature, long-wave radiation to nearby cooler surfaces and to the sky in proportion to its emissivity. The absorptivity and emissivity of dark-coloured matt finishes on brick, concrete and asphalt are both 0.9 and the long-wave radiation rate I_{l} is 93 $\mathrm{W/m^2}$ when the sky is cloudless at times of maximum solar heat gains to the building.

$$t_{\mathrm{eo}} = t_{\mathrm{ao}} + R_{\mathrm{so}} \, (0.9 \, I_{\mathrm{THd}} - 0.9 \, I_{\mathrm{l}}) \ {}^{\circ}\mathrm{C}$$

where R_{so} = external surface film thermal resistance, 0.07 $\mathrm{m^2 \, K/W}$

EXAMPLE 3.11

The peak I_{THd} of 850 $\mathrm{W/m^2}$ for a roof in southeast England is expected to occur at noon on 21 June when the outdoor dry bulb air temperature t_{ao} is 19 °C and the I_{l} is 93 $\mathrm{W/m^2}$. Calculate the sol-air temperature for the roof.

$$t_{\mathrm{eo}} = t_{\mathrm{ao}} + R_{\mathrm{so}} \, (0.9 \, I_{\mathrm{THd}} - 0.9 \, I_{\mathrm{l}}) \ {}^{\circ}\mathrm{C}$$

$$= 19 + 0.07 \times (0.9 \times 850 - 0.9 \times 93)$$

$$= 66.7 {}^{\circ}\mathrm{C}$$

EXAMPLE 3.12

The I_{TVd} of 375 W/m^2 for a south-facing vertical wall in southeast England occurs at 1500 h on 22 August when the outdoor dry-bulb air temperature t_{ao} is 21.5 °C and the I_l is 17 W/m^2. Calculate the sol-air temperature for the dark-coloured wall.

$$t_{\mathrm{eo}} = t_{\mathrm{ao}} + R_{\mathrm{so}}\,(0.9\,I_{\mathrm{TVd}} - 0.9\,I_\mathrm{l})\ °\mathrm{C}$$

$$= 21.5 + 0.07 \times (0.9 \times 375 - 0.9 \times 17)$$

$$= 44°\mathrm{C}$$

HEAT TRANSMISSION THROUGH GLAZING

Glass is formed from a molten mixture of 70% silica (SiO_2), 15% soda (Na_2O), 10% lime (CaO), 2.5% magnesia (MgO), 2.5% alumina (Al_2O_2) or other metallic elements to give it particular properties, such as gold (Au) and selenium (Se) to make it photosensitive. An oil-fired furnace liquefies the components at up to 1590 °C which are then floated on to a bath of molten tin and progressively cooled. This is the float-glass method in current use. Glass fibre and glass wool are made by passing the hot liquid through fine orifices (*Encyclopaedia Britannica*, 1980). Glass is manufactured in 4, 6, 10 and 12 mm thicknesses as clear float, modified float, toughened and laminated depending upon type, safety requirements and wind loading. Clear float glass is the most common but body-tinted grey, bronze, blue or green can be chosen for aesthetic, daylighting or thermal reasons. Colour pigments are added during the molten stage and the full thickness of the glass is coloured. Metallic particles added to the surface during the fluid stage of production form a reflective surface on one side of the glass. The reflected light may give a silver colour rendering but transmitted light may appear as bronze. The low-emissivity surface improves the thermal insulation properties of the glass.

Glass is transparent to high-temperature, high-frequency solar irradiance but is poor at passing long-wave radiation from low-temperature surfaces within the building, thus creating the greenhouse effect. The glass surface that has the reflective coating is often located on the inner face of double-glazed units for protection. Body-tinted glass is good at absorbing solar irradiance and reducing heat flow into the building; however, increased glass temperature and thermal expansion are caused. Freon gas can be used to fill the space within sealed double-glazing units as its reduced convection heat-transfer ability lowers the U value.

The area of glazing used may be limited by statutory regulations for energy-conservation reasons. Whether large or small areas are advantageous depends upon the annual balance between the adventitious solar heat gains, the internal heat generation from people, lights and equipment and the heat losses. The availability of natural internal illumination through the glazing is an important part of the energy balance, with reductions in the electrical energy used for artificial lighting. Whether large glass areas are advantageous or not can be analysed from the net energy consumption of the whole building on a monthly and annual basis.

Figure 3.14 shows the proportions of the solar energy flows through glass due to its three properties of transmissivity T, absorptivity A and reflectivity R. Glasses

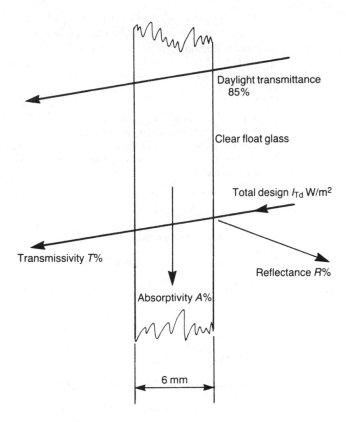

Fig. 3.14 Properties of glass.

have different transmittances for direct and diffuse irradiance. Some irradiance that is absorbed by the glass is radiated into the room and a total transmittance figure is given. These properties remain constant until the solar incidence angle exceeds about 45 ° when they rapidly diminish to zero at 90 ° incidence. Double glazing may consist of tinted or reflective glass in the exterior layer and a clear float inner pane. The outer glass is raised to a higher temperature by being insulated from the cooler interior of the building and sustains greater thermal expansion. The overall direct solar transmittance will be the multiplication of the two glass types such as 0.86 for the clear and 0.23 for the reflective, equalling 0.2. The outward view through tinted glass is not seriously affected and the effect may be hardly noticeable. Reflective glass provides a high degree of internal privacy for the occupants of the building.

Table 3.2 Performance data for glass

Glass type	Light	Direct solar irradiance			Total
	T	T	A	R	T
4 mm clear	0.89	0.82	0.11	0.07	0.86
6 mm tinted	0.5	0.46	0.49	0.05	0.62
6 mm reflective	0.1	0.08	0.6	0.32	0.23

Source: Pilkington Glass Ltd

The exception is at night when interior lighting and external darkness cause the viewing direction to be reversed. The interior is then clearly visible from the street. Typical values are given in Table 3.2.

EXAMPLE 3.13

A window facing southwest is 3 m long and 2.75 m high and has a shaded area of $0.2\,\text{m}^2$. Solar altitude is $43.5\,°$, solar azimuth is $66\,°$ W of S, direct solar irradiance normal to the sun is $832\,\text{W/m}^2$ and diffuse irradiance is $43\,\text{W/m}^2$. The transmissibilities of direct and diffuse irradiances are 0.85 and 0.8. Calculate the solar heat gain transmitted through the glass.

Glass azimuth is $45\,°$ W of S, so

$$\text{glass–solar azimuth} = 66° - 45°$$

$$= 21°$$

$$I_{\text{DVd}} = I_{\text{D}} \cos A \cos D$$

$$= 832\,\text{W/m}^2 \times \cos 43.5 \times \cos 21$$

$$= 563.4\,\text{W/m}^2$$

direct transmitted irradiance $Q_{\text{D}} = 0.85\,I_{\text{DVd}} \times \text{sunlit area W}$

diffuse transmitted irradiance $Q_{\text{d}} = 0.8 \times 43 \times \text{glass area W}$

$$Q_{\text{D}} = 0.85 \times 563.4 \times ((3 \times 2.75) - 0.2)\,\text{W}$$

$$= 3855.1\,\text{W}$$

$$Q_{\text{d}} = 0.8 \times 43 \times 3 \times 2.75\,\text{W}$$

$$= 283.8\,\text{W}$$

total transmitted heat gain $= Q_{\text{D}} + Q_{\text{d}}$

$$= 3855.1 + 283.8\,\text{W}$$

$$= 4138.9\,\text{W}$$

The absorbed solar irradiance causes the glass to rise in temperature above that of its surroundings. The air convection currents either side of the glazing remove the heat gained and, under steady-state conditions, the glass remains at a constant temperature. However, as the irradiance is cyclic, stable conditions may be brief. Figure 3.15 shows the temperature gradient through glazing and a heat balance can be made if it is assumed that the glass is at a higher temperature than the adjacent air. Consider $1\,\text{m}^2$ of glazing:

absorbed heat = heat lost to surroundings

$$A\,I_{\text{TVd}} = h_{\text{so}}\,(t_{\text{g}} - t_{\text{ao}}) + h_{\text{si}}\,(t_{\text{g}} - t_{\text{ai}})$$

Typically

$$R_{\text{so}} = 0.06\,\text{m}^2\,\text{K/W}, \quad \text{so} \quad h_{\text{so}} = 16.7\,\text{W/m}^2\,\text{K}$$

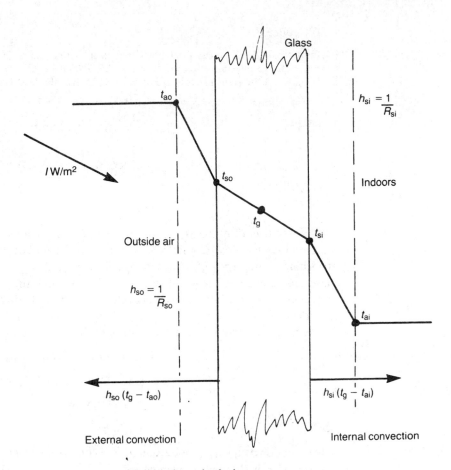

Fig. 3.15 Temperature gradient through glazing.

$$R_{si} = 0.12 \, m^2 \, K/W, \quad so \quad h_{si} = 8.3 \, W/m^2 \, K$$

(CIBSE, 1986a, Section A3)

$$A \, I_{TVd} = 16.7 \, (t_g - t_{ao}) + 8.3 \, (t_g - t_{ai})$$

Rearrange the equation to find the glass temperature:

$$A \, I_{TVd} = 16.7 \, t_g - 16.7 \, t_{ao} + 8.3 \, t_g - 8.3 \, t_{ai}$$

Gather the t_g terms together:

$$A \, I_{TVd} = 16.7 \, t_g + 8.3 \, t_g - 16.7 \, t_{ao} - 8.3 \, t_{ai}$$

$$A \, I_{TVd} = (16.7 + 8.3) \, t_g - 16.7 \, t_{ao} - 8.3 \, t_{ai}$$

$$A \, I_{TVd} = 25 \, t_g - 16.7 \, t_{ao} - 8.3 \, t_{ai}$$

$$A \, I_{TVd} + 16.7 \, t_{ao} + 8.3 \, t_{ai} = 25 \, t_g$$

$$t_g = \frac{A \, I_{TVd} + 16.7 \, t_{ao} + 8.3 \, t_{ai}}{25}$$

EXAMPLE 3.14

Find t_g for 6 mm clear float glass that is in a total solar irradiance of 640 W/m². External and internal air temperatures are 29 °C and 22 °C. Absorptivity A of the glass is 0.15 and the transmissibility T is 0.78. Calculate the heat flow into the room.

$$t_g = \frac{A\, I_{TVd} + 16.7\, t_{ao} + 8.3\, t_{ai}}{25}$$

$$= \frac{(0.15 \times 640) + (16.7 \times 29) + (8.3 \times 22)}{25}$$

$$= 30.5\,°C$$

It is worth noting that the glass temperature is higher than the mean of the air temperatures $(29 + 22)/2 = 25.5\,°C$ owing to the absorbed irradiance. Highly heat-absorbing 10 mm grey glass, where A is 0.71, would have a t_g of 44.9 °C. When the glass temperature is greater than that of the external air, there can be no U value convection and conduction heat gain into the room, only direct and diffuse radiation heat flow.

The heat flow rate into the room through the glass, Q, is

$$Q = h_{si}\, (t_g - t_{ai}) + T\, I_{TVd}$$

$$= 8.3 \times (30.5 - 22) + 0.78 \times 640$$

$$= 71 + 499\ \text{W/m}^2$$

$$= 570\ \text{W/m}^2$$

Q is a fraction of the available irradiance $(570/640) = 0.89$. Glass manufacturers may describe this as the total transmittance.

Further data on glass characteristics, combinations of glasses and the effect of internal and exterior shading devices are given in CIBSE (1986a) A5–14 and A5–15. Tabulated values of the plant-cooling load due to vertical glazing to maintain a constant internal dry resultant temperature in the UK are given in the CIBSE (1986a) Tables A9.14 and A9.15. These are for lightweight buildings, that are the most common modern design, and 10 h cooling plant operation in each 24 h period, used in commercial applications. The tables are for unshaded 6 mm clear glazing and the same with the intermittent use of light-coloured slatted blinds. Applications such as hospitals, manufacturing and containment buildings where continuous air conditioning is used, require other data.

HEAT GAINS THROUGH THE OPAQUE STRUCTURE

The opaque structural elements of the building, walls and roof, store solar heat gains for periods of up to 12 h prior to having a significant influence upon the internal resultant temperature. Their brick, concrete, steel and masonry gradually rise in temperature during prolonged periods of hot sunny weather, and remain warm. The time difference between the solar irradiance causing a heat gain to the outside surface of the structure and that same heat flow into storage, producing a corresponding heat flow into the interior, is important to the designer. The effect of cool evenings can be to produce an outflow of heat from the room, into the wall storage, at the time of peak solar heat gain into the room. The heat exchange between the room air and a dense wall at noon will be the result of heat loss from the outside wall surface to the external air at, perhaps, 9 h earlier, 0300 h.

The CIBSE method is to calculate the 24 h average heat flow through the structure and then the cyclic variation from the mean. The cyclic variation is calculated hourly and added to, or subtracted from, the mean heat flow. Thermal transmittance, U W/m² K, is used to calculate the mean flows and admittance, Y W/m² K, is used for the periodic oscillations. The mean conduction heat gains, or losses through an opaque structure Q_u are

$$Q_u = \frac{F_u}{F_v} A U (t_{eo} - t_c)$$

where

$$F_u = \frac{18 \, \Sigma A}{18 \, \Sigma A + \Sigma (A \, U)}$$

$$F_v = \frac{6 \, \Sigma A}{6 \, \Sigma A + 0.33 \, N \, V}$$

where t_{eo} = 24 h mean sol-air temperature, t_c = 24 h mean internal resultant temperature, A = surface area (m²), U = thermal transmittance (W/m² K), ΣA = summation of the room surface areas (m²), N = room air infiltration ventilation rate (air change per hour), and V = room volume (m³).

The cyclic variation from the mean heat flow $Q_{\tilde{u}}$ is

$$Q_{\tilde{u}} = \frac{F_y}{F_v} A U f t_{\tilde{eo}}$$

$$F_y = \frac{18 \, \Sigma A}{18 \, \Sigma A + \Sigma (A \, Y)}$$

where f = decrement factor, the ratio of the heat flow through the structure to the 24 h mean heat flow. Glazing is unity while dense concrete walls and roofs may be as low as 0.2 owing to their energy storage and release to outdoors and indoors. $t_{\tilde{eo}}$ = swing in the environmental temperature between that when the heat flow occurs at the external surface, and the 24 h mean environmental temperature. The swing will be negative when t_{eo} at the outdoor time of the heat gain is less than the 24 h mean t_{eo} and shows a loss of heat from the room. The time difference between these events is the time lag for the structure and this corresponds to the decrement factor reduction in the energy flows. Y = thermal admittance (W/m² K) is the rate of heat flow into the structure. It is the same as the U value for glass but higher than the U value for insulated, dense structures. F = surface factor is the ratio used for internal walls, floors and ceilings, and it relates to the variation in heat flows between rooms.

The net heat flow through the structure is the sum of the 24 h mean heat flow and the cyclic variation:

$$Q = Q_u + Q_{\tilde{u}} \text{ W}$$

EXAMPLE 3.15

An existing flat roof of $10\,\text{m} \times 8\,\text{m}$ is over an office that is to be maintained at a resultant temperature of 21 °C by an air-conditioning system. The thermal transmittance of the roof is $2.5\,\text{W/m}^2\,\text{K}$ and the admittance is $7\,\text{W/m}^2\,\text{K}$. The 24 h average outdoor environmental temperature on 21 June in Basingstoke is 26 °C. The roof decrement factor is 0.4 and its time lag is 7 h. The outdoor environmental temperature at 0500 h is 11.5 °C. The room height is 3 m. There are 1.5 air changes per hour due to the infiltration of outdoor air. Calculate the room cooling load through the roof at noon.

roof area $\Sigma A = 10 \times 8 \text{ m}^2$

$$= 80 \text{ m}^2$$

room volume $V = 10 \times 8 \times 3 \text{ m}^3$

$$= 240 \text{ m}^3$$

Mean 24 h cooling load Q_u:

$$Q_u = \frac{F_u}{F_v} A\, U\, (t_{eo} - t_c)$$

$$F_u = \frac{18\,\Sigma A}{18\,\Sigma A + \Sigma(A\,U)}$$

$$= \frac{18 \times 80}{18 \times 80 + \Sigma(80 \times 2.5)}$$

$$= \frac{1440}{1440 + 200}$$

$$= 0.88$$

$$F_v = \frac{6\,\Sigma A}{6\,\Sigma A + 0.33\,N\,V}$$

$$F_v = \frac{6 \times 80}{6 \times 80 + 0.33 \times 1.5 \times 240}$$

$$= \frac{480}{480 + 118.8}$$

$$= 0.8$$

$$Q_u = \frac{0.88}{0.8} \times 80 \times 2.5 \times (26 - 21)$$

$$= 1100\,\text{W}$$

The heat flow into the office at noon is due to the conditions 7 h earlier, at 0500 h. The swing in the environmental temperature between 0500 h and noon is

$$t_{\tilde{e}o} = 11.5 - 26$$

$$= -14.5\,°\text{C}$$

$$Q_{\tilde{u}} = \frac{F_y}{F_v} A U f t_{\tilde{eo}}$$

$$F_y = \frac{18 \times 80}{18 \times 80 + 80 \times 7}$$

$$= 0.72$$

$$Q_{\tilde{u}} = \frac{0.72}{0.8} \times 80 \times 2.5 \times 0.4 \times -14.5$$

$$= -1044 \text{ W}$$

net heat flow through the roof, $Q = Q_u + Q_{\tilde{u}}$ W

$$= 1100 - 1044$$

$$= 56 \text{ W}$$

PLANT COOLING LOAD

The cooling equipment in each room or building module is designed to remove the peak heat gain that occurs in that location. This peak load is found from a total of:

1. transmitted solar heat gain through the glazing;
2. U value convection and conduction heat gains through the glazing when appropriate;
3. the infiltration of warmer outdoor air directly into the room;
4. the net heat flows into the room through the opaque structure;
5. heat output from the occupants;
6. the electrical power input to lighting systems;
7. the electrical power input to office equipment;
8. the power input to electrical motors;
9. any heat output from machinery or industrial process equipment;
10. the heat content of high-temperature items that are transported into the air-conditioned room;
11. heat gains through the structure from adjacent warmer rooms;
12. ventilation air that enters from adjacent warmer rooms.

The transmitted heat gains from the direct and diffuse solar irradiances through the glazing are analysed as described earlier. Additional heat gains due to conduction and convection through the windows from the warmer outdoor air occur when the glass temperature is sufficiently low. The heat gain through the glazing is found from

$$Q_u = \frac{F_u}{F_v} A U (t_{ao} - t_c)$$

$t_{ao} = $ outdoor air temperature ($^\circ$C)

The decrement factor for glass, f, is 1, and the time lag is zero.

The infiltration of outdoor air can be avoided if the ducted air-conditioning system supplies more fresh air into the building than it mechanically extracts. This results in positive pressurization of the air within the building and causes leakage of air outwards. Under this circumstance, all the fresh outdoor air that is required for ventilation is supplied through the air-handling plant and is a cooling load upon the chilled-water coil there. This cooling duty is found from the psychrometric chart calculations for the air-handling plant. There is no fresh air cooling load within the room or its air-conditioning unit. The alternative is to extract more air from the building than is mechanically supplied to create a negative static air pressure within it. Warm outdoor air will leak inwards and create a ventilation cooling load on the room terminal. Such depressurization is used in containment buildings where it is essential to avoid the escape of contaminated air or particles and all the extract air is filtered to trap the dangerous or expensive matter. This is unlikely to be desirable for human-comfort air conditioning as untreated cold and hot air are continually drawn inwards. Where outdoor air is allowed to infiltrate the conditioned space, the instantaneous ventilation heat gain is calculated from

$$Q_v = 0.33 \, F_2 \, N \, V \, (t_{ao} - t_c)$$

$$F_2 = 1 + \frac{F_u \, \Sigma(A \, U)}{6 \, \Sigma A}$$

The heat output from the occupants is 90 W of sensible heat from an adult office worker in an air temperature of 22 °C (CIBSE, 1986a, Section A7) plus 50 W that is the latent heat equivalent of the evaporated moisture per person. Only the sensible heat is added to the cumulative gains for the room. The latent component is just moisture and does not influence the temperature of the room, only the person who is evaporating it. The designer needs to know how many occupants there will be and what their activity levels will be as sensible heat output varies strongly with metabolic rate. Heavy manual work produces up to 220 W sensible and 220 W latent heat outputs but these are likely to be for short periods of time. Hourly and daily schedules of occupancy heat gains are needed.

The electrical power input to lighting systems, photocopiers, computers, facsimile machines, vending machines and any other items, is emitted into the conditioned room as an equal amount of heat. This includes the visible light energy that is electro-magnetic radiation. Visible radiation is absorbed by the room surfaces by multiple reflections and an amount of energy is absorbed on each reflection depending upon the surface absorptivity. A proportion of the light energy may be radiated out of the building through windows and doorways. The control equipment of each luminaire emits heat and so does the wiring. Some or all of the lighting system heat output might be carried away from the room with ventilated luminaires that have room air directly extracted through them.

Fans, pumps, lift motors and other electric motor applications all dissipate heat into their surroundings and into the fluid being blown or pumped. Their efficiency is usually less than 50% and this proportion of their input kVA power is released into the conditioned room. Other machinery or industrial process equipment is similarly considered.

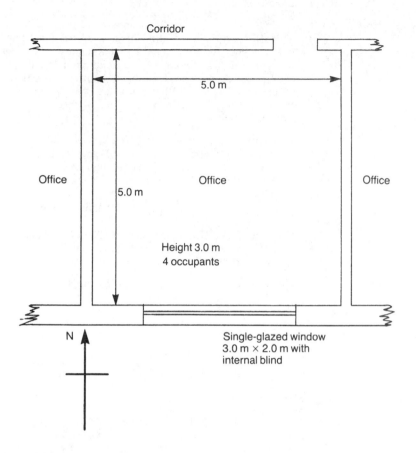

Fig. 3.16 Plan of office in Example 3.16.

High-temperature items may be transported into the air-conditioned room and their heat content that is above that of the room air temperature can be calculated from their mass and temperature. Other fuel-burning appliances such as gas-fired furnaces, water heaters or incinerators will have some of their input energy dissipated into the room though most of the waste heat will be flued.

Adjacent rooms that are at higher temperatures cause heat flows through the structure into the conditioned room and these are U value, area and temperature difference calculations hourly. Openings into adjacent rooms and corridors cause ventilation air movements that can lead to heat exchanges with the conditioned space. When static pressure differences exist across doorways and openings, the airflow rate is calculated from the area of opening and its flow coefficient depending upon its shape. The time duration of the opening allows calculation of the total quantity of air transferred. Otherwise, flow through the doorway is created by natural convection and an estimate can be made. Some flows between rooms will be deliberately created, such as from occupied rooms, through corridors, into toilet accommodation or kitchens and then extracted through ductwork.

EXAMPLE 3.16

A south-facing top-floor Southampton office is shown in Fig. 3.16. The exposure is normal. The office is to be maintained at a resultant temperature of 21 °C. There are four sedentary occupants and two continuously used computers having power consumptions of 200 W each. The air-conditioning plant operates for 10 h per day. There are 1.25 air changes per hour due to natural infiltration of outdoor air. The adjacent offices, corridor and office below are maintained at the same temperature as the example office. The constructional details and their CIBSE (1986a) table reference numbers are:

(a) double glazed window A3.14, clear float glass in an aluminium frame with thermal break and internal white-coloured venetian blinds;
(b) external wall A3.17, 27 (d) 100 mm heavyweight concrete block, 75 mm glass fibre, 100 mm lightweight concrete block, 13 mm lightweight plaster, dark exterior colour;
(c) internal walls A3.20, 5 (b) 100 mm medium weight concrete block, 13 mm lightweight plaster both sides;
(d) floor A3.21, 1(b) 50 mm screed, 150 mm cast concrete, 25 mm wood block floor;
(e) roof A3.19, 7(c) 19 mm asphalt, 13 mm fibreboard, 25 mm air gap, 75 mm glass fibre, 10 mm plasterboard, dark exterior colour.

Calculate the peak cooling load for the room air-conditioning unit.

The thermal data for each surface are shown in Table 3.3.

Table 3.3   Heat transfer data for Example 3.16

Surface	A (m^2)	U (W/m^2 K)	$A U$	Y (W/m^2 K)	$A Y$	f	Lag (h)
Glass	6	3.3	19.8	3.3	19.8	1	0
Ext. wall	9	0.33	3	2.4	21.6	0.35	9
Int. wall	45	1.7	0	3.5	157.5	0.72	1
Floor	25	1.5	0	2.9	72.5	0.7	1
Roof	25	0.4	10	0.7	17.5	0.99	1
$\Sigma A = 110$			$\Sigma (A U) = 32.8$		$\Sigma (A Y) = 288.9$		

Note that the internal walls and floor have no heat exchange with their adjacent rooms, so AU is zero. These internal surfaces do not have solar radiation decrement factors f or time lags, and surface factor F is used.

$$\text{office volume } V = 5 \times 5 \times 3 \text{ m}^3$$

$$= 75 \text{ m}^3$$

The cooling load due to the south-facing double clear float glazing for a lightweight building at 51.7 °N latitude with the intermittent use of internal slatted blinds is read from Table A9.15. South glazing always has a peak cooling load at noon sun time and the peak for the year occurs on 22 September and 21 March at 333 W/m^2. A correction factor of 0.74 applies when the blinds are closed. Notice the effects of different glasses, single glazing and the use and position of the blinds from the correction factors.

$$\text{solar gain through the window} = 333 \text{ W/m}^2 \times 0.74 \times 6 \text{ m}^2$$

$$= 1479 \text{ W}$$

Table 3.4 Sol-air data for Example 3.16

Surface	Lag (h)	24 h t_{eo}	24 h t_{ao}	Time (h)	t_{eo}	Swing $t_{\tilde{e}o}$
Ext. wall	9	25	15.5	0300	10.5	− 14.5
Roof	1	20	15.5	1100	38	18
					t_{ao}	$t_{\tilde{a}o}$
Window	0	—	15.5	1200	18.5	3

Find the 24 h mean sol-air temperatures and the sol-air temperatures at the time of the occurrence of the solar irradiance to the external surfaces from Table A2.33 (g). Table 3.4 shows the results. The time of irradiation of the wall is 9 h before noon, 0300 h and the t_{eo} then is 10.5 °C. The swing in the t_{eo} between the 24 h mean and the 0300 h value is − 14.5 °C. Similarly at 1100 h for the roof, the t_{eo} is 38 °C and this is 18 °C above the 24 h mean. The window is subject to air temperature heat transfers not sol-air temperatures and it has zero time lag. The outdoor air is at 18.5 °C at noon; that is a swing of 3 °C above the 24 h mean.

$$F_u = \frac{18\,\Sigma A}{18\,\Sigma A + \Sigma(A\,U)}$$

$$= \frac{18 \times 110}{18 \times 110 + 32.8}$$

$$= 0.98$$

$$F_v = \frac{6\,\Sigma A}{6\,\Sigma A + 0.33\,N\,V}$$

$$= \frac{6 \times 110}{6 \times 110 + 0.33 \times 1.25 \times 75}$$

$$= 0.96$$

$$F_y = \frac{18\,\Sigma A}{18\,\Sigma A + \Sigma(A\,Y)}$$

$$= \frac{18 \times 110}{18 \times 110 + 288.9}$$

$$= 0.87$$

$$F_2 = 1 + \frac{F_u\,\Sigma(A\,U)}{6\,\Sigma A}$$

$$F_2 = 1 + \frac{0.98 \times 32.8}{6 \times 110}$$

$$= 1.05$$

Calculate the 24 h mean conduction heat gains with

$$Q_u = \frac{F_u}{F_v}\,A\,U\,(t_{eo} - t_c)$$

For the external wall:

$$Q_u = \frac{0.98}{0.96} \times 9 \times 0.33 \times (25 - 21)$$

$$= 1.02 \times 9 \times 0.33 \times 4$$

$$= 12\ W$$

For the roof:

$$Q_u = 1.02 \times 25 \times 0.4 \times (20 - 21)$$

$$= -10\ W$$

For the window:

$$Q_u = 1.02\ A\ U\ (t_{ao} - t_c)$$

$$= 1.02 \times 6 \times 3.3 \times (15.5 - 21)$$

$$= -111\ W$$

The 24 h mean conduction gain $= 12 - 10 - 111\ W$

$$= -109\ W, \text{ a net loss.}$$

The cyclic variation from the mean heat flow $Q_{\tilde{u}}$, is

$$Q_{\tilde{u}} = \frac{F_y}{F_v}\ A\ U f\ t_{\tilde{eo}}$$

For the external wall:

$$Q_{\tilde{u}} = \frac{0.87}{0.96} \times 9 \times 0.33 \times 0.35 \times (-14.5)$$

$$= 0.91 \times 9 \times 0.33 \times 0.35 \times (-14.5)$$

$$= -14\ W$$

For the roof:

$$Q_{\tilde{u}} = 0.91 \times 25 \times 0.4 \times 0.99 \times 18$$

$$= 162\ W$$

For the window:

$$Q_{\tilde{u}} = 0.91\ A\ U f\ t_{\tilde{ao}}$$

$$= 0.91 \times 6 \times 3.3 \times 1 \times 3$$

$$= 54\ W$$

The net swing in conduction gains $= -14 + 162 + 54$

$$= 202\ W$$

The outdoor air that is at 18.5 °C at noon infiltrates the conditioned office causing a heat gain of

$$Q_v = 0.33\ F_2\ N\ V\ (t_{ao} - t_c)$$

$$= 0.33 \times 1.05 \times 1.25 \times 75 \times (18.5 - 21)$$

$$= -81\ W$$

The four occupants each emit 90 W:

$$Q = 4 \times 90 \text{ W}$$
$$= 360 \text{ W}$$

The two computers each emit 200 W:

$$Q = 2 \times 200 \text{ W}$$
$$= 400 \text{ W}$$

The total net heat gain Q to the office at noon on 22 September and 21 March is the sum of:

1. solar gain through the glazing, 1479 W;
2. 24 hour mean conduction through the structure, -109 W;
3. net swing in the conduction gain, 202 W;
4. ventilation air infiltration, -81 W;
5. occupancy gain, 360 W;
6. electrical equipment emission, 400 W.

$$Q = 1479 - 109 + 202 - 81 + 360 + 400 \text{ W}$$
$$= 2251 \text{ W}$$

A close approximation to this answer could have been acquired by ignoring the conduction gains through the structure and the cooling effect of the infiltration, 2239 W; however, this may not be appropriate in every case.

PEAK SUMMERTIME TEMPERATURE IN A BUILDING

During warm sunny weather, buildings without any method of lowering the internal temperature may become overheated and uncomfortable for normal work or habitation. The upper limit of acceptability may be as high as an internal environmental temperature of 27 °C. Such a choice is arbitrary and does not take account of the glaring effect of direct solar irradiance upon the person, their activity level, clothing, thermal insulation or temperature and speed of the air around them. An assessment of the peak summertime temperature within a building ought to be made before the decision to design a cooling system is made. The provision of low-cost cooling systems can be investigated.

The internal temperature will be the combination of 24 h mean heat gains, cyclic gains producing temperature swings about the mean and the heat loss from the internal environment. Some of this heat loss might be accomplished with some form of cooling. The final temperature reached is a balance between gains and losses, some of which are under the potential control of the occupier or engineer. Painting the exterior of the roof with white paint or spraying water on to the roof can reduce the heat gains. Some large areas of glass that were built during the 1950s and 1960s, when there was little thought given to energy economy, are known to have been painted white. Additional mechanical ventilation with outdoor air that is already at 25 °C to 30 °C or more may not produce human comfort conditions. It may be sufficient to avoid the overheating of hardware such as stored goods and operational electric motors. The increased air velocity around the personnel will alleviate discomfort and may produce tolerable conditions. Ideally there needs to be manual control over the direction and velocity of increased air circulation as weather conditions vary quickly.

The method of assessing the peak environmental temperature is to calculate the:

1. 24 h mean solar heat gains;
2. 24 h mean internal heat gains;
3. mean internal environmental temperature from the known gains and the 24 h mean external air temperature;
4. peak swing in heat gains above the 24 h mean;
5. swing in environmental temperature due to the swing in heat gains;
6. peak environmental temperature (the mean plus the swing values).

The mean heat gain from people, lights and equipment is found by multiplying their power by the hours of usage and dividing by 24 h. The 24 h mean solar gains come from the daily mean total irradiance (CIBSE, 1986a, Table A8.1) and correction factors for shading.

$$\text{mean gain } Q = \Sigma(A_g U_g)(t_{ei} - t_{ao}) + 0.33 N V (t_{ei} - t_{ao}) + \Sigma(A_f U_f)(t_{ei} - t_{eo})$$

where A_f = area of opaque fabric (m^2),
A_g = area of window (m^2),
U_f = thermal transmittance of opaque fabric (W/m^2 K),
U_g = thermal transmittance of window (W/m^2 K),
t_{ei} = mean internal environmental temperature (°C),
t_{eo} = mean external environmental temperature (°C),
t_{ao} = mean external air temperature (°C). The mean internal environmental temperature t_{ei} is found by rearranging the equation.

EXAMPLE 3.17

A south-facing Brighton office of 6 m × 6 m × 3 m high has single glazed clear float window openings of 10 m^2. The surrounding rooms are all similar. There are three occupants emitting 90 W and four electrical items of 150 W each. The office is used for 8 h in each 24 h. Its windows and door are shut at other times and the ventilation rate is 1 air change per hour. Use the data provided to estimate the 24 h mean internal environmental temperature. The peak solar irradiance on a south-facing vertical window is 710 W/m^2 at noon on 22 September in southeast England and the daily mean is 200 W/m^2. The solar gain correction factor for the glazing without blinds is 0.76. The thermal data are shown in Tables 3.5 and 3.6.

$$\text{mean solar gain} = 200 \text{ W/m}^2 \times 10 \text{ m}^3 \times 0.76$$

$$= 1520 \text{ W}$$

$$\text{mean internal gain} = \frac{(3 \times 90 \text{ W} \times 8 \text{ h}) + (4 \times 150 \text{ W} \times 8 \text{ h}) \, 24 \text{ h}}{24}$$

$$= 290 \text{ W}$$

$$\text{total mean gain } Q = 1520 + 290 \text{ W}$$

$$= 1810 \text{ W}$$

Table 3.5 Heat transfer data for Example 3.17

Surface	$A(m^2)$	$U(W/m^2 K)$	AU	$Y(W/m^2 K)$	AY	f	$Lag(h)$
Glass	10	5.7	57	57	57	1	0
Ext. wall	8	0.4	3.2	4.3	34.4	0.18	10
Int. wall	54	1.7	0	3.5	189	0.72	1
Floor	36	1.7	0	5.2	187.2	0.72	3
Ceiling	36	1.7	0	2.2	79.2	0.86	1

$$\Sigma(A\,Y) = 546.8$$

Table 3.6 Sol-air data for Example 3.17

Surface	$Lag(h)$	$24\,h\,t_{eo}$	$24\,h\,t_{ao}$	$Time(h)$	t_{eo}	$Swing\;t_{\tilde{e}o}$
Ext. wall	10	25	15.5	0200	11	-14
					t_{ao}	$t_{\tilde{a}o}$
Window	0	–	15.5	1200	18.5	3

$$Q = \Sigma(A_g\,U_g)\,(t_{ei} - t_{ao}) + 0.33\,N\,V\,(t_{ei} - t_{ao}) + \Sigma(A_f U_f)\,(t_{ei} - t_{eo})$$

$\Sigma(A_g\,U_g) = 57$ W/K

$\Sigma(A_f\,U_f) = 3.2$ W/K

$N = 1$ air change per hour

$V = 6 \times 6 \times 3$ m^3

$\quad = 108$ m^3

mean $t_{ao} = 15.5\,°C$

mean $t_{eo} = 25\,°C$

$1810 = 57 \times (t_{ei} - 15.5) + 0.33 \times 1 \times 108 \times (t_{ei} - 15.5) + 3.2 \times (t_{ei} - 25)$

$1810 = 57 \times t_{ei} - 883.5 + 35.6 \times t_{ei} - 552.4 + 3.2 \times t_{ei} - 80$

$1810 + 883.5 + 552.4 + 80 = 57 \times t_{ei} + 35.6 \times t_{ei} + 3.2 \times t_{ei}$

$3325.9 = 95.8\,t_{ei}$

$t_{ei} = 34.7\,°C$

The 24 h mean environmental temperature exceeds the recommended 27 °C and, unless the swing in heat gains is a large negative quantity, discomfort will result. A positive swing is expected owing to the additional irradiation through the glazing at noon. The swings about the mean of the environmental and air temperatures are shown in Table 3.6. An alternating solar gain factor is applied to the increase of the noon irradiance above the 24 h mean value from CIBSE (1986a) Table A8.6. This is 0.65 for clear single glazing in a lightweight building. The use of heat-absorbing or reflective glass, double glazing or blinds produces lower factors.

Calculate the total swing in the solar heat gains through the glazing, from the ventilation air, internal sources and the structure between the 24 h mean and the peak, $Q_{\tilde{i}}$. The swing, mean to peak, in the internal environmental temperature $t_{\tilde{e}i}$ is then found from

$$Q_{\tilde{i}} = [\Sigma(A\,Y) + 0.33\,N\,V]\,t_{\tilde{e}i}$$

The final room environmental temperature is the sum of the mean and swing figures.

EXAMPLE 3.18

Continue Example 3.17 to calculate the office peak environmental temperature. The alternating solar gain factor for the glass is 0.65.

$$\text{swing in solar irradiation gain} = 0.65 \times (710 - 200)\ \text{W/m}^2 \times 10\ \text{m}^2$$

$$= 3315\ \text{W}$$

$$\text{swing in S wall heat gain} = f A\, U\, t_{\widetilde{eo}}$$

$$= 0.18 \times 8\ \text{m}^2 \times 0.4\ \text{W/m}^2\,\text{K} \times (-14)\ \text{K}$$

$$= -8\ \text{W}$$

$$\text{swing in glass conduction} = f A\, U\, t_{\widetilde{ao}}$$

$$= 1 \times 10\ \text{m}^2 \times 5.7\ \text{W/m}^2\,\text{K} \times 3\ \text{K}$$

$$= 171\ \text{W}$$

$$\text{swing in ventilation gain} = 0.33\, N\, V\, t_{\widetilde{ao}}$$

$$= 0.33 \times 1 \times 108 \times 3\ \text{W}$$

$$= 107\ \text{W}$$

$$\text{swing in internal gains} = \text{peak gain} - \text{mean gains}$$

$$= (3 \times 90) + (4 \times 150) - 290\ \text{W}$$

$$= 580\ \text{W}$$

$$\text{total swing in gains } Q_{\widetilde{t}} = 3315 - 8 + 171 + 107 + 580\ \text{W}$$

$$= 4165\ \text{W}$$

From Table 3.5

$$\Sigma (A\, Y) = 546.8\ \text{W/K}$$

$$Q_{\widetilde{t}} = [\Sigma (A\, Y) + 0.33\, N\, V]\, t_{\widetilde{ei}}$$

$$4165 = (546.8 + 0.33 \times 1 \times 108) \times t_{\widetilde{ei}}$$

$$4165 = 582.4\, t_{\widetilde{ei}}$$

$$t_{\widetilde{ei}} = 7.2\ ^\circ\text{C}$$

$$\text{peak environmental temperature} = 34.7\ ^\circ\text{C} + 7.2\ ^\circ\text{C}$$

$$t_{ei} = 41.9\ ^\circ\text{C}$$

The result confirms the earlier conclusion and shows that a south-facing conservatory is unsuitable for work. Increased ventilation, heat-absorbing or reflecting glass and shading devices can be investigated.

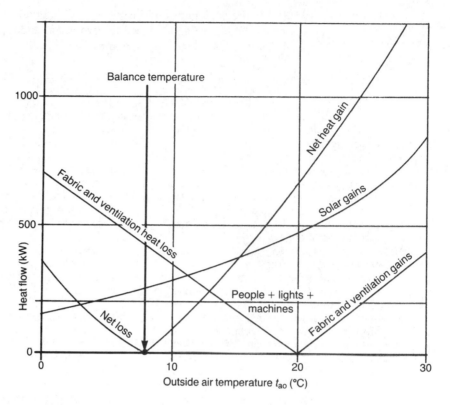

Fig. 3.17   Balance temperature for an air-conditioned building.

BALANCE TEMPERATURE

The balance temperature of a building is that outdoor air temperature at which there is no net heat loss or gain between the internal and external environments. At this condition, the heat losses to the cooler outdoor air are balanced by the solar and internal heat gains. A low balance temperature either means that the building is highly insulated and makes efficient use of the available heat gains, or it has very high internal heat generation from people and equipment. The heat flows into and out of the building need to be known for both the peak summer and winter conditions and at intermediate outdoor temperatures. If only the peak values are known, it may be assumed that the gains and losses are straight-line graphs between these points. Some inaccuracy will result, as solar gains have a sinusoidal variation.

Figure 3.17 shows the variation of heat losses and gains for an office building in London where:

(a) there is a constant heat gain of 200 kW from the occupants and equipment during the working day;
(b) thermal transmittance and ventilation heat losses occur up to an outdoor air temperature of 20 °C; above 20 °C these turn into heat gains;
(c) solar heat gains follow the curves shown.

Addition of the heat gains and losses at 2 °C intervals produces the net heat loss and net heat gain curves that balance to zero at an outdoor air temperature of 8 °C. This is the building's balance temperature.

Questions

Data from the *CIBSE Guide Vol. A* (CIBSE, 1986a) is provided for these questions and the reader should have access to it to verify the information and to become familiar with its use.

1. State the sources of heat gain that will affect the internal thermal environment in residences, offices, retail premises, an atrium in a shopping concourse, a pharmaceutical manufacturing building, an air-conditioning plantroom, a railway passenger carriage, a motor car, an aeroplane, and an entertainment theatre.

2. Explain how heat gains to an occupied building have intermittent characteristics.

3. Describe the angular position of the sun in relation to the four walls and the roof of a building. Define each term used.

4. Explain what is meant by the incidence of solar radiation upon a surface and define how it is found for vertical walls, horizontal roofs and sloping surfaces. State the reason for needing to know the incidence in the calculation of solar radiation heat gains to a surface.

5. A surveyor needed to assess the height of a tower and measured an angle of 52° from the ground to the top of the tower at a distance of 35 m from it. Calculate the building height.

6. A 19-storey office building was being built on a hillside and the surveyor needed to check the height of the structural steelwork during construction. The nearest point of the top of the structure was at an elevation of 40° from the surveyor's position and 55 m horizontally away. The furthest point of the steel frame down the hill was 75 m from the surveyor when measured down the slope of the hill and at a decline angle of 12°. Calculate the overall height of the frame and the average floor to floor height.

7. A building has a side 30 m long that slopes forwards from ground level at 78° from the horizontal. The roof line is at 66° from the ground at a distance of 32 m from the foot of the wall. Calculate the vertical height of the building, the length of the sloping face and the sloping face area.

8. A vertical wall in London faces south. At 1000 h sun time on 22 June the solar altitude is 55° and solar azimuth 128°. Calculate the wall solar azimuth and incidence angles.

9. Atrium glazing in Glasgow, latitude 55° 52'N, faces west and slopes 12° forwards from the vertical. The sun time is 1300 h on 21 August, the solar altitude is 45° and the solar azimuth is 201°. The intensity of the direct solar radiation normal to the sun is 840 W/m². Calculate the incidence angle and the solar intensity upon the glazing.

10. A flat-plate solar collector near Sydney at a latitude of 35° S slopes at 45° to the horizontal and faces north. At 1200 h sun time on 21 December the solar altitude is 78°, azimuth 0° and direct intensity normal to the sun is 918 W/m². Find the incidence angle by sketching the angles and not applying the general equation then calculate the solar intensity that is normal to the collector surface.

11. A solar collector in southeast England faces due S at an angle of 40° to the horizontal. Calculate the maximum solar irradiance and state when this will occur.

12. A window facing south is 4 m long and 3 m high and has a shaded area of 5 m². Solar altitude is 55°, solar azimuth is 12° E of S, direct solar irradiance normal to the sun is 540 W/m² and diffuse irradiance is 210 W/m². The glazing transmissibilities of direct and diffuse irradiances are 0.74 and 0.69. Calculate the solar heat gain transmitted through the glass.

13. The upper glazing on an atrium slopes at 60° to the horizontal and faces southeast. It is 2 m long and 2 m high and has a shaded area of 1 m². At 1000 h on 22 September the design direct solar irradiance on a horizontal surface, l_{DHd}, is 355 W/m². The design direct solar irradiance on a vertical surface, l_{DVd}, is 525 W/m². The design diffuse irradiance l_{dHd} is 160 W/m². The glazing transmissibilities of direct and diffuse irradiances are 0.46 and 0.4. Calculate the solar heat gain transmitted through the glass.

14. Find the glass temperature t_g, total heat flow into the room Q and the total transmittance T for 6 mm silver reflective float glass that is in a total solar irradiance of 625 W/m² at 0800 h on 21 June facing east. The external and internal air temperatures are 31 °C and 23 °C. The absorptivity A of the glass is 0.29, the transmissibility T is 0.43 and reflectance R is 0.28. The glass is low-emissivity and has values of $R_{si} = 0.3$ m² K/W and $R_{so} = 0.07$ m²K/W.

15. Find the glass temperature tg, direction of heat flow Q and the total transmittance T for 10 mm bronze-tinted heat-absorbing float glass that is in a total sol ar irradiance of 175 W/m² at 1300 h on 21 December facing west. The external and internal air temperatures are – 2 °C and 18 °C. The

absorptivity A of the glass is 0.67, the transmissibility T is 0.29 and reflectance R is 0.04. The glass is high-emissivity and has values of $R_{si} = 0.12 m^2$ K/W and $R_{so} = 0.03 m^2$K/W due to its exposed location.

16. A new light-coloured flat roof of $25 m \times 15 m$ is over a production area that is to be maintained at a resultant temperature of 19 °C by an air-conditioning system. The thermal transmittance of the roof is $0.4 W/m^2$ K and the admittance is $0.7 W/m^2$ K. The 24 h average outdoor environmental temperature for a roof on 23 July in Southampton is 22 °C. The roof decrement factor is 0.99 and its time lag is 1 h. The outdoor environmental temperature at 1300 h is 37.5 °C. The room height is 3 m and it has 0.5 air changes per hour due to the infiltration of outdoor air. Calculate the room-cooling load through the roof for 1400 h.

17. A south-facing office wall is 30 m long and 4 m high. The thermal transmittance of the wall is 0.33 W/m^2 K and the admittance is $2.4 W/m^2$ K. The office has a volume of $1440 m^3$ and is to be maintained at a resultant temperature of 20 °C by an air-conditioning system. The 24 h average outdoor environmental temperature for a dark-coloured wall on 23 July in Bournemouth is 26°C. The wall decrement factor is 0.35 and its time lag is 9 h. The outdoor environmental temperature at 0800 h is 25.5 °C. The room has 1.5 air changes per hour due to the infiltration of outdoor air. Calculate the room cooling load through the wall for 1700 h.

18. A southeast-facing top-floor London office is similar to that shown in Fig. 3.16. The exposure is normal. The office is to be maintained at a resultant temperature of 19 °C. There are two sedentary occupants and two continuously used computers having power consumptions of 250 W each. The air conditioning plant operates for 10 h/day. There are 1.5 air changes per hour due to natural infiltration of outdoor air. The adjacent offices, corridor and office below, are maintained at the same temperature as the example office. Room volume V is $600 m^3$. The glazing cooling load peak is $313 W/m^2$ at 1100 h on 22 September and the correction factor for glass and shading types is 0.55. Use the data provided and calculate the peak cooling load for the room air-conditioning unit.

The thermal data are shown in Tables 3.7 and 3.8.

Table 3.7 Heat transfer data for question 18

Surface	A (m^2)	U (W/m^2 K)	AU	Y (W/ m^2 K) A	Y	f	Lag (h)
Glass	25	3.3		3.3		1	0
Ext. wall	60	0.3		2.7		0.28	8
Int. wall	100	1.5		3		0.72	1
Floor	200	1.2		2.4		0.7	1
Roof	200	0.25		0.8		0.99	1
$\Sigma A =$		$\Sigma(AU) =$		$\Sigma(AY) =$			

Table 3.8 Sol-air data for question 18

Surface	Lag (h)	24 h t_{eo}	24 h t_{ao}	Time(h)	t_{eo}	Swing $t\tilde{e}o$
Ext. wall	8	23.5	15.5	0300	10.5	
Roof	1	20	15.5	1000	33.5	
					t_{ao}	$t\tilde{a}o$
Window	0	–	15.5	1100	17	

19. The top floor of an office building in Bristol is to be maintained at a resultant temperature of 20 °C. Use the data provided to calculate the peak cooling load for the top-floor office.

The air-conditioning plant operates for 10 h per day. There is 0.75 of an air change per hour due to natural infiltration of outdoor air and 0.25 of an air change per hour from uncooled air entering from the rooms on the floors below. Only the top floor of the building is cooled. The lower floors are expected to be at an air temperature of 26 °C at times of peak cooling load. The top floor is a rectangular open-plan general office 40 m long, 15 m wide and 3.5 m high with one long side facing south. There is $40 m^2$ of glass on the north and south sides but no glazing on the east and west walls. There are 40 occupants each emitting 90 W, 20 computers of 200 W each, one photocopier of 400 W and 10 fluorescent lamps permanently used of 65 W each.

The thermal data are shown in Tables 3.9 and 3.10.

Table 3.9 Heat transfer data for question 19

Surface A (m^2)	U (W/m^{2}K)	AU	Y (W/m^{2}K)A	f	Lag (h)
S glass	5.7		5.7	1	0
N glass	3.3		3.3	1	0
S wall	0.3		2.5	0.3	8
E wall	0.4		3	0.7	9
W wall	0.4		3	0.7	9
N wall	0.3		2.5	0.3	8
Floor	1.8		2.8	0.7	1
Roof	0.25		0.9	0.9	2
$\Sigma A =$	$\Sigma(AU) =$		$\Sigma(AY) =$		

It is expected that the peak cooling load for the top floor will occur at the time and date of the peak irradiance on the south glazing, that is noon sun time on 22 September and 21 March. This might not be the case and other times and dates could be analysed for comparison. Data from CIBSE (1986a) Table A9.15: south single glazing has a cooling load of $333 W/m^2$ and a correction factor of 0.77, the north double glazing has a cooling load of $104 W/m^2$ and a correction factor of 0.95.

Table 3.10 Sol-air data for question 19

Surface	Lag(h)	24 hteo	24 htao	Time (h)	teo	Swing $t\tilde{e}\tilde{o}$
S wall	8	25	15.5	0400	10.5	
E wall	9	20.5	15.5	0300	10.5	
W wall	9	20.5	15.5	0300	10.5	
N wall	8	17	15.5	0400	10.5	
Roof	2	20	15.5	1000	33.5	
					$t\tilde{a}\tilde{o}$	$t\tilde{a}\tilde{o}$
S glass	0	–	15.5	1200	18.5	
N glass	0	–	15.5	1200	18.5	

20. Calculate the net heat transfer through a west-facing single-glazed window for 22 April at 1400 h when the cooling load is 207 W/m^2 with a shading correction factor of 0.77, F_u/F_v is 0.9, F_u/F_y is 0.95, outside air temperature is 14.8°C, room resultant temperature is 21°C, 24 h mean external air temperature is 9°C, window U value is 5.7 W/m^2 and dimensions are 2.5 m × 15 m.

21. List the ways in which the south-facing office in Example 3.17 could be made comfortable with passive architectural changes and mechanical cooling methods.

22. A west-facing Plymouth office 15 m × 5 × 3 m high has double-glazed gold-coloured heat-reflecting window openings of 25 m^2. The surrounding rooms are all similar. There are six occupants emitting 90 W and five electrical items of 15 W each. The office is used for 8 h in each 24 h. Its windows and door are shut at other times and the ventilation rate is 1.5 air changes per hour. Use the data provided to estimate the 24 h mean and peak internal environmental temperatures. The peak solar irradiance on a west-facing vertical window is 625 W/m^2 at 1600 h on 21 June in south-east England and the daily mean is 185 W/m^2. The mean solar gain correction factor for the glazing without blinds is 0.25 and the alternating factor is 0.2. The thermal data are shown in Tables 3.11 and 3.12.

Table 3.11 Heat transfer data for question 22

Surface	A (m^2)	U (W/m^{2}W) AU	Y (W/m^{2}K) A Y	f	Lag (h)
Glass	3.3	3.3		1	0
Ext. wall	0.57	3.6		0.31	9
Int. wall	1	3.6		0.62	1
Floor	2	4.3		0.59	2
Ceiling	2	6		0.46	3
			Σ (AY) =		

Table 3.12 Sol-air data for question 22

Surface	Lag (h)	24 h $t$eo	24 h $t$ao	Time (h)	$t$eo	Swing $t\tilde{e}\tilde{o}$
Ext. wall	9	24.5	16.5	0700	15.5	
					$t$ao	$t\tilde{a}\tilde{o}$
Window	0	–	16.5	1600	22	

23. A vehicle production factory that was constructed in 1960 at Southampton is 100 m × 100 m × 10 m high and has no glazing. The structural steel frame is clad in corrugated metal sheet with no insulation for the walls and a lightweight flat roof, all painted white externally. The workforce of 200 occupies two 8 h shifts per 24 h emitting 110 W each. Heat-producing motors, lights, tools and processes generate 50 kW during the working periods. The 100% outdoor air mechanical ventilation systems operate for 24 h per day and 7 days per week and produce 1 air change per hour. The thermal data are shown in Tables 3.13 and 3.14. Use the data provided to estimate the peak internal environmental temperature.

Table 3.13 Heat transfer data for question 23

Surface A (m^2)	U (W/m^2 K) A U	Y (W/m^2 K) A Y	f	Lag (h)
S wall	5.7	5.7	1	0
E wall	5.7	5.7	1	0
W wall	5.7	5.7	1	0
N wall	5.7	5.7	1	0
Floor	1.7	5.2	0.72	3
Roof	1.1	1.2	0.99	1
	Σ (AU) =	Σ (AY) =		

Table 3.14 Sol-air data for question 23

Surface	Lag (h)	24 h $t$eo	24 h $t$ao	Time (h)	$t$eo	Swing $t\tilde{e}\tilde{o}$
S wall	0	22.5	19	1200	36	
E wall	0	22.5	19	1200	26.5	
W wall	0	22.5	19	1200	26.5	
N wall	0	20	19	1200	26	
Roof	1	22	19	1100	34.5	
					$t$ao	$t\tilde{a}\tilde{o}$
Air	0	–	19	1200	21.5	

24. Describe, with the aid of sketches, how shading devices and glass types are used to assist in the provision of thermal comfort within buildings and how they affect the natural illumination and cooling plant loads.

25. A window 3 m long and 2 m high is recessed 200 mm from the face of a building. The wall azimuth is 170° and the solar azimuth at 1300 h sun time is 190°. Calculate the shade width on the window.

26. A window 3.5 m long and 1.75 m high is recessed 150 mm from the face of the building. The solar altitude at 1500 h on 24 August is 37° and the solar azimuth is 240°. The wall azimuth is 275°. Calculate the unshaded area of the glass.

27. State the ways in which the shaded areas of buildings can be predicted and the use of this information.

28. Calculate the net heat transfer through a south-facing single-glazed window on 21 December at a latitude of 51.7°N at noon when the solar irradiance cooling load is 273 W/m^2, the outside air temperature is 5°C and the room resultant

temperature is 20°C. The 24 h mean outdoor air temperature is 2°C. The window is 2.5 m × 2.5 m and its thermal transmittance is 5.7 W/m^2 K.

29. Explain the use of the peak summertime internal environmental temperature in the analysis of the thermal comfort conditions. Include in your explanation when it is needed to be calculated, what importance it has to the building owner and the design engineer and what part it plays in the decisions to be made on the choice of air-conditioning system.

30. Explain why thermal admittance, decrement factor, surface factor, structural time lag and environmental temperature swing are used in preference to thermal transmittance in the assessment of summer internal conditions.

31. List the advice that can be given to the owner of a building that suffers from summer overheating. Explain why the use of additional outdoor air mechanical ventilation may be an unsuitable solution for some applications but correct for some buildings.

32. Draw a graph showing the balance temperature for a building from the following data:
    (a) constant heat gain from the occupants, lights and electrical equipment of 10 kW;
    (b) solar gains of 20 kW at an outdoor air temperature of 0°C, 30 kW at 10°C, 50 kW at 20°C and 70 kW at 30°C;
    (c) conduction and ventilation heat loss of 90 kW at 0°C, zero at 21°C and a heat gain of 30 kW at 30°C.

4 Fluid flow

INTRODUCTION

All air-conditioning systems rely upon the flow of fluid through pipes and ducts. The derived equations used here for airflow in ducts, hot and chilled water and methane in pipes, have advantages over solely using charts and tables, in that the student's own calculator and computer can reproduce published data with acceptable accuracy. The arrangements, formulae and data for spreadsheets for undertaking these calculations are provided.

The heat-carrying capacity of low-, medium- and high-temperature hot-water pipe and air-duct distributions are analysed and sizes compared. The performance of thermal storage arrangements is calculated. The pump and fan power consumption of distribution systems is calculated.

All the stages of calculations are shown once. Repeated parts are shown in edited form.

LEARNING OBJECTIVES

Study of this chapter will enable the user to:

1. calculate the flow rate of water, methane and air through pipes and ducts;

2. reproduce pipe and duct-size data published in the *CIBSE Guide* book C;

3. find pipe and duct sizes to satisfy heating and cooling loads;

4. write a pipe-sizing program into a calculator;

5. calculate pipe and duct sizes with computer spreadsheets;

6. compare the heat-carrying capacity of hot-water pipe and air-duct systems;

7. calculate the energy consumed by pump and fan heat-distribution systems;

8. understand how thermal storage systems are used;

9. apply thermal storage to heating and cooling systems.

PIPE AND DUCT EQUATION

The sizes of pipes and ducts are found from charts and extensive tabulated data in the Chartered Institution of Building Services Engineers *Guide Vol. C* (CIBSE, 1986c) and design engineers will have this within an arm's length at all times. Fluid flows are invariably fully developed turbulent, as laminar flows would be at very low velocity, require large pipe diameters and produce no benefit. The formula describing such flows is the D'Arcy equation. The Colebrook and White equation is used to find the design friction factor for the D'Arcy formula. Insertion of the appropriate fluid properties and pipe material factors produces a usable equation that can acceptably replicate published data.

Water flow at 10 °C

For water at 10 °C in copper pipe, Table X, BS 2871: Part 1:1971; the mass flow rate of water can be found from

$$M = -70.24 \, \Delta p^{0.5} \, d^{2.5} \log \left(\frac{4.05 \times 10^{-7}}{d} + \frac{7.33 \times 10^{-5}}{\Delta p^{0.5} \, d^{1.5}} \right)$$

where M = mass flow rate of water at 10°C (kg/s), Δp = pressure drop per metre of pipe (Pa/m), d = pipe internal diameter (m). The internal diameters of pipes are given in Table 4.1.

Table 4.1 Diameters of copper pipe Table X, B.S. 2877: Part 1

Nominal diameter (mm)	*Internal diameter* (m)
6.0	0.004 80
8.0	0.006 80
10.0	0.008 80
12.0	0.010 80
15.0	0.013 60
22.0	0.020 22
28.0	0.026 22
35.0	0.032 63
42.0	0.039 63
54.0	0.051 63

The equation is solved by inserting known or estimated values for Δp and d, evaluating M and checking to see if the flow rate exceeds the design requirement.

Table 4.2 HP-41C calculator listing for water flow at 10 °C in a Table X copper pipe

Step	Instruction	Explanation
01	LBL "TABLE X PIPE"	Program title
02	4.05 E − 7	4.05×10^{-7}
03	STOP	Enter d(m)
04	STO 01	Store d in store 01
05	/	Divide
06	7.33 E − 5	7.33×10^{-5}
07	RCL 01	Recall d from store 01
08	1.50	
09	Y^X	Evaluates $d^{1.5}$
10	/	Divide
11	STOP	Enter Δp (Pa/m)
12	STO 02	
13	SQRT	Square root of Δp
14	/	Divide
15	+	Add
16	LOG	Evaluate logarithm
17	RCL 01	
18	2.50	
19	Y^X	Evaluates $d^{2.50}$
20	×	Multiplication
21	RCL 02	
22	SQRT	Square root
23	×	Multiply
24	70.24	
25	×	Multiply
26	CHS	Change sign to −
27	END	Displays M (kg/s)

If not, a second estimate is tried. In the case of hot- and cold-water pipework for tap services, the pressure drop rate will be known from the available head and index circuit length, needing only diameter to be estimated. Table 4.2 shows the listing for a Hewlett-Packard HP-41C programmable calculator to solve this problem.

EXAMPLE 4.1

Calculate the flow rate of water at 10 °C that will flow through a 15.0 mm Table X copper pipe when the frictional resistance is 1500 Pa/m.

For this pipe, $d = 0.01360$ m and $\Delta p = 1500$ Pa/m. Insert values in:

$$M = -70.24 \, \Delta p^{0.5} \, d^{2.5} \log \left(\frac{4.05 \times 10^{-7}}{d} + \frac{7.33 \times 10^{-5}}{\Delta p^{0.5} \, d^{1.5}} \right)$$

First evaluate log bracket:

$$\text{value} = \log \left(\frac{4.05 \times 10^{-7}}{0.0136} + \frac{7.33 \times 10^{-5}}{1500^{0.5} \times 0.0136^{1.5}} \right)$$

$$= \log \left(0.000\,029\,7 + 0.001\,193\,3 \right)$$

$$= -2.9125$$

and

$$M = -70.24\,\Delta p^{0.5}\,d^{2.5}\,(-2.9125)$$
$$= -70.24 \times 1500^{0.5} \times 0.0136^{2.5} \times (-2.9125)$$
$$= 0.171\,\text{kg/s}$$

EXAMPLE 4.2

Find the pipe diameter required to carry 0.780 kg/s of water at 10 °C if the available pressure loss rate is 1000 Pa/m in Table X copper pipe.

Try a 28.0 mm pipe having $d = 0.026\,22$ m, $\Delta p = 1000$ Pa/m. From the flow equation, $M = 0.803$ kg/s; this is greater than the design requirement and is satisfactory.

A spreadsheet program to carry out this calculation greatly aids the designer. Enter the one in Table 4.3 into your computer. Do not type in the flow rates or velocity numbers. These are the results of the formulae that you put into these cells.

Table 4.3 Water flow at 10°C in Table X copper pipe

	B	C	D	E
3	File:	CHAP4001		
5	Water flow at 10°C in Table X copper pipe			
7	Data for Δ_p = 500 Pa			
9	pipe diameter		flow rate	velocity
11	nominal (mm) actual (m)	M (kg/s)	Q (m³/s)	v (m/s)
13	15 0.01360	0.0907	0.000091	0.625
15	22 0.02022	0.2680	0.000268	0.835
17	28 0.02622	0.5427	0.000543	1.005
19	35 0.03263	0.9803	0.000981	1.173
21	42 0.03963	1.6553	0.001656	1.342
23	54 0.05163	3.3693	0.003370	1.610

The spreadsheet has been set up to read the cell C7 containing the pressure loss rate Δp Pa, and then cell B13 containing the first pipe internal diameter in metres, calculate the water flow rate and put this into cell C13. This is repeated for each line of additional pipe sizes. The flow rate M kg/s is divided by the water density 999.7 kg/m³. This finds the volume flow rate Q m³/s that is put into cell D13. Cell D13 is then divided by the cross-sectional area of the pipe ($\pi d^2/4$) to find the water velocity v m/s, which is put into cell E13. Each spreadsheet package requires a slightly different equation format. Enter only the relevant part of the equations given. Omit the '$M = $' part of the equation given. Only the right-hand side of the equals sign is entered into the spreadsheet. Use the correct symbols for your software. The formulae are given.

Cell C13 contains:

$M = -70.24*\text{SQRT(C7)}*(\text{B13})\hat{}2.5*\text{LOG}(((4.05*10\,\hat{}-7)/\text{B13})$
$+ (7.33*10\,\hat{}-5)/\text{SQRT(C7)}/\text{B13}\hat{}1.5)$

Cell D13 contains: $Q = \text{C13}/999.7$

Cell E13 contains: $v = D\,13*4/PI/(B\,13)\wedge2$

Refer to Table 4.2 for the meanings of the symbols. * means multiplication. ^ means raise to the power: $2\wedge3 = 2^3 = 8$

Copy the formulae into the other cells and then edit them so that the correct cell references are used on each line.

The spreadsheet can then be used for any value of pressure drop rate Δp on an iterative basis. Other pipe sizes can be substituted or added to the list.

Save the spreadsheet in a file named as CHAP4001. Copy the file into files named CHAP4002, CHAP4003 and CHAP4004. Edit these files later so that they will be suitable for the further examples given in this chapter.

The density and viscosity of a fluid significantly influence the frictional resistance of a pipe and thus its mass flow rate. Values for these variables were included in the flow equation used.

Water flow at 75 °C

For Table X copper pipe carrying water at 75.0 °C the flow equation becomes:

$$M = -69.36\,\Delta p^{0.5}\,d^{2.5}\log\left(\frac{4.05\times10^{-7}}{d} + \frac{2.15\times10^{-5}}{\Delta p^{0.5}\,d^{1.5}}\right)$$

EXAMPLE 4.3

Calculate the flow rate of water at 75.0 °C that can be carried by a 42.0 mm Table X copper pipe when the allowable pressure loss rate due to friction is 220.0 Pa/m.

$$d = 0.03963\,\text{m}, \quad \Delta p = 220.0\ \text{Pa/m} \quad \text{and} \quad M = 1.194\ \text{kg/s}$$

For quiet operation of pipework systems, the water velocity should not exceed 1.0 m/s. Where turbulence-generated noise can be tolerated, then velocities of 3.0–6.0 m/s can be used. Such occasions will be where pipes are enclosed within concrete service ducts a long way from offices, or where the background noise level is high, such as in a factory. However, factory areas are increasingly high-technology facilities and such a choice must be carefully considered against all the relevant facts.

High water velocity also means high pressure-loss rate, large pump-pressure rise and high pump-running cost owing to the power consumption of the losses of energy in the pipework. Conversely, high water velocity means that smaller pipe diameters can be used that reduce capital costs of the pipe, pipe fittings, valves, thermal insulation materials, and service ducts, and reduce heat loss through smaller pipe surface areas. The pipeline designer optimizes all these factors in relation to the circumstances.

Water velocity can be calculated from the mass flow rate M kg/s, pipe internal diameter d m and water density, kg/m^3.

$$\text{density of water at } 10°C = 999.73\ \text{kg/m}^3$$

$$\text{density of water at } 75°C = 974.85\ \text{kg/m}^3$$

Using Q = fluid volume flow rate (m³/s), v = fluid velocity (m/s), ρ = fluid density (kg/m³), π = circle circumference/diameter, pipe cross-sectional area = $\pi\,d^2/4$ (m²):

$$\text{fluid velocity } v = \frac{Q \text{ m}^3/\text{s}}{\pi\,d^2/4 \text{ m}^2}$$

$$= \frac{4\,Q}{\pi\,d^2} \text{ m/s}$$

and as

$$Q = \frac{M \text{ kg/s}}{\rho \text{ kg/m}^3} \text{ m}^3/\text{s}$$

then

$$v = \frac{4\,M}{\rho\,\pi\,d^2} \text{ m/s}$$

EXAMPLE 4.4

Calculate the water velocity in a 28.0 mm Table X copper pipe carrying 0.519 kg/s at 75 °C.

For a 28.0 mm pipe, $d = 0.02622$ m, $\rho = 974.85$ kg/m³, $\pi = 3.1416$.

$$v = \frac{4 \times 0.519}{974.85 \times 3.1416 \times 0.02622^2} \text{ m/s}$$

$$= 0.986 \text{ m/s}$$

The spreadsheet to carry out this calculation is the copy in the file named as CHAP4002 and needs to be edited (Table 4.4)

Table 4.4 Water flow at 75 °C in Table X copper pipe

	B		C	D	E
3	File:	CHAP4002			
5	Water flow at 75 °C in Table X copper pipe				
7		Data for $\Delta p = 300$ Pa/m			
9	pipe diameter		flow rate		velocity
11	nominal (mm)	actual (m)	M (kg/s)	Q (m³/s)	v (m/s)
13	15	0.01360	0.0801	0.000082	0.525
15	22	0.02022	0.2336	0.000240	0.746
17	28	0.02622	0.4696	0.000482	0.892
19	35	0.03263	0.8437	0.000865	1.035
21	42	0.03963	1.4182	0.001455	1.179
23	54	0.05163	2.8702	0.002944	1.406

Cell C13 contains:

$$M = -69.36*\text{SQRT}(C7)*(B13)^{\wedge}2.5*\text{LOG}(((4.05*10^{\wedge}-7)/B13)$$

$$+ (2.15*10^{\wedge}-5)/\text{SQRT}(C7)/B13^{\wedge}1.5$$

Cell D13 contains: $Q = C13/974.85$

Cell E13 contains: $v = D13*4/\text{PI}/(B13)^{\wedge}2$

Flow of methane in pipes

The equation for Table X copper pipe carrying methane gas during turbulent flow, is

$$Q = -2.695\,\Delta p^{0.5}\,d^{2.5}\log\left(\frac{4.05\times10^{-7}}{d} + \frac{2.30\times10^{-5}}{\Delta p^{0.5}\,d^{1.5}}\right)$$

where Q = methane volume flow rate (m³/s), Δp pressure loss rate (Pa/m), d = pipe internal diameter (m).

EXAMPLE 4.5

Calculate the flow rate of methane in a 35.0 mm Table X copper pipe when the pressure loss rate is 6.0 Pa/m. From Table 4.1, $d = 0.032\,63$ m.

$$Q = -2.695\times6^{0.5}\times0.03263^{2.5}\times\log\left(\frac{4.05\times10^{-7}}{0.03263} + \frac{2.30\times10^{-5}}{6^{0.5}\times0.03263^{1.5}}\right)$$

$$= 0.003\,55\ \mathrm{m^3/s}$$

The spreadsheet to carry out this calculation is the copy in the file named as CHAP4003 and needs to be edited (Table 4.5)

Table 4.5 Flow of methane in Table X copper pipe

		B	C	D	E
3	File:	CHAP4003			
5	Flow of methane in Table X copper pipe				
7	Data for Δp = 3 Pa/m				
9	pipe diameter		flow rate		velocity
11	nominal (mm)	actual (m)	Q (m³/s)	Q (l/s)	v (m/s)
13	15	0.01360	0.000209	0.209	1.439
15	22	0.02022	0.000633	0.633	1.972
17	28	0.02622	0.001300	1.300	2.408
19	35	0.03263	0.002374	2.374	2.840
21	42	0.03963	0.004044	4.044	3.279
23	54	0.05163	0.008321	8.321	3.975

Cell C13 contains:

$Q = -2.695*\mathrm{SQRT}(C7)*(B13)^{\wedge}2.5*\mathrm{LOG}(((4.05*10^{\wedge}-7)/B13)$

$+ (2.3*10^{\wedge}-5)/\mathrm{SQRT}(C7)/B13^{\wedge}1.5)$

Cell D13 contains: $Q = C\,13*1000$.
Cell E13 contains: $v = C13*4/\mathrm{PI}/(B13)^{\wedge}2$

Airflow in ducts

The flow of air at 20 °C d.b., 43% percentage saturation and a barometric pressure of 1013.25 mb through clean galvanized sheet metal ducts having joints made in accordance with good practice, is given by

$$Q = -2.0278\,\Delta p^{0.5}\,d^{2.5}\,\log\left(\frac{4.05 \times 10^{-5}}{d} + \frac{2.933 \times 10^{-5}}{\Delta p^{0.5}\,d^{1.5}}\right)$$

where, Q = airflow rate (m^3/s), Δp = pressure loss rate (Pa/m), d = duct internal diameter (m).

EXAMPLE 4.6

A 500 mm internal diameter galvanized sheet steel duct is carrying air at 20 °C d.b. at a design pressure loss rate of 1.0 Pa/m. Calculate the air volume flow rate being passed and the air velocity.

$$d = 0.50\,\text{m}, \quad \Delta p = 1.0\,\text{Pa/m}$$

$$Q = -2.0278 \times 1.0^{0.5} \times 0.50^{2.5} \times \log\left(\frac{4.05 \times 10^{-5}}{0.50} + \frac{2.933 \times 10^{-5}}{1.0^{0.5} \times 0.50^{1.5}}\right)$$

$$= 1.357\,\text{m}^3/\text{s}$$

$$\text{air velocity, } v = \frac{1.357\,\text{m}^3/\text{s}}{\pi \times 0.50^2/4\,\text{m}^3}$$

$$= 6.91\,\text{m/s}$$

The spreadsheet to carry out this calculation is the copy in the file named as CHAP4004 and needs to be edited (Table 4.6).

Table 4.6 Flow of air at 20 °C in galvanized sheet steel ducts

	B	C	D	E
3	File:	CHAP4004		
5	Flow of air at 20 °C in glavanized sheet steel ducts			
7	Data for Δp = 1 Pa/m			
9	duct diameter		flow rate	velocity
11	nominal (mm) actual (m)		Q (m^3/s) Q (l/s)	v (m/s)
13	150	0.15	0.0550 55.0	3.1
15	300	0.3	0.3502 350.2	5.0
17	500	0.5	1.3569 1356.9	6.9
19	600	0.6	2.1963 2196.3	7.8
21	800	0.8	4.6874 4687.3	9.3
23	1000	1.0	8.4275 8427.5	10.7

Cell C13 contains:

$Q = -2.027*\text{SQRT}(C7)*(B13)^{\wedge}2.5*\text{LOG}(((4.05*10^{\wedge}-5)/B13)$

$\qquad + (2.933*10^{\wedge}-5)/\text{SQRT}(C7)/B13^{\wedge}1.5)$

Cell D13 contains: $Q = C13*1000$
Cell E13 contains: $v = C13*4/\text{PI}/(B13)^{\wedge}2$

EXAMPLE 4.7

Find a suitable diameter for a galvanized sheet steel duct that is to carry 3250 l/s of air at 20 °C d.b. at a maximum velocity of 5.0 m/s when the pressure loss rate is not to exceed 1.0 Pa/m. Duct diameters are in increments of 50 mm.

An iterative procedure is needed. Try $d = 1.0$ m.

Air flow $Q = 3.250$ m^3/s

air velocity $v = \dfrac{3.250 \text{ m}^3/\text{s}}{\pi \times 1.0^2/4 \text{ m}^2}$

$= 4.14$ m/s, which is less than the limit

Try $d = 0.950$ m for Q = 3.250 m^3/s.

$$v = \frac{3.250}{\pi \times 0.95^2/4} \text{ m/s}$$

$$= 4.59 \text{ m/s}$$

Trying $d = 0.900$ m produces $v = 5.11$ m/s, which is too high. Thus a 950 mm diameter duct will be used, provided that the frictional pressure loss rate does not exceed $\Delta p = 1.0$ Pa/m. Insert $\Delta p = 1.0$ Pa/m and $d = 0.950$ m into duct formula and calculate the maximum carrying capacity, irrespective of limiting velocity.

$$Q = -2.0278 \times 1.0^{0.5} \times 0.95^{2.5} \times \log\left(\frac{4.05 \times 10^{-5}}{0.95} + \frac{2.933 \times 10^{-5}}{1.0^{0.5} \times 0.95^{1.5}}\right)$$

$$Q = 7.365 \text{ m}^3/\text{s and as } 1.0 \text{ m}^3 = 1000.0 \text{ l},$$

$Q = 7365$ l/s, which is greater than required, so the actual pressure loss rate will be less than 1.0 Pa/m, around 0.22 Pa/m.

HEAT-CARRYING CAPACITY

The engineering services designer uses the most cost-effective fluid to transfer heating or cooling power to the occupied space. This may mean using a high-temperature hot-water pipework distribution when there are long distances between the energy-conversion source and the target location. It is more economical to transfer heated or cooled water than conditioned air, as smaller-diameter pipes are used and pumping costs are less than those for fan and air-duct networks. The circulation of air is limited to the smallest practical area with the minimum lengths and diameters of air ductwork.

Table 4.7 gives physical data on the heat transfer fluids used to enable a comparison to be made between their relative transportation capacities.

To compare the heating or cooling carrying capacities of water and air, find the flow rates required for 1.0 kW of heat transfer, from

heat flow = mass flow × SHC × allowable temperature drop

$$M \text{ kg/s} = \frac{\text{heat flow kW}}{\text{SHC} \times \text{allowable drop}}$$

Table 4.7 Fluid physical data

Fluid	Mean temp. (°C)	Allowable drop (K)	SHC (kJ/kg K)	Density (kg/m³)
Water	10	4	4.193	999.7
	75	12	4.194	974.9
	120	30	4.248	943.1
Air	10	4	1.012	1.242
	30	8	1.012	1.160
	40	12	1.012	1.123

EXAMPLE 4.8

Compare the mass flow rates of water needed to transfer 1.0 kW of heating power with low-temperature hot water (LTHW) and medium-temperature hot water (MTHW) heating systems.

From Table 4.7 mean water temperature for LTHW = 75 °C and HTHW = 120 °C, requiring the water to be pressurized to avoid boiling.

$$M \text{ kg/s} = \frac{\text{heat flow kW}}{\text{SHC} \times \text{allowable drop}}$$

(a) LTHW, heat flow = 1.0 kW, SHC = 4.194 kJ/kg K

$$\text{allowable drop} = 12.0 \text{ K}$$

$$M \text{ kg/s} = \frac{1.0}{4.194 \times 12.0} = 0.020 \text{ kg/s}$$

(b) MTHW, heat flow = 1.0 kW, SHC = 4.248 kJ/kg K

$$\text{allowable drop} = 30.0 \text{ K}$$

$$M \text{ kg/s} = \frac{1.0}{4.248 \times 30.0} = 0.008 \text{ kg/s}$$

The advantage of MTHW and HTHW is that larger temperature drops between flow and return can be used, reducing the water flow rate, while providing a high mean water temperature at the heat emitter, rapid heat transfer and small terminal units further away from the occupants, all to economize on the pipework costs.

EXAMPLE 4.9

Find suitable Table X copper pipe diameters for LTHW and MTHW heating systems to transfer 12 kW from a boiler to a convector if the pressure loss rate is not to exceed 500 Pa/m and water velocity is not to exceed 1 m/s.

(a) LTHW:

$$M = \frac{12}{4.194 \times 12} \text{ kg/s} = 0.238 \text{ kg/s}$$

Try a 22 mm pipe whose $d = 0.020\,22$ m.

$$\text{water volume flow rate } Q = \frac{0.238\,\text{kg/s}}{974.9\,\text{kg/m}^3}$$

$$= 2.441 \times 10^{-4}\,\text{m}^3\text{/s}$$

$$v = \frac{2.441 \times 10^{-4}\,\text{m}^3\text{/s}}{\pi \times 0.020\,22^2/4\,\text{m}^2}\,\text{m/s}$$

$$= 0.76\,\text{m/s}$$

Calculate the carrying capacity of the 22 mm pipe for $\Delta p = 500\,\text{Pa/m}$, $d = 0.020\,22\,\text{m}$ using

$$M = -69.36\,\Delta p^{0.5}\,d^{2.5}\log\left(\frac{4.05 \times 10^{-7}}{d} + \frac{2.15 \times 10^{-5}}{\Delta p^{0.5}\,d^{1.5}}\right)$$

$$\log(\ ) = \left(\frac{4.05 \times 10^{-7}}{0.020\,22} + \frac{2.15 \times 10^{-5}}{500^{0.5} \times 0.020\,22^{1.5}}\right)$$

$$= -3.45$$

$$M = -69.36 \times 500^{0.5} \times 0.020\,22^{2.5} \times (-3.45)\,\text{kg/s}$$

$$= 0.311\,\text{kg/s}$$

This is in excess of the design requirement. The actual pressure loss rate will be less than 500 Pa/m. The pipe size is satisfactory. The reader should try 15 mm.

(b) MTHW:

$$M = \frac{12}{4.248 \times 30}\,\text{kg/s} = 0.094\,\text{kg/s}$$

Try a 15 mm pipe whose $d = 0.0136$ m.

$$\text{water volume flow rate } Q = \frac{0.094\,\text{kg/s}}{943.1\,\text{kg/m}^3}$$

$$= 9.984 \times 10^{-5}\,\text{m}^3\text{/s}$$

$$v = \frac{9.984 \times 10^{-5}\,\text{m}^3\text{/s}}{\pi \times 0.0136^2/4\,\text{m}^2}\,\text{m/s}$$

$$= 0.69\,\text{m/s}$$

This is within the design maximum. Calculate the carrying capacity of the 15 mm pipe for $\Delta p = 500\,\text{Pa/m}$, $d = 0.0136\,\text{m}$ using

$$\log(\ ) = \log\left(\frac{4.05 \times 10^{-7}}{0.0136} + \frac{2.15 \times 10^{-5}}{500^{0.5} \times 0.0136^{1.5}}\right)$$

$$= -3.197$$

$$M = -69.36 \times 500^{0.5} \times 0.0136^{2.5} \times (-3.197)\,\text{kg/s}$$

$$= 0.107\,\text{kg/s}$$

This is in excess of the design requirement. The actual pressure drop rate will be less than the design maximum.

EXAMPLE 4.10

A mechanical ventilation system is to be used to heat a theatre having a total heat loss of 60 kW. The boiler plant room is 30 m from the theatre. Compare the diameters of LPHW pipes and warm-air ducts that can be used, stating your preference and reasons.

(a) LPHW system:

$$M = \frac{60}{4.194 \times 12} \text{ kg/s} = 1.192 \text{ kg/s}$$

Try a 42 mm Table X copper pipe whose $d = 0.039\,63$ m.

$$\text{water volume flow rate } Q = \frac{1.192 \text{ kg/s}}{974.9 \text{ kg/m}^3}$$

$$= 1.223 \times 10^{-3} \text{ m}^3/\text{s}$$

$$v = \frac{1.223 \times 10^{-3} \text{ m}^3/\text{s}}{\pi \times 0.03963^2/4 \text{ m}^2} \text{ m/s}$$

$$= 0.991 \text{ m/s}$$

Estimate the actual pressure loss rate by calculating the carrying capacity of the 42 mm pipe for $\Delta p = 220$ Pa/m, $d = 0.039\,63$ m, using

$$M = -69.36\,\Delta p^{0.5}\, d^{2.5} \log\left(\frac{4.05 \times 10^{-7}}{d} + \frac{2.15 \times 10^{-5}}{\Delta p^{0.5} \times d^{1.5}}\right)$$

$$\log(\ ) = \log\left(\frac{4.05 \times 10^{-7}}{0.039\,63} + \frac{2.15 \times 10^{-5}}{220^{0.5} \times 0.039\,63^{1.5}}\right)$$

$$= -3.712$$

$$M = -69.36 \times 220^{0.5} \times 0.039\,63^{2.5} \times (-3.712) \text{ kg/s}$$

$$= 1.194 \text{ kg/s}$$

This is slightly in excess of the design requirement. The pipe size is satisfactory.

(b) For the air-duct system, warm supply air would leave the plant room at 40 °C, travel through galvanized steel ducts, entèr the theatre at 38 °C, heat and ventilate the theatre, leave the theatre at the room-air temperature of 22 °C and arrive back in the plant room at 20 °C, having lost some heat to the cold external air.

A heater battery in the plant room heats the air from 20 °C to 40 °C with LPHW from the boiler.

$$\text{supply air temperature change in the room} = (38 - 22) \text{ K}$$

$$= 16 \text{ K}$$

$$M = \frac{60}{1.012 \times 16} \text{ kg/s} = 3.706 \text{ kg/s}$$

Try an 800 mm diameter duct, $d = 0.8$ m.

$$\text{water volume flow rate } Q = \frac{3.706\,\text{kg/s}}{1.1233\,\text{kg/m}^3}$$

$$= 3.3\,\text{m}^3/\text{s}$$

$$v = \frac{3.3\,\text{m}^3/\text{s}}{\pi \times 0.8^2/4\,\text{m}^2}\,\text{m/s}$$

$$= 6.6\,\text{m/s}$$

Estimate the actual pressure loss rate by calculating the carrying capacity of the 800 mm duct for $\Delta p = 0.7$ Pa/m, $d = 0.8$ m, using

$$M = -2.0278\,\Delta p^{0.5}\,d^{2.5}\log\left(\frac{4.05 \times 10^{-5}}{d} + \frac{2.93 \times 10^{-5}}{\Delta p^{0.5}\,o^{1.5}}\right)$$

$$\log(\ ) = \log\left(\frac{4.05 \times 10^{-5}}{0.8} + \frac{2.93 \times 10^{-5}}{0.7^{0.5} \times 0.8^{1.5}}\right)$$

$$= -4.002$$

$$M = -2.0278 \times 0.7^{0.5} \times 0.8^{2.5} \times (-4.002)\,\text{kg/s}$$

$$= 3.89\,\text{kg/s}$$

This is slightly in excess of the design requirement. The duct size is satisfactory.

The choice facing the designer is between a 42 mm LPHW flow and return pipes and 800 mm supply and recirculation air ducts. The more concentrated form of energy, water, is used for long- distance heat transportation. A smaller underground service duct would be needed and less thermal insulation material would be used.

PUMP AND FAN POWER CONSUMPTION

The power expended in overcoming frictional resistance in a pipe or duct is:

$$\text{power, } P\,\text{W} = Q\,\text{m}^3/\text{s} \times H\,\text{Pa}$$

Verify this is correct by checking the units using: $1\,\text{Pa} = 1\,\text{N/m}^2$ and $1\,\text{Nm/s} = 1\,\text{W}$; $H =$ total pressure drop overcome by the pump or fan (Pa); $H = (\Delta p\,\text{Pa/m} \times l\,\text{m}) +$ pressure drop in fittings; $l =$ length (m)

For a pump or fan,

$$\text{input power} = \frac{\text{fluid power consumption}}{\text{overall efficiency}}$$

The overall efficiency will incorporate the electrical power losses in the motor, the mechanical energy absorbed in bearings and drive systems plus the fluid energy losses due to casing friction and changes in flow direction. Typically, overall efficiency will be 60%

EXAMPLE 4.11

Calculate the pump and fan energy consumptions for the alternative systems in Example 4.10. The overall electromechanical efficiency of each machine is 60% The resistance of pipe fittings is 30% of that for straight pipes. The resistance of air-duct fittings is 200% of that for straight ducts.

The theatre is 30m from the plant room, so 60 m of pipe or duct would be used.

(a) LPHW pipework resistance $H = (220 \, \text{Pa/m} \times 60 \, \text{m}) \times 1.3$

$$= 17\,160 \, \text{Pa}$$

The 1.3 represents an addition of 30% for pipe fittings.

$$\text{fluid power consumption} = 1.223 \times 10^{-3} \, \text{m}^3/\text{s} \times 17\,160 \, \text{Pa}$$

$$= 21 \, \text{W}$$

$$\text{input power to pump motor} = \frac{21}{0.6} \, \text{W} = 35 \, \text{W}$$

(b) air duct resistance $H = (0.7 \, \text{Pa/m} \times 60 \, \text{m}) \times 3$

$$= 126 \, \text{Pa}$$

The 3 represents an addition of 200% for duct fittings.

$$\text{fluid power consumption} = 3.3 \, \text{m}^3/\text{s} \times 126 \, \text{Pa}$$

$$= 415.8 \, \text{W}$$

$$\text{input power to motor} = \frac{415.8}{0.6} \, \text{W} = 693 \, \text{W}$$

This demonstrates that a fan and air-duct system absorbs several times the electrical power of a hot-water system to deliver the same amount of heating energy.

THERMAL STORAGE

The storage of heat in a water tank, rocks or bricks, is used to connect an intermittent demand for energy with an intermittent source. The demand for heat may come from the heating system of a building or the use of hot water by sanitary appliances.

Low-temperature heat storage with a chilled water tank is used between a refrigeration compressor and the air-conditioning system it serves. It is necessary to limit the on–off operation of the refrigeration compressor to six starts per hour to minimize the temperature and wear of the starting equipment.

Typical demands for heating and cooling are cyclically variable during each 24 h period and follow weather patterns. The economical supply of energy to meet such a demand has to be equally variable. Simple on–off and multistep controls are designed for the convenience of the boiler or refrigeration compressor and not for the thermal comfort of the occupants of the building.

Intermittent supplies of energy come from night-time electrical use when the building is unoccupied, the on–off operation of boilers and refrigeration compressors, solar collectors and sources of waste heat. Some examples of thermal storage arrangements are as follows.

Hot-water storage cylinder for sanitary services

(a) A low-power-output fossil-fuel-burning appliance heats the cylinder during long running periods at 100% output and maximum efficiency.
(b) An array of solar collector panels or concentrators on the roof is used to preheat the cold water supplied to the storage cylinder.
(c) Off-peak night-time electrical immersion heaters replace the hot water used during the day.

Space heating

(a) Off-peak electrical immersion heaters are used to raise the temperature of a storage tank or cylinder to 80 °C overnight to meet the building heat demand for the following day.
(b) Off-peak electrical storage radiators that have brick, concrete or cast iron heat stores.
(c) Passive architecture utilizes solar heating of exposed masonry walls to absorb heat energy that is released up to 10 h later. The heat output can be partly controlled by increasing air circulation with fans.
(d) Passive architecture uses glazing to allow solar gains into the building. Shading is used to control heat input. A heat-storage wall is often placed close to the glazing to avoid short-term overheating and allow control of heat release over a period of hours.
(e) Phase-change salts and chemicals can absorb energy from waste heat sources or off-peak electricity in order to change phase. Heat release from the store is achieved by allowing a reversal of the phase change when needed.

Refrigeration

A chilled-water storage tank is cooled by on–off refrigeration compressors. Compressors can be run at full load for long periods to chill the water and then remain off for an hour or more. This is particularly advantageous when the cooling demand within the building is low during mild weather. Compressors are not to be allowed to start more than six times per hour as the starting current is around twice the running current and motor overheating is to be avoided.

Table 4.8 gives the physical data needed for thermal storage calculations.

Table 4.8   Thermal storage data

Material	Density (kg/m^3)	SHC (kJ/kg K)
Water at 75 °C	974.9	4.194
Water at 10 °C	999.7	4.193
Cast iron	7000	480
Concrete	2300	840
Granite	2650	1000

To calculate the quantity of heat stored:

heat stored, $kW = mass\,kg \times SHC\,kJ/kg\,K \times (t_1 - t_2)\,K$

mass of storage material = volume $m^3 \times$ density kg/m^3

t_1 = mean storage temperature °C

t_2 = lowest usable storage temperature °C

EXAMPLE 4.12

A house has an average rate of heat loss of 6.4 kW between 0700 h and 2300 h. A hot-water storage tank is to be raised to 80 °C between 2300 h and 0700 h by off-peak electrical immersion heaters. Calculate a suitable design for the system.

---

heat demand period = (2300 − 0700) h = 16 h

electrical charge period = 8 h

heat storage required = $6.4\,kW \times 16\,h = 102.4\,kWh$

The heat store will lose energy through its thermal insulation. The rate of heat loss can be accurately calculated when the tank dimensions are known. An estimate of 15% can be added initially.

Add 15% heat loss:

heat storage = $(102.4 \times 1.15)\,kWh$

$= 117.76\,kWh$

heat stored overnight = heater power kW × charge period h

117.76 kWh = heater power kW = 8 h

heater power = 14.7 kW

Five stages of 3 kW immersion heaters, 15 kW, would be suitable.
Calculate the volume and size of the water storage tank:

heat stored = water mass $\times$ SHC $\times (t_1 - t_2)$

suitable lowest storage temperature $t_2 = 60$ °C, $t_1 = 80$ °C,

water density = $974.9\,kg/m^3$, SHC = $4.194\,kJ/kg\,K$

heat stored = $15\,kW \times 8\,h = 120\,kWh$

$= 120 \times 3600\,kWs = 120 \times 3600\,kJ$

Also,

heat stored = mass kg $\times 4.194\,kJ/kg\,K \times (80 - 60)\,K$

$= mass \times 4.194 \times 20\,kJ$

so

$$120 \times 3600 = mass \times 4.194 \times 20$$

and

$$\text{mass} = \frac{120 \times 3600}{4.194 \times 20} \, \text{kg} = 5150 \, \text{kg}$$

Find the storage volume:

$$5150 \, \text{kg} = \text{volume m}^3 \times 974.9 \, \text{kg/m}^3$$

$$\text{volume} = \frac{5150}{974.9} = 5.28 \, \text{m}^3$$

The storage tank will be no higher than 1 m, so its plan dimensions will be $(5.28)^{0.5}$ m = 2.3 m..

One tank of 2.3 m × 2.3 m × 1 m, with the economic thickness of thermal insulation, would be needed.

EXAMPLE 4.13

A 150 kW refrigeration compressor system chills water from 10 °C to 6 °C in a building's air-conditioning system. The compressor is not to cycle on–off more than six times per hour. During mild weather, the building cooling load is 25 kW. Heat gains to the stored chilled water amount to 10% of the cooling energy used. Calculate a suitable design for an intermediate chilled water storage tank.

$$\text{building cooling load in 1 h} = 25 \, \text{kW} \times 1.10 \times 1 \text{h} = 27.5 \, \text{kWh}$$

If the refrigeration system operates at 150 kW for 12 minutes:

$$\text{cooling output} = 150 \, \text{kW} \times \frac{12}{60} \, \text{h} = 30 \, \text{kWh}$$

So the compressors can run for 12 min each hour to lower the storage tank to 6 °C, satisfying the building cooling load. In order to store 30 kWh, water storage will be:

$$\text{mass} = \frac{30 \times 3600}{4.193 \times 4} \, \text{kg} = 6440 \, \text{kg}$$

$$\text{water volume} = \frac{6440}{999.7} = 6.44 \, \text{m}^3$$

$$\text{tank size} = 2.54 \text{m} \times 2.54 \text{m} \times 1 \text{m}$$

Questions

1. State the engineering and economic objectives of heat-distribution systems. Give examples of good practice. Outline the economic considerations needed.
2. Calculate the maximum carrying capacity of a 28 mm Table X copper pipe at a pressure loss rate of 1200 Pa/m when the water temperature is 75 °C.
3. Find the Table X copper pipe size appropriate to a cold-water flow rate of 0.123 kg/s and a pressure loss rate of 850 Pa/m.
4. A low-temperature hot-water heating system that serves air-conditioning heater batteries has a two-pipe circuit that is to be sized on the basis of a constant pressure loss rate of 500 Pa/m. Calculate the maximum flow capacity of Table X copper pipes from 15 mm to 54 mm nominal diameters to assist the designer.

5. Calculate the water velocity in a 42 mm Table X copper pipe carrying 1.52 kg/s at 75 °C.

6. Find the flow rate of natural gas, methane, in a 15 mm Table X copper pipe at a pressure loss rate of 12.5 Pa/m.

7. The pressure loss rate available for a gas supply installation is 8 Pa/m. Table X copper pipe is to be used. Calculate the gas volume flow capacity of pipes from 22 mm to 54 mm nominal diameter.

8. A 300 mm internal diameter galvanized sheet steel duct is to carry air at 20 °C d.b. with a pressure loss rate of 2 Pa/m. Calculate the air flow rate that can be carried.

9. Calculate the air velocity in a 450 mm internal diameter galvanized sheet steel duct when the air temperature is 20 °C d.b. and the pressure loss rate is 0.6 Pa/m.

10. An air-conditioning system is to have galvanized sheet metal ducts carrying air at 20 °C d.b. at a maximum velocity of 7.5 m/s. The maximum allowable frictional resistance of straight duct is to be 3 Pa/m.

    The airflow rates to be carried in different sections of duct are 0.12, 0.3, 0.6, 1 and 2 m^3/s.

    Ducts are manufactured in sizes from 100 mm internal diameter upwards in 50 mm increments.

    Calculate the most economical size for each duct.

11. Calculate the water mass flow rates needed for LTHW and MTHW heating systems in order to transfer 3.5 kW and 420 kW. State which system would be used for each heat load.

12. Find a suitable Table X copper pipe diameter for a LTHW heating system to transfer 25 kW from the boiler to a system of heat emitters if the pressure loss rate is not to exceed 750 Pa/m and the water velocity is not to be above 1 m/s. State the actual pressure loss rate and water velocity.

13. District heating is to have a total connected head load of 550 kW using a MTHW system having a flow temperature of 150°C and a return of 120°C. Water velocity can be up to 3 m/s in the underground distribution mains. The pressure-loss rate is not to exceed 750 Pa/m. If Table X copper pipe is to be used, what would be the appropriate size, actual pressure loss rate and water velocity?

14. The air-conditioning system in a 20-storey office block is to comprise air-handling plants and a basement boiler room. A total connected load of 100 kW was calculated.

    Compare the diameters of LTHW heating system pipes with air ducts that would be needed to transport the heating capacity vertically through the building. State how the design engineer would configure the heating service to minimize the spaces occupied by the distribution services. Maximum water velocity is to be 1 m/s. Air velocity is not to exceed 8 m/s.

15. Calculate the pump and fan energy consumptions to operate LTHW or a ducted-air heating system where the boiler plant is 75 m from the heat load of 75 kW. Overall electromechanical efficiency of the pump and fan is 65% Water velocity can be up to 2 m/s and air velocity up to 10 m/s.

    The frictional resistance of the pipeline fittings amounts to 25% of the length of straight pipe and that for the air duct is 75% of the duct length.

16. Discuss why thermal storage is needed in heating and cooling services installations and list the principal methods used.

17. A living room has an average rate of heat loss of 2.2 kW during a winter day while occupied between 0700 h and 2300 h. An electrically charged off-peak storage heater is to be installed to maintain comfort conditions. The charge period is 2300 h to 0700 h. Calculate the input power and the total energy storage required for the heater.

18. A building having an average rate of daily heat loss of 15 kW between 0700 h and 230 h is to be heated from a hot-water storage tank operating between 85 °C and 60 °C. The tank has electric immersion heaters operating between 2300 h and 0700 h. Calculate the storage tank size needed, heater power and quantity of heat stored in MJ.

19. A 230 kW refrigeration compressor system chills water from 12 °C to 7 °C in a building's air-conditioning system. The compressor is not to cycle on–off more than four times per hour. During mild weather, the building cooling load is 100 kW. Heat gains to the stored chilled water amount to 15% of the cooling energy used. Calculate a suitable design for an intermediate chilled water storage tank.

5 System design

INTRODUCTION

System design brings together the constituent parts of an air-conditioning system from the other chapters. The fresh air inlet quantity is explained in terms of its dilution property for safety and human comfort. The airflow quantity supplied to a conditioned space is calculated from the heating and cooling loads on the building. This forms the basis for air-conditioning design.

Sensible and latent heat gains and losses are used. Equations are derived from basic principles. An unknown supply air temperature can be calculated from a knowledge of the other factors. Calculation examples are constructed from simple principles into full design applications. Data that are appropriate to different parts of the world are used.

Psychrometric charts are used for complete system designs for both winter and summer states. Schematic duct layouts are introduced to account for all the air flows through a building. Practical problem-solving is used for realistic cases. Designers are faced with a fresh challenge by each new application and there are often several correct solutions. The answers shown are illustrative and will encourage the user to apply their principles.

LEARNING OBJECTIVES

Study of this chapter will enable the user to:

1. know the reasons for ventilation;
2. identify suitable quantities for ventilation air;
3. calculate the air volume and mass flow rate needed to control sensible heat gains;
4. calculate air density;
5. calculate room air change rate;
6. calculate winter supply-air temperature;
7. decide suitable combinations of quantity and temperature for supply-air systems;
8. manipulate the design formulae;
9. validate the accuracy of design calculations;

10. calculate the moisture content of supply air used to control latent heat gains;

11. use psychrometric data;

12. use the sensible-to-total-heat ratio line for a room;

13. find the airflows for the distribution ductwork;

14. know how to create zones of different static air pressure;

15. use schematic logic diagrams for airflows through air-handling duct systems;

16. calculate air flows between rooms;

17. carry out air quantity and condition calculations for complete systems;

18. plot psychrometric processes;

19. calculate and use psychrometric data;

20. construct psychrometric chart processes from incomplete data;

21. calculate heater and cooler battery loads;

22. undertake design calculations for different countries;

23. identify where condensation occurs.

Key terms and concepts

air density	138	occupational exposure limit	136
air mass flow rate	143	outdoor air	135
air volume flow rate	137	plant air flow	146
condensation	159	psychrometric processes	153
cooler battery load	153	recirculated air	136
dual-duct system	165	room air change rate	142
ductwork schematic	147	sensible heat	137
evaporation	143	sensible-to-total-heat ratio	145
exhaust air	148	single-duct system	142
fan-coil units	155	static air pressure	146
fresh air	135	supply-air moisture content	144
heat balance	137	supply-air temperature	137
heater battery load	155	system logic	147
latent heat	143	transfer between rooms	147
lower explosive limit	136	upper explosive limit	136
mixed air	153	ventilation	135
moisture	143	warm-air heating system	141

VENTILATION REQUIREMENTS

Ventilation air is needed to sustain safe and comfortable conditions for the occupants. The quantity of outdoor air is related to human activity level and the need to remove or dilute atmospheric pollutants. Occupational exposure limits (OEL) are stipulated by the Health and Safety Executive (HSE, 1990) based on an average for an 8 h working day, or a peak 10 min period. The carbon dioxide (CO_2) limit is 5000 parts per million (ppm) or 0.5%.

$$5000\,\text{ppm} = \frac{5000\,\text{parts}\,CO_2}{1 \times 10^6\,\text{parts air}} \times 100\%$$

$$= 0.5\%$$

Other pollutants have much lower 8 h OEL. The ammonia (NH_3) limit is 25 ppm, carbon monoxide (CO) 50 ppm, hydrogen sulphide (H_2S) 10 ppm, nitrogen dioxide (NO_2) 2 ppm and sulphur dioxide (SO_2) 2 ppm. Their threshold of noticeable smell is often even lower: NH_3 is 5 ppm, H_2S 0.1 ppm and chlorine 0.02 ppm.

Gas and vapour in air can form explosive mixtures. The lower explosive limit (LEL) for methane (CH_4) is 5%, meaning that below 5% methane in air, the mixture is not combustible. The upper explosive limit, (UEL) for CH_4 is 15%, so an explosive mixture is created between 5% and 15%. To create the explosion, the mixture has to be exposed to a surface, spark or flame that is at its ignition temperature of 538 °C. Gas detectors and alarms can be installed in the building or in service ducts. However, upon a gas or pollutant leak, fresh air ventilation is relied upon to disperse the fuel and lower the local temperature.

The recommended outdoor air supply rates for air-conditioned spaces are given in Table 5.1.

Table 5.1 Recommended outdoor air supply rates for air conditioned spaces (CIBSE, 1986a, Table A1.5)

Type of space	l/s per person	l/s per m² floor area
Factory	8	0.8
Open-plan office	8	1.3
Shop	8	3
Theatre	8	–
Private office	12	1.3
Conference room	18–25	–
Residence	12	–
Bar	18	–
Dining room	18	–
Heavy smoking	25	6
Corridor	–	1.3
House kitchen	–	10
Restaurant kitchen	–	20
Toilets	–	10

Where the polluted air in a room cannot be recirculated to other areas, such as from kitchens and toilets, it is exhausted directly to the outdoor atmosphere. Heat exchangers can be used for energy recovery. Such areas are 100% fresh-air-ventilated. Most comfort applications will utilize the recirculation of air from conditioned spaces to retain the already correct temperature and humidity. The fresh air will be typically 5% to 25% of the total air in circulation.

AIR-HANDLING EQUATIONS

The air-handling system is normally designed to remove the sensible and latent heat gains to the conditioned space. These are calculated from the Chapter 3 equations or dedicated computer software. Their extraction should be at the same rate as their occurrence. When this happens, the conditioned space remains at constant temperature and humidity. The designer calculates the maximum heating, cooling and moisture plant loads. These determine the plant size. The control engineer arranges for the plant to react to maintain the desired conditions within an acceptable band of values.

SH = sensible heat gain, or loss, to the space (kW); LH = latent heat gain, or loss, to the space (kW); t_R = room-air temperature (°C d.b.); t_S = supply-air temperature (°C d.b.); SHC = specific heat capacity of air (1.0048 kJ/kg K at 20 °C); ρ = density of air (1.1906 kg/m³ at 20 °C d.b. 50% saturation).

Figure 5.1 shows the arrangement of conditioned supply air entering a room at t_S, absorbing the sensible and latent heat gains and leaving through the extract duct at the room air temperature t_R. While either the supply-air temperature or quantity is appropriate to the gains, the room condition remains stable. Any change in the weather, solar or internal gains or occupancy will result in a changed room-air temperature. The room-air temperature detector may be located in the extract air duct or in the conditioned space. This instigates a change in the fluid flow rate through the heating or cooling coil when a variable-temperature system is used, or a variation to the supply air flow rate in a VAV system.

A heat balance can be made for summer cooling:

heat gain to room = heat gained by air flow through the room
SH kW = air mass flow rate × increase in specific enthalpy
SH kW = M kg/s × $(h_R - h_S)$ kJ/kg

The sensible heat gain SH kW and the room-air specific enthalpy h_R are known but both the quantity M and condition h_S of the supply air are to be found. It is the volume flow rate Q m³/s and temperature t_S °C d.b. of the supply air that are more useful to the designer. Q and t_S are interrelated and have implications for the thermal comfort of the occupants and for the type of air terminal device to be used. There is no easy answer to the formulation of a suitable design and an iterative, or trial and error, routine can be necessary. SH kW = air mass flow rate × specific heat capacity × air temperature rise.

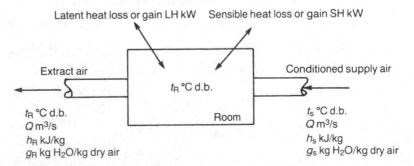

Fig. 5.1 Basic data for airflow design.

$$SH\,kW = M\,kg/s \times SHC\,kJ/kg\,K \times (t_R - t_S)\,K$$

$$M\,kg/s = Q\,m^3/s \times \rho\,kg/m^3$$

$$SH\,kW = Q\,m^3/s \times \rho\,kg/m^3 \times SHC\,kJ/kg\,K \times (t_R - t_S)\,K$$

The quantity of supply air is to be calculated but density depends upon temperature. From the general gas equation,

$$PV = MRT$$

where P = absolute pressure of gas (N/m^2), V = volume of gas (m^3), M = mass of gas (kg), R = gas constant (287.1 J/kg K), T = absolute temperature of gas (K):

$$\rho = \frac{M\,kg}{V\,m^3} = \frac{P}{R\,T}$$

The gas constant R, humidity and pressure should be constants along the supply air duct, so

$$\rho_1 \approx \frac{1}{T}$$

If standard air density = ρ_1 kg/m^3, and supply air density = ρ_2 kg/m^2:

$$\rho_1 \approx \frac{1}{T_1}$$

$$\rho_2 \approx \frac{1}{T_2}$$

Dividing and inverting the equations gives

$$\frac{\rho_2}{\rho_1} = \frac{T_1}{T_2}$$

$$\rho_2 = \rho_1 \frac{T_1}{T_2}$$

The density of air can be corrected for other supply-air temperatures:

$$\rho_2 = 1.1906 \times \frac{273 + 20}{273 + t_S}\,kg/m^3$$

Substitute this into the heat balance formula:

$$SH\,kW = Q\,m^3/s \times \rho_2\,kg/m^3 \times SHC\,kJ/kg\,K \times (t_R - t_S)\,K$$

$$= Q \times 1.1906 \times \frac{273 + 20}{273 + t_S} \times 1.0048 \times (t_R - t_S)$$

$$= Q \times 351 \times \frac{t_R - t_S}{273 + t_S}$$

Rearrange to find Q:

$$Q = \frac{SH\,kW \times (273 + t_S)}{351\,(t_R - t_S)}\,m^3/s \text{ for summer cooling}$$

This is the general formula that will be used to find the combination of supply-air temperature t_S °C and volume flow rate Q m³/s that will satisfy the room-cooling load.

EXAMPLE 5.1

An office has a sensible heat gain of 10 kW when the room-air temperature is 20 °C d.b. Calculate the necessary volume flow rate of supply air to maintain the room at the design temperature when the supply-air temperature can be 10 °C d.b.

$$Q = \frac{SH \text{ kW} \times (273 + t_S)}{351 \, (t_R - t_S)} \text{ m}^3/\text{s}$$

$$= \frac{10 \text{ kW} \times (273 + 10)}{351 \times (20 - 10)} \text{ m}^3/\text{s}$$

$$= 0.806 \text{ m}^3/\text{s}$$

EXAMPLE 5.2

A lecture theatre is 12 m × 12 m × 6 m. It is maintained at 20 °C d.b. and has four air changes per hour of cooled outdoor air supplied at 15 °C d.b. Calculate the maximum cooling load that the equipment can meet.

$$Q = \frac{N \text{ air changes}}{h} \times V \text{ m}^3 \times \frac{1 \text{ h}}{3600 \text{ s}}$$

$$= 4 \times 12 \times 12 \times 6/3600 \text{ m}^3/\text{s}$$

$$= 0.96 \text{ m}^3/\text{s}$$

$$Q = \frac{SH \text{ kW} \times (273 + t_S)}{351 \, (t_R - t_S)} \text{ m}^3/\text{s}$$

$$SH = \frac{Q \times 351 \times (t_R - t_S)}{273 + t_S} \text{ kW}$$

$$= \frac{0.96 \times 351 \times (20 - 15)}{273 + 15}$$

$$= 5.85 \text{ kW}$$

The summer supply air temperature that is needed for fixed values of sensible heat gain and volume flow rate can be found by rearrangement.

$$Q = \frac{SH \text{ kW} \times (273 + t_S)}{351 \, (t_R - t_S)} \text{ m}^3/\text{s}$$

$$Q \times 351 \, (t_R - t_S) = SH \, (273 + t_S)$$

$$Q \times 351 \, t_R - Q \times 351 \, t_S = SH \times 273 + SH \times t_S$$

$$Q \times 351 \, t_R - SH \times 273 = Q \times 351 \, t_S + SH \times t_S$$

$$Q \times 351\, t_R - SH \times 273 = t_S\, (Q \times 351 + SH)$$

$$t_S = \frac{Q \times 351\, t_R - SH \times 273}{Q \times 351 + SH}$$

EXAMPLE 5.3

An open-plan office is 20 m × 15 m × 3 m. It is maintained at 22 °C d.b. and has eight air changes per hour of cooled supply air. The cooling plant load is 26 kW. Calculate the supply-air temperature.

$$Q = \frac{N \text{ air changes}}{h} \times V \text{ m}^3 \times \frac{1\,h}{3600\,s}$$

$$= 8 \times 20 \times 15 \times 3/3600 \text{ m}^3/s$$

$$= 2 \text{ m}^3/s$$

$$t_S = \frac{Q \times 351\, t_R - SH \times 273}{Q \times 351 + SH}$$

$$= \frac{2 \times 351 \times 22 - 26 \times 273}{2 \times 351 + 26}$$

$$= 11.5 \,°C$$

Check the accuracy of the new formula by substituting for t_S.

$$Q = \frac{SH \text{ kW} \times (273 + t_S)}{351\,(t_R - t_S)} \text{ m}^3/s$$

$$= \frac{26 \times (273 + 11.5)}{351 \times (22 - 11.5)} \text{ m}^3/s$$

$$= 2.007 \text{ m}^3/s \quad \text{correct to within 0.4\% due to rounding}$$

The same procedure is adopted for winter heating, the difference being that the supply-air temperature is higher than that of the room air; $t_S > t_R$.

$$Q = \frac{SH \text{ kW} \times (273 + t_S)}{351\,(t_S - t_R)} \text{ m}^3/s \text{ for winter heating}$$

EXAMPLE 5.4

A room has a winter sensible heat loss of 12 kW when the room-air temperature is 21 °C d.b. Calculate the necessary volume flow rate of supply air to maintain the room at the design temperature when the supply-air temperature is 35 °C.

$$Q = \frac{SH \text{ kW} \times (273 + t_S)}{351\,(t_S - t_R)} \text{ m}^3/s$$

$$= \frac{12 \times (273 + 35)}{351 \times (35 - 21)} \text{ m}^3/s$$

$$= 0.752 \text{ m}^3/s$$

EXAMPLE 5.5

A ducted warm-air heating system serves a public hall of 18 m × 11 m × 4 m. It is maintained at 20 °C d.b. and has 1.5 air changes per hour of recirculated air supplied at 30 °C d.b. Calculate the maximum heat loss from the building that the system can meet.

$$Q = \frac{N \text{ air changes}}{h} \times V \text{ m}^3 \times \frac{1 \text{ h}}{3600 \text{ s}}$$

$$= 1.5 \times 18 \times 11 \times 4/3600 \text{ m}^3/\text{s}$$

$$= 0.33 \text{ m}^3/\text{s}$$

$$Q = \frac{SH \text{ kW} \times (273 + t_S)}{351 (t_S - t_R)} \text{ m}^3/\text{s}$$

$$SH = \frac{Q \times 351 (t_S - t_R)}{273 + t_S} \text{ kW}$$

$$= \frac{0.33 \times 351 \times (30 - 20)}{273 + 30}$$

$$= 3.823 \text{ kW}$$

It may be assumed that the winter Q m³/s is the same as the summer Q m³/s in a constant-volume system. The same air-conditioning system is providing both conditions and operating continuously throughout the year. The error in this assumption is that the summer and winter supply-air temperatures and densities are different. The higher winter supply temperature increases the air volume flow rate. The winter supply-air temperature that is needed for fixed values of sensible heat loss and volume flow rate can be found by rearrangement.

$$Q = \frac{SH \text{ kW} \times (273 + t_S)}{351 (t_S - t_R)} \text{ m}^3/\text{s}$$

$$Q \times 351 (t_S - t_R) = SH (273 + t_S)$$

$$Q \times 351 t_S - Q \times 351 t_R = SH \times 273 + SH \times t_S$$

$$Q \times 351 t_S - SH \times t_S = Q \times 351 t_R + SH \times 273$$

$$t_S (351 Q - SH) = 351 Q t_R + 273 SH$$

$$t_S = \frac{351 Q t_R + 273 SH}{351 Q - SH}$$

EXAMPLE 5.6

An open-plan office is 20 m × 15 m × 3 m. It is maintained at 18 °C d.b. and has eight air changes per hour of supply air for both summer and winter. The room heat loss is 38 kW. Calculate the supply-air temperature.

$$Q = 2 \text{ m}^3/\text{s from Example 5.3}$$

$$t_S = \frac{351 \, Q \, t_R + 273 \, SH}{351 \, Q - SH}$$

$$= \frac{351 \times 2 \times 18 + 273 \times 38}{351 \times 2 - 38}$$

$$= 34.7 \,^{\circ}\text{C}$$

Check the accuracy of the new formula by substituting for t_S.

$$Q = \frac{SH \text{ kW} \times (273 + t_S)}{351 \, (t_S - t_R)} \text{ m}^3/\text{s}$$

$$= \frac{38 \times (273 + 34.7)}{351 \times (34.7 - 18)} \text{ m}^3/\text{s}$$

$$= 1.995 \text{ m}^3/\text{s} \text{ which is correct to within 0.3\% due to rounding.}$$

The air mass flow rate will be needed for heater and cooler battery load calculation. Either the air density or specific volume is used.

EXAMPLE 5.7

A single-duct air-conditioning system serves a public hall of 960 m^3 that has a sensible heat gain of 20 kW and a winter heat loss of 18 kW. It is maintained at 21 °C d.b. during winter and summer. The summer supply-air temperature is 13 °C d.b. The vapour pressure of the supply air is 1023 Pa in summer and winter. The supply-air fan is fitted upstream of the heater and cooler batteries and operates in an almost constant air temperature that is close to that of the room air. Calculate the mass flow rate of the supply air in summer, room air-change rate, the winter supply-air temperature and comment upon the volume flow rate under winter heating conditions.

In summer:

$$Q = \frac{SH \text{ kW} \times (273 + t_S)}{351 \, (t_R - t_S)} \text{ m}^3/\text{s}$$

$$= \frac{20 \times (273 + 13)}{351 \times (21 - 13)} \text{ m}^3/\text{s}$$

$$= 2.037 \text{ m}^3/\text{s}$$

$$N = \frac{Q \text{ m}^3}{\text{s}} \times \frac{3600 \text{ s}}{1 \text{ h}} \times \frac{1 \text{ air change}}{V \text{ m}^3}$$

$$= \frac{2.037 \text{ m}^3}{\text{s}} \times \frac{3600 \text{ s}}{1 \text{ h}} \times \frac{1 \text{ air change}}{960 \text{ m}^3}$$

$$= 7.6 \text{ air changes/h}$$

From Chapter 2, the specific volume of humid air at standard atmospheric pressure is

$$v = \frac{287.1 \times (273 + 13)}{101325 - 1023} \, \text{m}^3/\text{kg}$$

$$= 0.8186 \, \text{m}^3/\text{kg}$$

The summer supply air mass flow rate

$$M = \frac{Q}{v}$$

$$= \frac{2.037 \, \text{m}^3}{\text{s}} \times \frac{1 \, \text{kg}}{0.8186 \, \text{m}^3}$$

$$= 2.489 \, \text{kg/s}$$

2.037 m³/s is assumed to be the supply-air volume flow rate in winter. This may not be strictly accurate owing to changes in air density with the supply-air temperature. The winter supply-air temperature is found from

$$t_S = \frac{351 \, Q \, t_R + 273 \, SH}{351 \, Q - SH}$$

$$= \frac{351 \times 2.037 \times 21 + 273 \times 18}{351 \times 2.037 - 18}$$

$$= \frac{15015 + 4914}{697}$$

$$= 28.6 \, ^\circ\text{C d.b.}$$

The winter supply air specific volume

$$v = \frac{287.1 \times (273 + 28.6)}{101\,325 - 1023} \, \text{m}^3/\text{kg}$$

$$= 0.863 \, \text{m}^3/\text{kg}$$

The winter supply-air mass flow rate will be the same as the summer value as the fan is handling constant-temperature air. The winter supply-air volume flow rate will be

$$Q = M \, \text{kg/s} \times v \, \text{m}^3/\text{kg}$$

$$= 2.489 \times 0.863 \, \text{m}^3/\text{s}$$

$$= 2.148 \, \text{m}^3/\text{s}$$

Substitute 2.148 m³/s for 2.037 m³/s and recalculate the winter supply air temperature to be 28.2 °C rather than 28.6 °C.

The quantity of air that is supplied to offset the sensible heat transfer is also utilized to balance the latent heat exchange. There is normally a latent heat gain to the conditioned space from the occupants owing to their respiration and moisture evaporation through the skin. The supply air absorbs the moisture produced by the occupants and leaves the room at the design room-moisture content and percentage saturation. Some industrial applications may require the supply air to add moisture to the conditioned space as paper or textiles may be absorbing moisture and removing water from the air.

The latent heat equivalent of the evaporated moisture is used for convenience but the psychrometric cycle requires the use of air moisture content g kg H_2O/kg dry air.

g_r = room-air moisture content kg H_2O/kg dry air

g_s = supply-air moisture content kg H_2O/kg dry air

Q = supply-air volume flow rate from the thermal load m³/s

h_{fg} = latent heat of vaporization of water at 0 °C = 2453.61 kJ/kg

A mass balance can be made between the moisture evaporated in the room and the moisture absorbed by the supply air flow:

moisture evaporated in room = moisture added to supply air
latent heat gain to the room LH kW
= air mass flow rate M kg/s × moisture added $(g_r - g_s)$ kg H_2O/kg dry air × latent heat of vaporization of moisture h_{fg} 2453.61 kJ/kg H_2O

$$LH = Q\,\frac{m^3}{s} \times 1.1906\,\frac{kg}{m^3} \times \frac{(273+20)}{(273+t_S)} \times (g_r - g_s)\,\frac{kg\ H_2O}{kg\ air} \times 2453.61\,\frac{kJ}{kg\ H_2O} \times \frac{kW\ s}{kJ}$$

$$LH = Q \times 1.1906 \times \frac{273+20}{273+t_S} \times (g_r - g_s) \times 2453.61\,kW$$

$$LH = Q \times \frac{855\,932}{273+t_S} \times (g_r - g_s)\,kW$$

$$Q = \frac{LH\,kW \times (273+t_S)}{(g_r - g_s) \times 860\,000}\,m^3/s$$

The 855 932 can be written as 860 000 with an insignificant 0.5% error. In cases such as a single-person office, the difference between g_r and g_s will be too small to calculate, measure with a sensing instrument or control with a humidifier. A 5% change in room percentage saturation is likely to be unimportant to the occupants, so calculation accuracy can be relaxed.

EXAMPLE 5.8

An office has a sensible heat gain of 12 kW and four occupants each having a latent heat output of 40 W when the room air condition is 20 °C d.b., 50% saturation. Calculate the necessary volume flow rate of supply air and its moisture content to maintain the room at the design state when the supply-air temperature can be 12 °C d.b. The room-air moisture content is 0.007 376 kg/kg.

Calculate the supply-air quantity from the sensible heat load only:

$$Q = \frac{SH\,kW \times (273+t_S)}{351\,(t_R - t_S)}\,m^3/s$$

$$= \frac{12\,kW \times (273+12)}{351 \times (20-12)}\,m^3/s$$

$$= 1.218\,m^3/s$$

latent heat gain $LH = 4$ people $\times$ 40 W/person

$$= 160 \times 10^{-3} \text{kW}$$

$$= 0.16 \text{kW}$$

Use the latent heat equation to find the supply-air moisture content that is needed.

$$Q = \frac{LH\,\text{kW} \times (273 + t_s)}{(g_r - g_s) \times 860\,000} \text{ m}^3/\text{s}$$

$$(g_r - g_s) = \frac{LH\,\text{kW} \times (273 + t_s)}{Q \times 860\,000}$$

$$= \frac{0.16 \times (273 + 12)}{1.218 \times 860\,000}$$

$$= 0.044 \times 10^{-3} \text{ kg/kg}$$

$$g_s = g_r - 0.044 \times 10^{-3} \text{ kg/kg}$$

$$= 0.007\,376 - 0.044 \times 10^{-3} \text{ kg/kg}$$

$$= 0.007\,332 \text{ kg/kg}$$

The sensible-to-total-heat ratio S/T can be used on the psychrometric chart during the design of the heating, cooling and humidity control processes.

$$\text{S/T} = \frac{\text{sensible heat gain or loss}}{\text{total heat gain or loss}}$$

The ratio is used for heating and cooling plant operations. Total heat transfer is the aggregate of $(SH + LH)$ kW, either of which may be positive or negative. The ratio is always less than unity: $\text{S/T} \leq 1$. S/T ratio is measured on a quadrant on a psychrometric chart. Draw a straight line from the centre of the quadrant radius to the perimeter. This produces a sloping line that connects the room- and supply-air conditions at the peak heating or cooling load. The lower quadrant is used when cooling the room and the upper quadrant when the supply air is heating the conditioned space. Transfer the sloping line on to the chart with drawing instruments. Figure 5.2 demonstrates the use of S/T heat ratio lines. Plotting points on the chart can often be aided with the S/T line when calculation of the air condition cannot be easily made.

EXAMPLE 5.9

A room has a sensible heat gain of 38 kW and a latent heat gain of 2 kW. The room-air condition is 20 °C d.b. and 50% saturation. The supply-air temperature is 12 °C d.b. Plot the room and supply conditions on a psychrometric chart.

$$\text{S/T ratio} = \frac{38}{38 + 2}$$

$$= 0.95$$

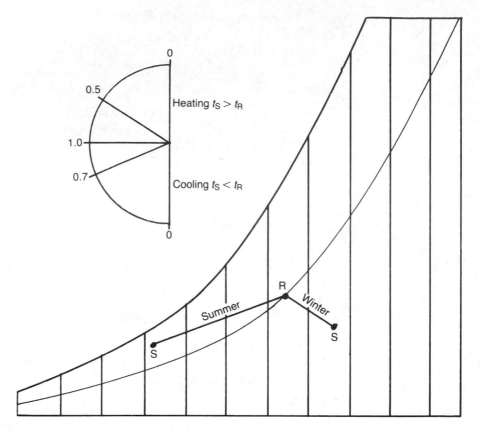

Fig. 5.2   Sensible-heat-to-total-heat ratio line on a psychrometric chart.

The intersection of the S/T ratio line and 12 °C d.b. occurs at the supply-air condition and is shown in Fig. 5.3.

PLANT AIRFLOW DESIGN

A schematic drawing of the air ductwork logic aids comprehension. The quantity of airflow is specified for each section of duct for sizing purposes and these data form an integral part of the overall system design. All airflows need to be specified, particularly where transfers take place between rooms. Accuracy down to a single litre per second (l/s) is required. The designer's accuracy may not be matched by reality when the commissioning tolerance of 5% is allowed. The effect of air transfers between rooms can be to create draughts around doors, whistling noise through gaps and suction forces on doors. Air-transfer grilles in walls or doors may be acceptable but they also pass speech and other sounds. Air ductwork that connects rooms can transfer speech and noise unless they have acoustic attenuation. The commissioning engineer uses the design airflow data, and a current record of all the information is essential. Design changes to airflows are used to update the records.

   The static air pressure that is to be maintained in a space is a function of the supply and extract air-duct pressures and airflows into and out from the space. The

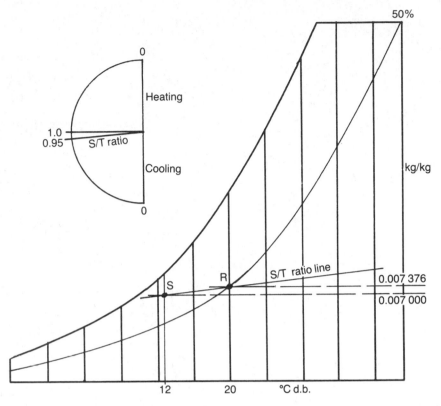

Fig. 5.3   Solution to Example 5.9.

design total airflow into a space must balance the quantity extracted. Temperature differences in the various parts of the air-circulation network cause density changes and volume corrections are applied.

Figure 5.4 is the schematic logic diagram of the airflow quantities for a single-duct system serving an office area. Ignore the figures in brackets. They are used later. Some conditioned air is exhausted to atmosphere through the toilet accommodation. Notice that all the flows of air are accounted for and that movement is deliberately in the direction of the toilets by restricting the extract grille quantities in the office and corridor. The fresh air inlet quantity is based upon the occupancy requirement, say, 50 people at 15 l/s per person, or 750 l/s. The building is assumed to be air-sealed from the outdoor environment. There is no continuous airflow through the external perimeter, doors or windows. This is unlikely in practice.

EXAMPLE 5.10

Calculate the summer airflows for all the ducts shown in Fig. 5.4 using the following data:

Room	Volume(m³)	SH (kW)	Occupants
Office	2000	20	40
Corridor	270	3	5
Toilet	150	2	4

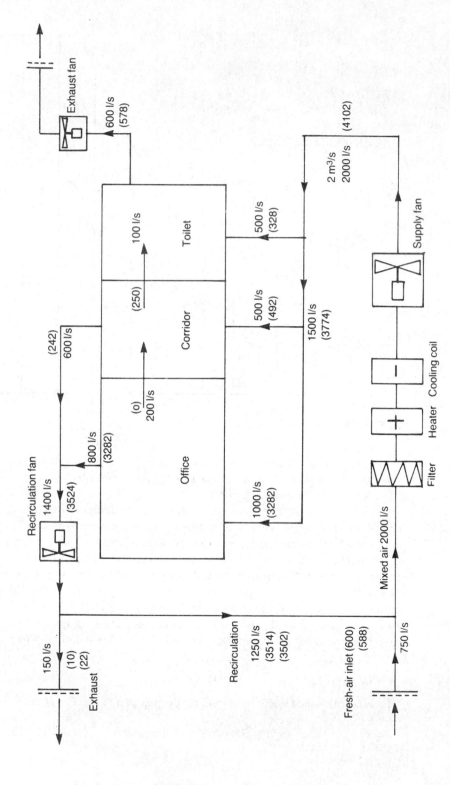

Fig. 5.4 Schematic logic diagram of airflows to rooms. (Solution to Example 5.10 is in brackets.)

Supply-air temperature = 15 °C d.b.
Room-air temperature = 20 °C d.b.
Toilet to have a minimum of three air changes per hour
Fresh air requirement = 12 l/s per person
The office and corridor air static pressures are equal.
The toilet door has a grille of free area of 0.1 m² into the corridor. The maximum air velocity through the grille is to be 2.5 m/s.

For each room, use

$$Q = \frac{SH \, kW \times (273 + t_S)}{351 \, (t_R - t_S)} \, m^3/s$$

For the office:

$$Q = \frac{20 \, kW \times (273 + 15)}{351 \times (20 - 15)} \, m^3/s$$

$$= 3.282 \, m^3/s$$

For the corridor:

$$Q = 0.492 \, m^3/s$$

For the toilet:

$$Q = 0.328 \, m^3/s$$

Air-change rate for the toilet:

$$N = \frac{Q \, m^3}{s} \times \frac{3600 \, s}{1 \, h} \times \frac{1 \, air \, change}{V \, m^3}$$

$$= \frac{0.328 \, m^3}{s} \times \frac{3600 \, s}{1 \, h} \times \frac{1 \, air \, change}{150 \, m^3}$$

$$= 7.9 \text{ air changes/h. This exceeds the minimum.}$$

Total supply air:

$$Q = 3.282 + 0.492 + 0.328 \, m^3/s$$

$$= 4.102 \, m^3/s$$

Fresh air quantity for each room:

$$\text{office } Q_f = 40 \text{ occupants} \times 12 \, l/s$$

$$= 480 \, l/s$$

$$\text{corridor } Q_f = 5 \text{ occupants} \times 12 \, l/s$$

$$= 60 \, l/s$$

$$\text{toilet } Q_f = 4 \text{ occupants} \times 12 \, l/s$$

$$= 48 \, l/s$$

$$\text{total fresh air inlet } Q = 480 + 60 + 48 \text{ l/s}$$

$$= 588 \text{ l/s}$$

$$\text{airflow through toilet door} = 0.1 \text{ m}^2 \times 2.5 \text{ m/s}$$

$$= 0.25 \text{ m}^3/\text{s}$$

$$= 250 \text{ l/s}$$

The airflows are shown in brackets in Figure 5.4. Check that the correct fresh airflow rate is supplied into each room. The office is the most important room for fresh air. It requires $(480/3282) \times 100\% = 14.6\%$ fresh air. This is higher than the average for all the rooms, $(588/4102) \times 100\% = 14.3\%$. The fresh air proportion of the office supply air is $14.3\% \times 3282 \text{ l/s} = 470 \text{ l/s}$ instead of 480 l/s. The office will be deficient in fresh air supply. The fresh air intake to the plant should be increased to $14.6\% \times 4102 \text{ l/s} = 600 \text{ l/s}$ when the decimal places are included in calculations. The updated airflows are shown in Figure 5.4.

COORDINATED SYSTEM CALCULATIONS

The essential principles of system design can now be applied to practical problems. The difference between exercises that are structured for learning purposes and realistic designs is that there may be several possible correct solutions in the real cases. Selection of data and assumptions is open to interpretation. The solutions provided are considered appropriate but the reader may arrive at unique answers that can be correct.

A range of applications is analysed by example and questions. Use the CIBSE psychrometric chart and data tables to find any missing data. There is no need to calculate data from first principles if it is published. Make any necessary assumptions to find a correct solution.

EXAMPLE 5.11

The office module shown in Fig. 5.5 is to have a single-duct air-conditioning system. There is to be a roof-mounted air-handling and refrigeration plant. The building has five floors of offices and each floor has eight similar layouts. Use a psychrometric chart and the data provided to find the peak summer and winter design loads and air conditions.

    Summer room air 22 °C d.b., 50% saturation.
    Summer outdoor air 30 °C d.b., 23 °C w.b
    Winter room air 18 °C d.b., 50% saturation.
    Winter outdoor air -2 °C d.b., -3 °C w.b.
    The office has six occupants, 90 W sensible, 50 W latent.
    The fresh air provision is 12 l/s per person.
    There are four computers of 150 W.
    There is no infiltration of outdoor air into the office.
    The peak cooling load through the south-facing glazing is 333 W/m^2 at noon on 22 September.
    A solar-gain correction factor of 0.77 applies when the light-coloured slatted blinds are used.

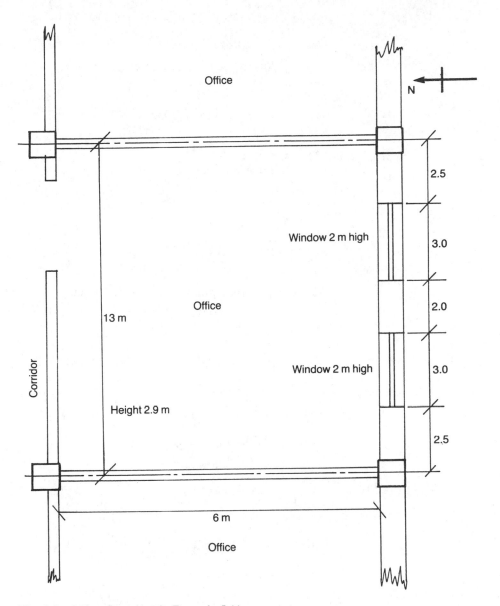

Fig. 5.5 Office floor plan in Example 5.11.

Ignore heat gains through the structure.
Surrounding rooms have the same conditions.
Glazing U value is 5.7 W/m^2 K.
External wall U value is 0.4 W/m^2 K.
Use the simplest forms of heat gain and loss calculation to obtain approximate plant loads.

Peak summer cooling load:

$$\text{glazing load} = 2 \times 3 \times 2 \text{ m}^2 \times 333 \text{ W/m}^2 \times 0.77$$

$$= 3077 \text{ W}$$

$$\text{internal gain} = 6 \text{ occupants} \times 90 \text{ W} + 4 \text{ PCs} \times 150 \text{ W}$$

$$= 1140 \text{ W}$$

$$\text{peak summer gain} = 3077 + 1140 \text{ W}$$

$$= 4217 \text{ W}$$

$$\text{office volume} = 13 \text{ m} \times 6 \text{ m} \times 2.9 \text{ m}$$

$$= 226.2 \text{ m}^3$$

$$\text{heat gain per m}^3 = \frac{4217 \text{ W}}{226.2 \text{ m}^3}$$

$$= 18.6 \text{ W/m}^3$$

This will be typical for similar offices in the UK facing south. Heat gains and losses vary from 5 W/m^3 to 30 W/m^3, depending upon orientation, thermal insulation, glazing area and internal heat generation from people, lights and equipment.

Winter heat loss:

$$\text{glazing loss} = 2 \times 3 \text{ m} \times 2 \text{ m} \times 5.7 \text{ W/m}^2 \text{ K} \times (18 - -2) \text{ K}$$

$$= 1368 \text{ W}$$

$$\text{wall loss} = (13 \text{ m} \times 2.9 \text{ m} - 2 \times 3 \text{ m} \times 2 \text{ m}) \times 0.4 \text{ W/m}^2 \text{ K} \times 20 \text{ K}$$

$$= 206 \text{ W}$$

$$\text{design heat loss} = 1368 + 206 \text{ W}$$

$$= 1574 \text{ W}$$

In summer, try $t_S = 13$ °C d.b.

$$Q = \frac{SH \text{ kW} \times (273 + t_S)}{351 \, (t_R - t_S)} \text{ m}^3/\text{s}$$

$$= \frac{4.217 \times (273 + 13)}{351 \times (22 - 13)} \text{ m}^3/\text{s}$$

$$= 0.382 \text{ m}^3/\text{s}$$

$$= 382 \text{ l/s}$$

$$N = \frac{Q \text{ m}^3}{\text{s}} \times \frac{3600 \text{ s}}{1 \text{ h}} \times \frac{1 \text{ air change}}{V \text{ m}^3}$$

$$= \frac{0.382 \text{ m}^3}{\text{s}} \times \frac{3600 \text{ s}}{1 \text{ h}} \times \frac{1 \text{ air change}}{226.2 \text{ m}^3}$$

$$= 6.1 \text{ air changes/h}$$

This is within the acceptable range of 4–20 air changes per hour for most rooms. It would be possible to use a higher supply-air temperature, 15 °C d.b. at Q of 0.494m³/s and N of 7.9 if the lower supply temperature would cause draughts.

latent heat gain $LH = 6$ people $\times 50$ W/person

$$= 300 \times 10^{-3} \, \text{kW}$$

$$= 0.3 \, \text{kW}$$

$$Q = \frac{LH \, \text{kW} \times (273 + t_{\text{s}})}{(g_{\text{r}} - g_{\text{s}}) \times 860\,000} \, \text{m}^3/\text{s}$$

$$(g_{\text{r}} - g_{\text{s}}) = \frac{LH \, \text{kW} \times (273 + t_{\text{s}})}{Q \times 860\,000}$$

$$= \frac{0.3 \times (273 + 13)}{0.382 \times 860\,000}$$

$$= 0.261 \times 10^{-3} \, \text{kg/kg}$$

From psychrometric data, summer $g_{\text{r}} = 0.008\,366$ kg/kg.

$$g_{\text{s}} = g_{\text{r}} - 0.261 \times 10^{-3} \, \text{kg/kg}$$

$$= 0.008\,366 - 0.261 \times 10^{-3} \, \text{kg/kg}$$

$$= 0.008\,105 \, \text{kg/kg}$$

fresh air supply $= 12 \, \text{l/s} \times 6$ occupants

$$= 72 \, \text{l/s}$$

fresh air proportion $= \dfrac{72}{382} \times 100\%$

$$= 19\%$$

To find the mixed air condition entering the cooling coil, proportion the outdoor air and recirculated room air flows:

$$t_{\text{m}} = 22 + 0.19 \times (30 - 22) \, ^\circ\text{C d.b.}$$

$$= 23.5 \, ^\circ\text{C d.b.}$$

Plot the mixed air conditions on the psychrometric chart, a sketch of which is shown in Fig. 5.6: t_{m} 17.2 °C w.b., specific enthalpy h_{m} 48 kJ/kg, 0.853 m³/kg. Plot the supply-air condition on the chart at 13 °C d.b., 0.0081 kg/kg, 11.8 °C w.b., h_{s} is 33.5 kJ/kg.

$$\text{cooling coil load} = \frac{0.382 \, \text{m}^3}{\text{s}} \times (48 - 33.5) \, \frac{\text{kJ}}{\text{kg}} \times \frac{\text{kg}}{0.853 \, \text{m}^3} \times \frac{\text{kW s}}{\text{kJ}}$$

$$= 6.494 \, \text{kW}$$

Find the winter supply-air temperature for $Q = 0.382 \, \text{m}^3/\text{s}$:

$$t_{\text{s}} = \frac{351 \, Q \, t_{\text{R}} + 273 \, SH}{351 \, Q - SH}$$

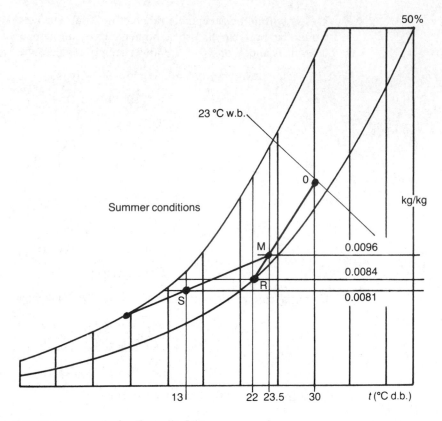

Fig. 5.6 Summer cycle for Example 5.11.

$$= \frac{351 \times 0.382 \times 18 + 273 \times 1.574}{351 \times 0.382 - 1.574}$$

$$= 21.5 \,^{\circ}\text{C d.b.}$$

Plot the winter room and outdoor air conditions on the psychrometric chart. Join them with a straight line and calculate the mixed-air state.

$$t_{\mathrm{m}} = 18 - 0.19 \times (18 - - 2) \,^{\circ}\text{C d.b.}$$

$$= 14.2 \,^{\circ}\text{C d.b.}$$

From the earlier calculation, ignoring the change in t_{s}:

$$(g_{\mathrm{r}} - g_{\mathrm{s}}) = 0.261 \times 10^{-3} \,\text{kg/kg}$$

From psychrometric data, winter $g_{\mathrm{r}} = 0.006 \ 492$ kg/kg.

$$g_{\mathrm{s}} = 0.006 \ 492 - 0.261 \times 10^{-3} \,\text{kg/kg}$$

$$= 0.006 \ 231 \,\text{kg/kg}$$

A sketch of the winter heating cycle is shown in Fig. 5.7: t_{m} 14.2 °C w.b., specific enthalpy h_{m} is 30 kJ/kg, 0.823 m^3/kg. Plot the supply-air condition on the chart at 21.5 °C d.b., 0.006 231 kg/kg, 13.5 °C w.b., h_{s} is 37.5 kJ/kg.

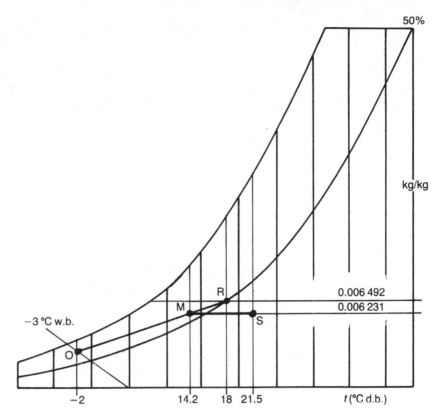

Fig. 5.7 Winter cycle for Example 5.11.

$$\text{heating coil load} = \frac{0.382 \, \text{m}^3}{\text{s}} \times (37.5 - 30) \, \frac{\text{kJ}}{\text{kg}} \times \frac{\text{kg}}{0.823 \, \text{m}^3} \times \frac{\text{kW s}}{\text{kJ}}$$

$$= 3.481 \, \text{kW}$$

Figure 5.8 shows the duct air flows for one room of this type.

Nearly all air-conditioning systems are in the single-duct air-handling category. The principal variation is the addition of terminal heating and cooling coils in fan-coil, variable air volume or induction systems. The use of air volume flow-rate regulation (VAV systems) does not affect the psychrometric peak loads, only the part load operation. Where terminal heating or cooling coils are used, the central air-handling plant may only be supplying conditioned fresh air. Recirculated air remains within the room.

The fresh air may provide some sensible cooling to the room or it may be at the desired room temperature. The fresh air can be designed to provide all the humidity control. It is common to do sensible cooling only with recirculation units in computer halls. The computer hall fresh air supply plant controls room humidity to ±5% of the desired value. In all applications, the supply-air condition will be determined from the room or zone having the greatest demand upon it. This will be the space requiring the most extreme central plant supply temperature and moisture content. The other rooms that are connected to the same duct will use their terminal coils to provide locally correct conditions.

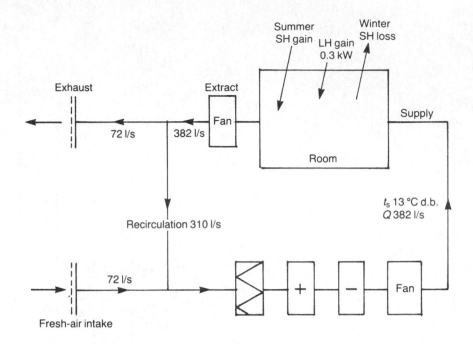

Fig. 5.8 Air-handling schematic for Example 5.11.

EXAMPLE 5.12

A Sydney building has two north-facing office floors served from a single-duct air-conditioning system with fan-coil units. Use a psychrometric chart and the data provided to find the peak summer and winter design loads, air flows and conditions.

  Summer room air 21 °C d.b., 50% saturation.
  Summer outdoor air 35 °C d.b., 24 °C w.b.
  Winter room air 21 °C d.b., 50% saturation.
  Winter outdoor air 6 °C d.b., 3 °C w.b.
  Office 1 has 60 m of perimeter length, north-facing glazing area of 100 m², is 7 m wide and 3 m high. It has 40 occupants and 25 computers of 120 W each.
  Office 2 has 30 m of perimeter length, glazing of 60 m², is 5 m wide and 4 m high. It has 15 occupants and 10 computers of 150 W each.
  The occupancy heat gains are 90 W sensible and 50 W latent.
  The fresh air provision is 12 l/s per person.
  There is no infiltration of outdoor air into the office.
  The peak cooling load through the north-facing glazing is 450 W/m² at noon on 21 February.
  A solar-gain correction factor of 0.55 applies to the glass type and shading blinds.
  Ignore heat gains through the structure.
  Surrounding rooms have the same conditions.
  Glazing U value is 2.8 W/m² K.
  External wall U value is 0.6 W/m² K.
  Use the simplest forms of heat gain and loss calculation to obtain approximate plant loads.
  Recommend a suitable design for the fan coil units.

Office 1

Peak summer cooling load:

$$\text{glazing load} = 100\,\text{m}^2 \times 450\,\text{W/m}^2 \times 0.55$$

$$= 24.75\,\text{kW}$$

$$\text{internal gain} = 40\,\text{occupants} \times 90\,\text{W} + 25\,\text{PCs} \times 120\,\text{W}$$

$$= 6.6\,\text{kW}$$

$$\text{peak summer gain} = 31.35\,\text{kW}$$

$$\text{office volume} = 60\,\text{m} \times 7\,\text{m} \times 3\,\text{m}$$

$$= 1260\,\text{m}^3$$

$$\text{heat gain per m}^3 = \frac{31\,350\,\text{W}}{1260\,\text{m}^3}$$

$$= 24.9\,\text{W/m}^3$$

Winter heat loss:

$$\text{glazing loss} = 100\,\text{m}^2 \times 2.8\,\text{W/m}^2\,\text{K} \times (21 - 6)\,\text{K}$$

$$= 4.2\,\text{kW}$$

$$\text{wall loss} = (60\,\text{m} \times 3\,\text{m} - 100\,\text{m}^2) \times 0.6\,\text{W/m}^2\,\text{K} \times 15\,\text{K}$$

$$= 0.72\,\text{kW}$$

$$\text{design heat loss} = 4.2 + 0.72\,\text{kW}$$

$$= 4.92\,\text{kW}$$

Office 2

Peak summer cooling load:

$$\text{glazing load} = 60\,\text{m}^2 \times 450\,\text{W/m}^2 \times 0.55$$

$$= 14.85\,\text{kW}$$

$$\text{internal gain} = 15\,\text{occupants} \times 90\,\text{W} + 10\,\text{PCs} \times 150\,\text{W}$$

$$= 2.85\,\text{kW}$$

$$\text{peak summer gain} = 17.7\,\text{kW}$$

$$\text{office volume} = 30\,\text{m} \times 5\,\text{m} \times 4\,\text{m}$$

$$= 600\,\text{m}^3$$

$$\text{heat gain per m}^3 = \frac{17\,700\,\text{W}}{600\,\text{m}^3}$$

$$= 29.5\,\text{W/m}^3$$

Winter heat loss:

$$\text{glazing loss} = 60\,\text{m}^2 \times 2.8\,\text{W/m}^2\,\text{K} \times (21 - 6)\,\text{K}$$

$$= 2.52\,\text{kW}$$

$$\text{wall loss} = (30\text{ m} \times 4\text{ m} - 60\text{ m}^2) \times 0.6\text{ W/m}^2\text{ K} \times 15\text{ K}$$

$$= 0.54\text{ kW}$$

$$\text{design heat loss} = 2.52 + 0.54\text{ kW}$$

$$= 3.06\text{ kW}$$

In summer, try $t_S = 12\,°\text{C}$ d.b. for office 1:

$$Q = \frac{SH\text{ kW} \times (273 + t_S)}{351(t_R - t_S)}\text{ m}^3/\text{s}$$

$$= \frac{31.35 \times (273 + 12)}{351 \times (21 - 12)}\text{ m}^3/\text{s}$$

$$= 2.828\text{ m}^3/\text{s}$$

$$N = \frac{Q\text{ m}^3}{\text{s}} \times \frac{3600\text{ s}}{1\text{ h}} \times \frac{1\text{ air change}}{V\text{ m}^3}$$

$$= \frac{2.828\text{ m}^3}{\text{s}} \times \frac{3600\text{ s}}{1\text{ h}} \times \frac{1\text{ air change}}{60 \times 7 \times 3\text{ m}^3}$$

$$= 8\text{ air changes/h}$$

$$\text{latent heat gain } LH = 40\text{ people} \times 50\text{ W/person}$$

$$= 2\text{ kW}$$

$$(g_r - g_s) = \frac{LH\text{ kW} \times (273 + t_S)}{Q \times 860\,000}$$

$$= \frac{2 \times (273 + 12)}{2.828 \times 860\,000}$$

$$= 0.234 \times 10^{-3}\text{ kg/kg}$$

From psychrometric data, summer $g_r = 0.007\,857$ kg/kg.

$$g_s = 0.007\,857 - 0.234 \times 10^{-3}\text{ kg/kg}$$

$$= 0.007\,622\text{ kg/kg}$$

Now calculate the summer supply-air conditions for office 2 using $t_S = 12\,°\text{C}$ d.b.

$$Q = 1.6\text{ m}^3/\text{s}$$

$$N = 9.6\text{ air changes/h}$$

$$LH = 0.75\text{ kW}$$

$$(g_r - g_s) = 0.155 \times 10^{-3}\text{ kg/kg}$$

$$g_s = 0.007\,857 - 0.155 \times 10^{-3}\text{ kg/kg}$$

$$= 0.007\,702\text{ kg/kg}$$

The supply-air moisture content required for office 1 is lower than that for office 2, so a g_s of 0.007 622 kg/kg will be used for the air that leaves the central air-handling plant. This means that the humidity control of office 2 will not be correct; however, the error will not be noticeable and probably not measurable.

Figure 5.9 shows the air-handling schematic for the two offices. It is likely that several fan-coil units would be installed in each office to provide local control of each part of the room. Manufacturers' literature is used to find units that deliver suitable airflow rates. If each unit can handle the supply air for 6 m of perimeter, then ten are needed for office 1 and five for office 2. The cooling coil in the central air-handling plant will remove the excess heat from the fresh air intake and the fan-coil units will take out the local heat gains. The mixing of conditioned fresh air and recirculated room air takes place within each fan coil unit.

Find the mixed air condition entering an office 1 unit:

$$\text{office 1 fresh air supply} = 12 \, \text{l/s} \times 40 \text{ occupants}$$

$$= 480 \, \text{l/s}$$

$$= 0.48 \, \text{m}^3/\text{s}$$

$$\text{total office 1 supply air } Q = 2.828 \, \text{m}^3/\text{s}$$

$$\text{fresh air proportion} = \frac{0.48}{2.828} \times 100\%$$

$$= 17\%$$

$$\text{each fan-coil unit will pass } Q = \frac{2.828}{10} \, \text{m}^3/\text{s}$$

$$= 0.283 \, \text{m}^3/\text{s}$$

$$= 283 \, \text{l/s}$$

17% of this will be from the fresh air inlet duct:

$$Q_f = 17\% \times 283 \, \text{l/s}$$

$$= 48 \, \text{l/s}$$

The summer cycle can be drawn on the psychrometric chart (Fig. 5.10), and unknown points or lines are found by experimental construction. Plot the known outside O, room R and supply air S conditions. Construct a horizontal line from O to the room-air temperature 21 °C d.b., but 19.9 °C w.b. This is condition C. It is the fresh air state that leaves the central air-handling plant. Maintaining C at the room-air temperature means that the ducted-air system is at room temperature and will have negligible heat gains or losses through the building.

Construct a vertical line from C to R. Recirculated room air R and fresh air C mix in the fan-coil unit along this line. The mixture M must contain 17% fresh air and 83% recirculated room air. This mixing process takes place along line CR such that CM will be 83% of CR. The fresh air proportion MR will be 17% of CR. The ratio can be calculated along the moisture content axis as it has equal increments.

$$g_m = 0.007\,857 + 0.17 \times (0.014\,15 - 0.007\,857) \, \text{kg/kg}$$

$$= 0.008\,927 \, \text{kg/kg}$$

Condensation will not take place in the mixing chamber of the fan-coil unit as the outside air dew point is not reached. This mixed air M passes to the terminal unit

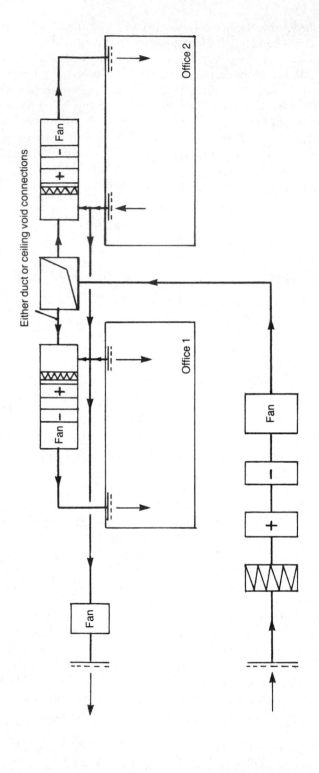

Fig. 5.9 Air-handling plant schematic for the fan-coil unit system in Example 5.12.

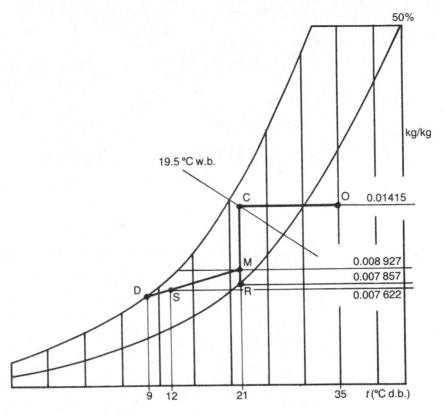

Fig. 5.10 Summer cycle for the fan-coil unit system in Example 5.12.

cooling coil where it is further cooled and dehumidified to the correct supply condition S.

Construct a straight line from M, through S, to the saturation curve at 9 °C. This is the room-coil dew point D. The supply air leaves the room-cooling coil at 12 °C d.b., 10.8 °C w.b., S. The data for the two cooling coils are

(a) at O, 35 °C d.b., 24 °C w.b., 71.2 kJ/kg, 0.8923 m³/kg, 0.014 15 kg/kg;
(b) at C, 21 °C d.b., 19.9 °C w.b., 57.2 kJ/kg, 0.014 15 kg/kg;
(c) at M, 21 °C d.b., 15.8 °C w.b., 43.74 kJ/kg, 0.845 m³/kg, 0.008 927 kg/kg;
(d) at S, 12 °C d.b., 10.8 °C w.b., 31.23 kJ/kg, 0.007 622 kg/kg.

$$SH \text{ kW} = \text{air-mass flow rate} \times \text{decrease in specific enthalpy}$$

$$= M \text{ kg/s} \times (h_R - h_S) \text{ kJ/kg}$$

$$= \frac{Q \text{ m}^3}{\text{s}} \times (h_R - h_S) \frac{\text{kJ}}{\text{kg}} \times \frac{\text{kg}}{v \text{ m}^3} \times \frac{\text{kW s}}{\text{kJ}}$$

The room unit cooling coil load M to S is

$$\text{load} = \frac{0.283 \text{ m}^3}{\text{s}} \times (43.74 - 31.23) \frac{\text{kJ}}{\text{kg}} \times \frac{\text{kg}}{0.845 \text{ m}^3} \times \frac{\text{kW s}}{\text{kJ}}$$

$$= 4.19 \text{ kW}$$

There are ten room units in office 1.

The fresh air plant cooling coil load O to C is

$$\text{load} = \frac{2.828\,\text{m}^3}{\text{s}} \times (71.2 - 57.2)\,\frac{\text{kJ}}{\text{kg}} \times \frac{\text{kg}}{0.8923\,\text{m}^3} \times \frac{\text{kW s}}{\text{kJ}}$$

$$= 44.371\,\text{kW}$$

Find the winter supply air temperature for $Q = 0.283\,\text{m}^3/\text{s}$.

$$t_S = \frac{351Qt_R + 273SH}{351Q - SH}$$

$$= \frac{351 \times 0.283 \times 21 + 273 \times 3.06}{351 \times 0.283 - 3.06}$$

$$= 30.3\,^\circ\text{C d.b.}$$

The supply air moisture content in winter is the same as during summer:

$$g_s = 0.007\,622\,\text{kg/kg}$$

Plot the winter outdoor air condition O on the psychrometric chart. Raise the outdoor air supply to the room-air temperature of 21 °C d.b. but at 10.5 °C w.b., condition H. Draw a vertical line from H to R. This represents the heated fresh air mixing with the recirculated room air that enters the fan-coil unit. M is the mixed air and 17% of it comes from H and 83% from R. This mixing process takes place along line HR such that HM will be 83% of HR. The ratio can be calculated along the moisture content axis as it has equal increments.

$$g_m = 0.007\,857 - 0.17 \times (0.007\,857 - 0.003\,432)\,\text{kg/kg}$$

$$= 0.007\,105\,\text{kg/kg}$$

Draw a horizontal line from M to the supply air temperature S at 30.3 °C d.b., but 0.007 105 kg/kg. M to S is the heating process through the coil in the room terminal unit. The supply air leaves the room-heating coil at 30.3 °C d.b., 17.5 °C w.b. and 0.007 105 kg/kg. This is below the required supply-air moisture content of 0.007 622 kg/kg. Estimate the room-air percentage saturation that will be produced, g_r, by adding the difference $(g_r - g_s)$, 0.234×10^{-3} kg/kg to g_s.

$$g_r = 0.007\,105 + 0.234 \times 10^{-3}\,\text{kg/kg}$$

$$= 0.007\,339\,\text{kg/kg}$$

Plot the expected room air condition 21 °C d.b., 0.007 339 kg/kg and find that it will be at around 47% saturation. This is sufficiently close to the desired 50% for humidification to be avoided. Any value above 40% is likely to be acceptable for comfort purposes.

The data for the two heating coils are:

(a) at O, 6 °C d.b., 3 °C w.b., 14.66 kJ/kg, 0.7947 m³/kg, 0.003 432 kg/kg;
(b) at H, 21 °C d.b., 10.5 °C w.b., 29.91 kJ/kg, 0.003 432 kg/kg
(c) at M, 21 °C d.b., 14 °C w.b., 38.95 kJ/kg, 0.8423 m³/kg, 0.007 105 kg/kg;
(d) at S, 30.3 °C d.b., 17.5 °C w.b., 48.5 kJ/kg, 0.007 105 kg/kg.

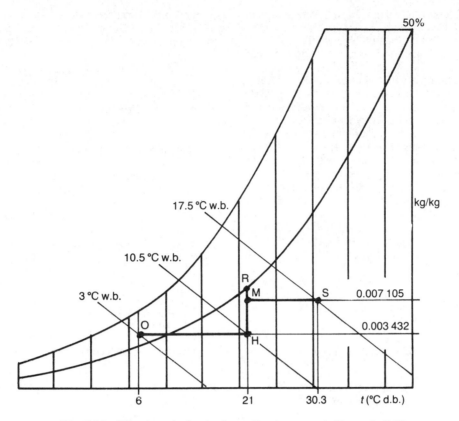

Fig. 5.11 Winter cycle for the fan-coil unit system in Example 5.12.

A sketch of the winter heating cycle is shown in Fig. 5.11.
The heating load on each fan coil unit is:

$$\text{heating coil load} = \frac{0.283 \text{ m}^3}{\text{s}} \times (48.5 - 38.95) \frac{\text{kJ}}{\text{kg}} \times \frac{\text{kg}}{0.8423 \text{ m}^3}$$

$$= 3.209 \text{ kW}$$

There are 10 room units in office 1.
    The fresh air plant heating coil load O to H is

$$\text{load} = \frac{2.828 \text{ m}^3}{\text{s}} \times (29.91 - 14.66) \frac{\text{kJ}}{\text{kg}} \times \frac{\text{kg}}{0.7947 \text{ m}^3}$$

$$= 54.268 \text{ kW}$$

Questions

1. An office has a sensible heat gain of 22 kW when the room-air temperature is 23°C d.b. Calculate the necessary volume flow rate of supply air to maintain the room at the design temperature when the supply-air temperature can be 14°C d.b.

2. A lecture theatre is 18 m x 10 m x 4 m. It is maintained at 21°C d.b. and has six air changes per hour of cooled outdoor air supplied at 16°C d.b. Calculate the maximum cooling load that the equipment can meet.

3. An hotel lounge is 15 m × 15 m × 4 m. It is maintained at 22 °C d.b. and has nine air changes per hour of cooled supply air. The cooling plant load is 19 kW. Calculate the required supply air temperature.

4. An exhibition hall has a winter sensible heat loss of 96 kW when the room-air temperature is 18 °C d.b. Calculate the necessary volume flow rate of supply air to maintain the room at the design temperature when the supply air temperature is 40 °C.

5. A ducted warm-air heating system serves a shop of 30 m × 20 m × 4 m. It is maintained at 18 °C d.b. and has three air changes per hour of recirculated air supplied at 25 °C d.b. Calculate the maximum heat loss from the building that the system can meet.

6. An open-plan office is 30 m × 12 m × 2.8 m. It is maintained at 19 °C d.b. and has six air changes per hour of supply air for both summer and winter. The room heat loss is 27 kW. Calculate the required supply air temperature.

7. A single-duct air-conditioning system serves a theatre of 1500 m³ that has a sensible heat gain of 52 kW and a winter heat loss of 39 kW. It is maintained at 22 °C d.b. during winter and summer. The summer supply-air temperature is 14 °C d.b. The vapour pressure of the supply air is 1044 Pa in summer and winter. The supply-air fan is fitted upstream of the heater and cooler batteries and operates in an almost constant air temperature that is close to that of the room air. Calculate the mass flow rate of the supply air in summer, room air change rate, the winter supply-air temperature and comment upon the volume flow rate under winter heating conditions.

8. An office has a sensible heat gain of 16 kW and seven occupants each having a latent heat output of 50 W when the room air condition is 23 °C d.b., 50% saturation. Calculate the necessary volume flow rate of supply air and its moisture content to maintain the room at the design state when the supply-air temperature can be 14 °C d.b. The room-air moisture content is 0.008 905 kg/kg.

9. A room has a sensible heat gain of 66 kW and a latent heat gain of 3 kW. The room-air condition is 21 °C d.b. and 50% saturation. The supply-air temperature is 13 °C d.b. Plot the room and supply conditions and sensible-to-total-heat ratio line on a psychrometric chart.

10. Calculate the summer air flows for all the ducts shown in Figure 5.4 using the following data;

Room	Volume (m³)	SH (kW)	Occupants
Office	3500	65	70
Corridor	400	12	6
Toilet	550	3	4

Supply air temperature = 14 °C d.b.
Room air temperatures = 19 °C d.b.
Toilet to have a minimum of six air changes per hour
Fresh air requirement = 12 l/s per person

The office and toilet doors each have a grille of free area of 0.25 m² into the corridor. The maximum air velocity through the grille is to be 2.75 m/s.

11. A space has sensible heat gains of 60 kW and 3 kW latent. The space condition is 20 °C d.b., 50% saturation. The supply-air temperature is 15 °C d.b. Calculate the supply-air quantity and moisture content needed.

12. A room 5 m × 12 × 3 m is to be air-conditioned. The maximum sensible heat gain is 7 kW when the latent heat gain is 1 kW. The room condition is to be maintained at 21 °C d.b., 50% saturation by a ventilation rate of eight air changes per hour. Calculate the required supply air condition.

13. A shop has solar heat gains of 8 kW and an internal air condition of 20 °C d.b., 50% saturation. There will be 25 occupants emitting 110 W of sensible heat and 50 W of latent heat each. There are 2 kW of heat gains from lighting and 1 kW of heat output from refrigerated display cabinets. If the supply air can be at 16 °C d.b., calculate its quantity and moisture content.

14. An office floor is to have a single-duct air-conditioning system. Use a psychrometric chart and the data provided to find the peak summer and winter design loads and air conditions.
    Summer room air 23 °C d.b., 50% saturation.
    Summer outdoor air 29 °C d.b., 22 °C w.b.
    Winter room air 19 °C d.b., 50% saturation.
    Winter outdoor air – 1 °C d.b., – 2 °C w.b.
    Summer supply-air temperature 15 °C d.b.
    Office volume 900 m³.
    Glazing area 45 m².
    External wall area 100 m².
    The office has 25 occupants, 90 W sensible, 50 W latent.
    The fresh air provision is 12 l/s per person.
    There are 12 computers of 180 W.
    There is one air change per hour of infiltration by the outdoor air into the office.
    The peak cooling load through the west-facing glazing is 314 W/m² at 1600 h on 21 June.
    A solar-gain correction factor of 0.48 applies to the reflective glazing.
    Ignore heat gains through the structure.
    Surrounding rooms are at the same conditions.
    Glazing U value is 2.7 W/m² K..
    External wall U value is 0.6 W/m² K.
    Use the simplest forms of heat gain and loss calculation to obtain approximate plant loads.

15. A Melbourne hotel has a dual-duct air-conditioning system. Use a psychrometric chart and the data provided to find the peak summer and winter design loads, air conditions and supply airflows from the hot and cold ducts for one room.
    Summer room air 22 °C d.b., 50% saturation.

Summer outdoor air 39 °C d.b., 23 °C w.b.
Winter room air 20 °C d.b., 50% saturation.
Winter outdoor air 3 °C d.b., 2 °C w.b.
Summer supply-air temperature 16 °C d.b.
Air leaves the plant-cooling coil in summer at 13 °C d.b.
Air leaves the plant-heating coil in winter at 25 °C d.b.
Glazing area is $12 m^2$ and external wall area is $30 m^2$.
The room has two occupants, 90 W sensible, 50 W latent.
The fresh air provision is 12 l/s per person.

There is no infiltration of outdoor air.
The peak cooling load through the north-facing glazing is $347 W/m^2$ at noon on 21 January.
A solar-gain correction factor of 0.66 applies to the shading.
Ignore heat gains through the structure.
Surrounding rooms have the same conditions.
Glazing U value is $5.7 W/m^2 K$.
External wall U value is $0.8 W/m^2 K$.
Room volume is $150 m^3$.
Use the simplest forms of heat gain and loss calculation to obtain approximate plant loads.

6 Ductwork design

INTRODUCTION

The design procedure for air-duct systems is shown from the use of Bernoulli's equation through to a spreadsheet that can be used for duct routes. Air-pressure terms and methods of measurement are explained. Detailed pressure changes at duct fittings and fans are calculated. The equations for the flow of air in ducts that were developed in Chapter 4 are used for duct sizing and have been used for the production of a duct-sizing chart for a limited range of flows.

LEARNING OBJECTIVES

Study of this chapter will enable the user to:
1. understand the meaning of static, velocity and total air pressure terms and units;
2. apply Bernoulli's theorem to airflow in ducts;
3. know how to measure air pressure in ducts;
4. calculate the air pressures used in ductwork design;
5. analyse the changes in pressure when air flows through changes in duct size;
6. calculate and plot graphs of the changes of air pressure that occur at fans;
7. define the fan pressure terms;
8. design air-duct systems using CIBSE data;
9. enter a spreadsheet program into a computer and carry out duct-sizing calculations (tuition in spreadsheet use is not provided);
10. understand the methods used for the measurement of air flows in duct systems.

Key terms and concepts

AIR PRESSURES IN A DUCT

Air flowing through a duct exerts two types of pressure on its surroundings. The first is dynamic pressure due to its motion or kinetic energy; this is velocity pressure, p_v. The second is the bursting pressure of the air trying to escape from the enclosure or of the surrounding air trying to enter the enclosing duct. This is static pressure p_s and it acts in all directions.

The sum of these two pressures is the total pressure p_t.

$$\text{total pressure} = \text{static pressure} + \text{velocity pressure}$$

$$p_t = p_s + p_v$$

Losses of pressure due to the frictional resistance of the duct, its bends, branches, changes in cross-section, dampers, filters and air-heating or cooling coils, cause losses of total pressure.

Bernoulli's theorem states that for a fluid flowing from position 1 to position 2, the total pressure energy at 1 equals the total pressure energy at 2 plus the loss of energy due to friction. All energy lost in friction is dissipated as heat and in some cases can cause a noticeable rise in the air temperature.

The balance of air pressures between 1 and 2 is

$$p_{t1} = p_{t2} + \Delta p$$

where p_{t1} = total pressure at 1, p_{t2} = total pressure at 2, Δp = pressure drop due to friction.

Figure 6.1 shows how the total, velocity and static pressures are measured. A water U-tube manometer or electronic micromanometer is connected to a probe in the airway. Static pressure is measured from a tapping in the side of the duct. In this case, the duct static air pressure is above the atmospheric air pressure, p_a. When this pressure is measured on the inlet side of a fan, static pressure may be below atmospheric.

The total pressure is found from the probe facing the air stream as this records both the dynamic and static pressures. The difference between the total and static pressures is the velocity pressure.

From Bernoulli

$$p_t = p_s + p_v$$

so

$$p_v = p_t - p_s$$

$$\text{velocity pressure } p_v = 0.5 \, \rho \, v^2 \text{ Pa}$$

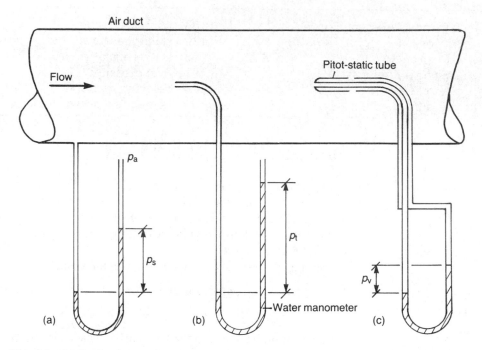

Fig. 6.1 Methods of pressure measurement in an air duct: (a) static pressure measurement tapping; (b) total pressure tune; (c) velocity pressure.

Air density ρ is $1.2\,\text{kg/m}^3$ at 20 °C d.b., 43% saturation and 101 325 Pa, which is the standard atmospheric pressure at sea level. The density at 101 325 Pa but at other temperatures or percentage saturations is found by calculation or by reading the specific volume from the psychrometric chart and inverting it. Density at any other condition is found from the atmospheric pressure p_a Pa and air temperature t °C d.b.:

$$\rho = 1.2 \times \frac{p_a}{101\,325} \times \frac{273 + 20}{273 + t} \frac{\text{kg}}{\text{m}^3}$$

The pressure p_a includes any deviation from the surrounding atmospheric pressure due to the static air pressure within an enclosure or duct p_s:

$$p_a = 101\,325 + p_s \text{ Pa}$$

Note that p_s can be positive or negative. Pressures measured with water manometers are converted into Pascals by

$$p_s = 9.807 \times H \text{ mmH}_2\text{O Pa}$$

The air velocity in a circular duct is found from the volume flow rate Q m³/s and duct internal diameter d m:

$$v = \frac{4Q}{\pi d^2} \text{ m/s}$$

EXAMPLE 6.1

Calculate the total pressure of air flowing at 0.2 m³/s in a 250 mm internal diameter duct if the air temperature is 22 °C d.b., the static pressure of the air in the duct is 25 mm water gauge above the atmospheric pressure of 101 450 Pa.

The atmospheric pressure within the duct,

$$p_a = 101\,450 + p_s \text{ Pa}$$

where

$$p_s = 9.807 \times 25 \text{ mmH}_2\text{O Pa}$$

$$p_s = 245.2 \text{ Pa}$$

and

$$p_a = 101\,450 + 245.2 = 101\,695.2 \text{ Pa}$$

It is reasonable to ignore decimal parts of Pascal pressures, so $p_s = 245$ Pa and $p_a = 101\,695$ Pa.

$$\rho = 1.2 \frac{p_a}{101\,325} \times \frac{273 + 20}{273 + t} \text{ kg/m}^3$$

$$= 1.2 \times \frac{101\,695}{101\,325} \times \frac{293}{273 + 22} \text{ kg/m}^3$$

$$= 1.196 \text{ kg/m}^3$$

$$\text{air velocity } v = \frac{4 \times 0.2}{\pi \times 0.25^2} \text{ m/s}$$

$$= 4.074 \text{ m/s}$$

$$p_v = 0.5 \times 1.196 \times 4.074^2 \text{ Pa}, \quad p_v = 9.925 \text{ Pa} = 10 \text{ Pa}$$

$$\text{total pressure } p_t = p_s + p_v$$

$$p_t = 245 + 10 \text{ Pa}$$

$$= 255 \text{ Pa}$$

VARIATIONS OF PRESSURE ALONG A DUCT

Pressure gradients are produced along the length of a duct as shown in Fig. 6.2. The gradient caused along the constant-diameter duct is calculated from the pressure drop rate due to friction calculated from the methods used in Chapter 4. The gradients of the total and static pressure lines are equal. Friction losses can be calculated as total or static pressure drops. It is convenient to calculate all pressure drops as reductions in the available total pressure.

Duct fittings, such as the reducer shown, cause loss of pressure through surface friction and additional turbulence because of the shape of the enclosure. Such losses

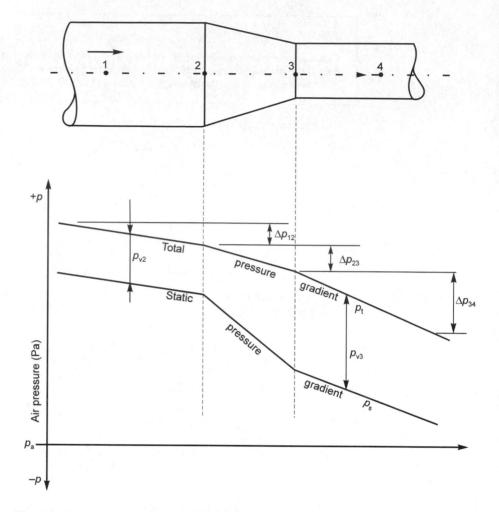

Fig. 6.2   Variation of air pressures through a reducer.

are found from experiment and require the use of unique factors that are multiplied by the velocity pressure. Data for commonly used duct fittings are given in Figs 6.3 and 6.4 from CIBSE (1986c).

To produce a pressure-gradient diagram similar to Figure 6.2, draw the ductwork to be analysed directly above a graph as shown. Clearly mark each point where a change of section takes place and identify each with a number. The sequence is as follows.

1. Find the total pressure at 1, p_{t1} Pa, and plot it to an appropriate scale above, or below, atmospheric pressure p_a Pa.
2. Calculate the pressure drop between 1 and 2 in the straight duct, Δp_{12}. Note that Δp_{12} means the pressure drop due to friction between nodes 1 and 2.
3. Subtract Δp_{12} from p_{t1} to find the total pressure at point 2, p_{t2} Pa.
4. Draw the total pressure straight line from p_{t1} to p_{t2}.
5. Find the velocity pressure-loss factor for the duct fitting, k, from Figs 6.3 or 6.4 or the reference.

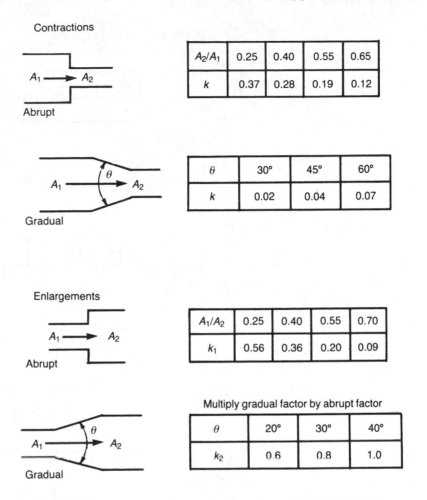

Fig. 6.3 Velocity pressure loss factors for air duct fittings (CIBSE, 1986c). Factors are multiplied by the velocity pressure in the smaller area: $\Delta p_{12} = k_1 k_2 p_{v1}$

6. Find the velocity pressure to be used for the frictional resistance calculation for the fitting. In this case, it is the velocity pressure in the smaller duct p_{v3}.
7. Multiply the fitting k factor by the velocity pressure p_{v3} to calculate the pressure drop through the fitting, Δp_{23}.
8. Subtract pressure drop Δp_{23} from the total pressure p_{t2} to find the total pressure at 3, p_{t3}.
9. Plot p_{t3} and connect the total pressure straight line from p_{t2}.
10. Calculate the pressure drop between 3 and 4 in the straight duct, Δp_{34}.
11. Subtract Δp_{34} from p_{t3} to find the total pressure at point 4, p_{t4} Pa.
12. Draw the total pressure straight line from p_{t3} to p_{t4}.
13. Calculate the velocity pressure at 1, p_{v1} Pa. Note that $p_{v1} = p_{v2}$ as the velocity remains constant in an equal diameter duct along its length.
14. Subtract p_{v1} from both p_{t1} and p_{t2}. The differences are the static pressures at 1 and 2, p_{s1} and p_{s2}.
15. Plot p_{s1} and p_{s2} and connect them with a straight line.

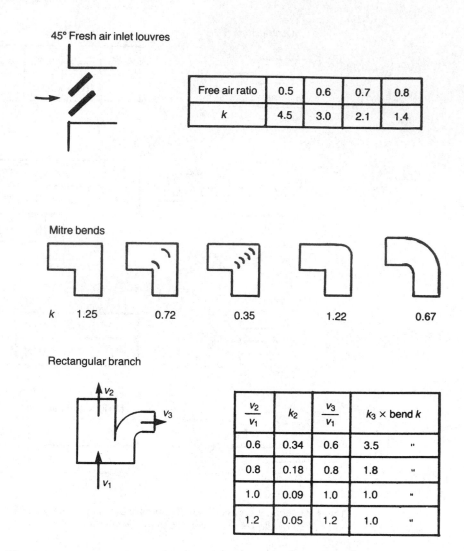

45° Fresh air inlet louvres

Free air ratio	0.5	0.6	0.7	0.8
k	4.5	3.0	2.1	1.4

Mitre bends

| k | 1.25 | 0.72 | 0.35 | 1.22 | 0.67 |

Rectangular branch

$\dfrac{v_2}{v_1}$	k_2	$\dfrac{v_3}{v_1}$	$k_3 \times$ bend k	
0.6	0.34	0.6	3.5	"
0.8	0.18	0.8	1.8	"
1.0	0.09	1.0	1.0	"
1.2	0.05	1.2	1.0	"

Fig. 6.4   Dynamic pressure loss factors for air duct fittings (CIBSE, 1986c).

16. Subtract p_{v3} from p_{t3} and find static pressure at 3, p_{s3}.
17. Draw a straight line from p_{s2} to p_{s3}.
18. Subtract p_{v3} from p_{t4} and find p_{s4}. Note that $p_{v3} = p_{v4}$.
19. Draw a straight line from p_{s3} to p_{s4}.
20. Mark the upper gradient as the total pressure line.
21. Mark the lower line as the gradient of static pressure.
22. Mark the differences between the two gradients as velocity pressures.
23. Identify each of the corresponding points of the duct diagram and the graph.
24. Mark on the graph all the calculated pressures.

This procedure applies to other air-duct cases and to the changes of pressure across a fan.

Notice that for this reduction in diameter of the duct, the static pressure reduces. This is because the velocity pressure has increased. The available total pressure has also fallen owing to friction losses. If the duct diameter increases between 2 and 3, the static pressure may rise owing to a fall in the velocity pressure. Such a regain of static pressure depends upon how much frictional pressure drop occurs. Static regain can also take place at branches owing to a reduction in the airflow quantity and velocity. Such regains of static pressure may be usefully employed in overcoming the resistance of downstream ducts or terminal equipment such as grilles, dampers or heater coils.

EXAMPLE 6.2

Calculate the pressures that occur when 4 m³/s flow through a 5 m long, 1 m diameter duct that then reduces to 600 mm diameter and remains at 600 mm for 5 m. The air total pressure at the commencement of the 1 m duct is 300 Pa above atmospheric. The reducer is the 45° concentric type shown in Figure 6.3. The air density is 1.2 kg/m³. The frictional pressure loss rates are 0.22 Pa/m in the 1 m diameter and 3 Pa/m in the 600 mm diameter ducts.

Using the node numbers from Figure 6.2 and the sequence numbers listed:
1. $p_{t1} = 300$ Pa. Plot graph axes.
2. Along the 1 m duct that is 5 m long:

$$\Delta p_{12} = 0.22\ \text{Pa/m} \times 5\ \text{m}$$

$$= 1\ \text{Pa}$$

3. Find p_{t2}, p_{s2} and p_{v3}

$$p_{t2} = p_{t1} - \Delta p_{12}$$

$$= 300 - 1\ \text{Pa}$$

$$= 299\ \text{Pa}$$

4. Draw the total pressure line 1 to 2.
5. The reducer has a velocity pressure loss factor of 0.04 and this is multiplied by the smaller duct velocity pressure, p_{v3}.

6.
$$v_3 = \frac{4\ \text{m}^3/\text{s}}{\pi \times 0.6^2/4\ \text{m}^2}$$

$$= 14.2\ \text{m/s}$$

$$p_{v3} = 0.5 \times 1.2 \times 14.2^2\ \text{Pa}$$

$$= 121\ \text{Pa}$$

7. Pressure drop through reducer;

$$\Delta p_{23} = \text{facator} \times p_{v3}$$

$$= 0.04 \times 121\ \text{Pa}$$

$$= 5\ \text{Pa}$$

8.
$$p_{t3} = p_{t2} - \Delta p_{23}$$
$$= 299 - 5$$
$$= 294 \text{ Pa}$$

9.   Plot total pressure line 2 to 3.

10.   Pressure drop through 600 mm duct:
$$\Delta p_{34} = 5 \text{ m} \times 3 \text{ Pa/m}$$
$$= 15 \text{ Pa}$$

11.
$$p_{t4} = p_{t3} - \Delta p_{34}$$
$$= 294 - 15$$
$$= 279 \text{ Pa}$$

12.   Draw total pressure line 3 to 4.

13.
$$v_1 = \frac{4 \text{ m}^3/\text{s}}{\pi \times 1^2/4 \text{ m}^2} = 5.1 \text{ m/s}$$
$$p_{v1} = 0.5 \times 1.2 \times 5.1^2 \text{ Pa}$$
$$= 16 \text{ Pa}$$
$$p_{v2} = 16 \text{ Pa}$$

14.
$$p_{t1} = p_{s1} + p_{v1}$$
$$p_{s1} = p_{t1} - p_{v1}$$
$$p_{s1} = 300 - 16 \text{ Pa}$$
$$= 284 \text{ Pa}$$
$$p_{s2} = p_{t2} - p_{v2}$$
$$p_{v2} = p_{v1}$$
$$p_{s2} = 299 - 16 \text{ Pa}$$
$$= 283 \text{ Pa}$$

15.   Plot static pressure 1 and 2 and draw line.

16.
$$p_{s3} = p_{t3} - p_{v3}$$
$$= 294 - 121$$
$$= 173 \text{ Pa}$$

17.   Draw static pressure line from 2 to 3.

18.
$$p_{s4} = p_{t4} - p_{v4}$$
$$p_{v3} = p_{v4}$$
$$p_{s4} = 279 - 121$$
$$= 158 \text{ Pa}$$

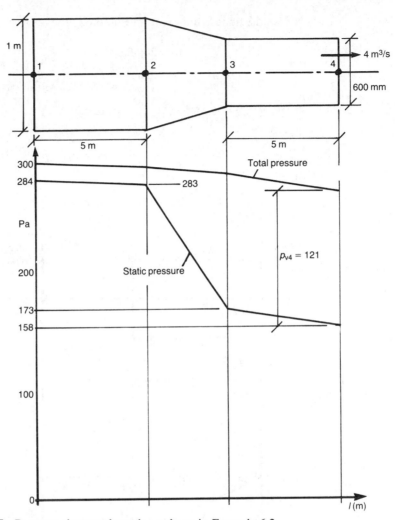

Fig. 6.5 Pressure changes through a reducer in Example 6.2.

19. Draw static pressure line from 3 to 4.

20–24. Figure 6.5 shows these answers plotted to scale.

The solutions to the following example are not listed in strict accordance with the sequence. The reader may follow the sequence and validate the answer with the results shown.

EXAMPLE 6.3

Calculate the static regain and pressure changes that occur when 2 m³/s flow through a 12 m long, 500 mm diameter duct that then enlarges to 800 mm diameter and remains at 800 mm for 10 m. The air total pressure at the commencement of the 800 mm duct is 200 Pa above atmospheric. The enlarger is the 20° concentric type shown in Figure 6.3. The air density is 1.2 kg/m³. The frictional pressure loss rates are 2 Pa/m in the 500 mm diameter and 0.2 Pa/m in the 800 mm diameter ducts.

Use the node numbers from Fig. 6.6.

$$p_{t1} = 200 \text{ Pa}, \quad v_1 = \frac{2 \text{ m}^3/\text{s}}{\pi \times 0.5^2/4 \text{ m}^2} = 10.2 \text{ m/s}$$

$$p_{v1} = 0.5 \times 1.2 \times 10.2^2 \text{ Pa},$$

$$= 62 \text{ Pa}$$

From

$$p_{t1} = p_{s1} + p_{v1}$$

then

$$p_{s1} = p_{t1} - p_{v1}$$

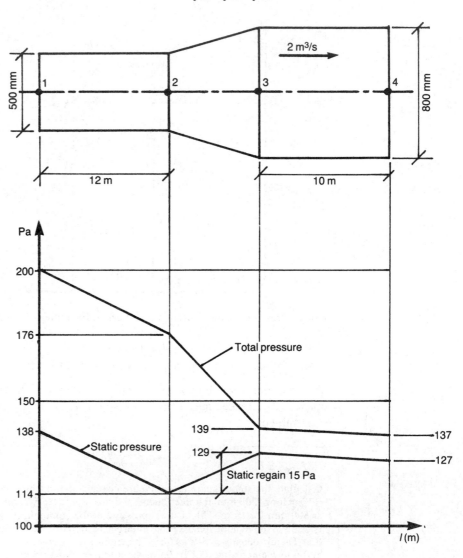

Fig. 6.6   Pressure changes through an enlargement in Example 6.3.

$$= 200 - 62 \, \text{Pa}$$

$$= 138 \, \text{Pa}$$

Along the 500 mm duct that is 12 m long:

$$\Delta p_{12} = 2 \, \text{Pa/m} \times 12 \, \text{m}$$

$$= 24 \, \text{Pa}$$

Find p_{t2}, p_{s2} and p_{v3}.

$$p_{t2} = p_{t1} - \Delta p_{12}$$

$$= 200 - 24 \, \text{Pa},$$

$$= 176 \, \text{Pa}$$

$$p_{s2} = p_{t2} - p_{v2}, \quad p_{v2} = p_{v1}, \quad p_{s2} = 176 - 62 \, \text{Pa}$$

$$= 114 \, \text{Pa}$$

$$v_3 = \frac{2 \, \text{m}^3/\text{s}}{\pi \times 0.8^2/4 \, \text{m}^2}$$

$$= 4 \, \text{m/s}$$

$$p_{v3} = 0.5 \times 1.2 \times 4^2 \, \text{Pa},$$

$$= 10 \, \text{Pa}$$

The enlarger has a velocity pressure loss factor of 0.6 and this is multiplied by the smaller duct velocity pressure, p_{v2}. Pressure drop through reducer:

$$\Delta p_{23} = \text{factor} \times P_{v2}$$

$$= 0.6 \times 62.4 \, \text{Pa}$$

$$= 37 \, \text{Pa}$$

So

$$p_{t3} = p_{t2} - \Delta p_{23}$$

$$= 176 - 37$$

$$= 139 \, \text{Pa}$$

$$p_{s3} = p_{t3} - p_{v3}$$

$$= 139 - 10$$

$$= 129 \, \text{Pa}$$

Pressure drop through 800 mm duct:

$$\Delta p_{34} = 10 \, \text{m} \times 0.2 \, \text{Pa/m}$$

$$= 2 \, \text{Pa}$$

$$p_{t4} = p_{t3} - \Delta p_{34}$$

$$= 139 - 2$$

$$= 137 \, \text{Pa}$$

$$p_{s4} = p_{t4} - p_{v4}, \text{ as } p_{v3} = p_{v4}$$

$$= 137 - 10$$

$$= 127 \, \text{Pa}$$

The regain of static pressure:

$$SR = p_{s3} - p_{s2}$$

$$= 129 - 114$$

$$= 15 \, \text{Pa}$$

Fig. 6.6 shows these answers plotted to scale.

EXAMPLE 6.4

An air duct branch is shown in Fig. 6.7. Use the data provided to calculate all the duct pressures at the nodes. 1 m diameter duct 1–2 is 15 m long and carries 6 m^3/s. 800 mm diameter duct 3–4 is 20 m long and carries 3 m^3/s. 600 mm diameter duct 5–6 is 10 m long and carries 3 m^3/s. The branch offtake is a short-radius bend. The straight-through contraction has an angle of 30°. Pressure drop rates are: duct 1–2, 0.52 Pa/m; duct 3–4, 0.72 Pa/m; duct 5–6, 0.8 Pa/m. The total pressure at node 1 is 500 Pa.

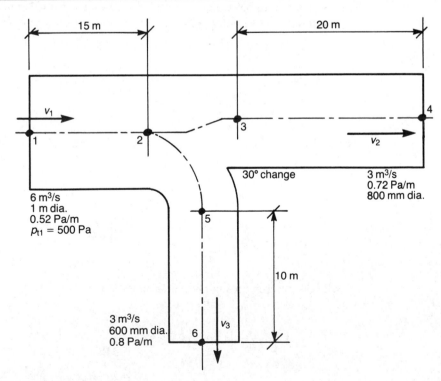

Fig. 6.7 Air-duct branch for Example 6.4.

$$p_{t1} = 500 \text{ Pa}, \ v_1 = \frac{6 \text{ m}^3/\text{s}}{\pi \times 1^2/4 \text{ m}^2} = 7.6 \text{ m/s}$$

$$p_{v1} = 0.5 \times 1.2 \times 7.6^2 \text{ Pa}$$

$$= 35 \text{ Pa}$$

From

$$p_{t1} = p_{s1} + p_{v1}$$

then

$$p_{s1} = p_{t1} - p_{v1}$$

$$= 500 - 35 \text{ Pa}$$

$$= 465 \text{ Pa}$$

Along the 1 m duct that is 15 m long:

$$\Delta p_{12} = 0.52 \text{ Pa/m} \times 15 \text{ m}$$

$$= 8 \text{ Pa}$$

Find p_{t2}, p_{s2} and p_{v3}.

$$p_{t2} = p_{t1} - \Delta p_{12}$$

$$= 500 - 8 \text{ Pa}$$

$$= 492 \text{ Pa}$$

$$p_{s2} = p_{t2} - p_{v2}$$

$$p_{v2} = p_{v1}$$

$$p_{s2} = 492 - 35 \text{ Pa}$$

$$= 457 \text{ Pa}$$

$$v_3 = \frac{3 \text{ m}^3/\text{s}}{\pi \times 0.8^2/4 \text{ m}^2}$$

$$= 6 \text{ m/s}$$

$$p_{v3} = 0.5 \times 1.2 \times 6^2 \text{ Pa}$$

$$= 22 \text{ Pa}$$

The straight-through part of the branch has a velocity ratio $v_3/v_2 = 6/7.6 = 0.8$. Refer to Fig. 6.4 to find the velocity pressure loss factor of 0.18; this is multiplied by the smaller duct velocity pressure, p_{v3}.

Pressure drop through reducer:

$$\Delta p_{12} = \text{factor} \times p_{v3}$$

$$= 0.18 \times 22 \text{ Pa}$$

$$= 4 \text{ Pa}$$

So

$$p_{t3} = p_{t2} - \Delta p_{23}$$

$$= 492 - 4$$

$$= 488 \text{ Pa}$$

$$p_{s3} = p_{t3} - p_{v3}$$

$$= 488 - 22$$

$$= 466 \text{ Pa}$$

Pressure drop through 800 mm duct:

$$\Delta p_{34} = 20 \text{ m} \times 0.72 \text{ Pa/m}$$

$$= 14 \text{ Pa}$$

$$p_{t4} = p_{t3} - \Delta p_{34}$$

$$= 488 - 14$$

$$= 474 \text{ Pa}$$

$$p_{s4} = p_{t4} - p_{v4}, \, as \, p_{v3} = p_{v4}$$

$$= 474 - 22$$

$$= 452 \text{ Pa}$$

The regain of static pressure:

$$SR = p_{s3} - p_{s2}$$

$$= 466 - 457$$

$$= 9 \text{ Pa}$$

Find the air velocity in the branch:

$$v_5 = 3 \text{ m}^3/\text{s}/(\pi \times 0.6^2/4 \text{ m}^2) = 10.6 \text{ m/s}$$

$$p_{v5} = 0.5 \times 1.2 \times 10.6^2 \text{ Pa}$$

$$= 67 \text{ Pa}$$

The bend part of the branch has a velocity ratio:

$$\frac{v_5}{v_2} = \frac{10.6}{7.6} = 1.4$$

Refer to Fig. 6.4 to find the velocity pressure loss factor of 1; this is multiplied by the bend factor 0.67 and the branch duct velocity pressure, p_{v5}.

Pressure drop through branch:

$$\Delta p_{25} = \text{factors} \times p_{v5}$$

$$= 1 \times 0.67 \times 67 \text{ Pa}$$

$$= 45 \text{ Pa}$$

So

$$p_{t5} = p_{t2} - \Delta p_{25}$$

$$= 492 - 45$$

$$= 447 \text{ Pa}$$

$$p_{s5} = p_{t5} - p_{v5}$$

$$= 447 - 67$$

$$= 380 \text{ Pa}$$

Pressure drop through 600 mm duct:

$$\Delta p_{56} = 10 \text{ m} \times 0.8 \text{ Pa/m}$$

$$= 8 \text{ Pa}$$

$$p_{t6} = p_{t5} - \Delta p_{56}$$

$$= 447 - 8$$

$$= 439 \text{ Pa}$$

$$p_{s6} = p_{t6} - p_{v6}, \text{ as } p_{v5} = p_{v6}$$

$$= 439 - 67$$

$$= 372 \text{ Pa}$$

PRESSURE CHANGES AT A FAN

The fan produces a rise of total pressure in the whole system. This rise is the fan total pressure, FTP Pa. The fan total pressure rise is equal to the drop of total pressure due to friction in the ductwork system. The fan total pressure is calculated from the total pressure in the fan outlet duct minus the total pressure in the fan inlet duct. Figure 6.8 shows the pressure changes across a centrifugal fan and duct system.

$$FTP = p_{t2} - p_{t1} \text{ Pa}$$

Fan velocity pressure, FVP, is defined as the velocity pressure in the discharge area from the fan, p_{v2}.

$$FVP = p_{v2} \text{ Pa}$$

Fan static pressure, FSP, is defined as fan total pressure minus fan velocity pressure. Note that this is not necessarily the same as the change in static pressure across the fan connections.

$$FSP = FTP - FVP \text{ Pa}$$

The procedure is as follows.
1. Find air density.
2. Calculate air velocity and velocity pressure in each duct.
3. Convert all pressures to pascals.

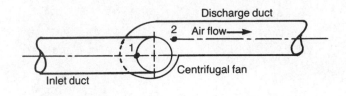

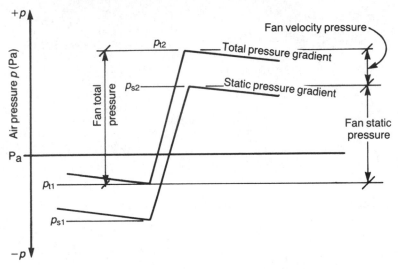

Fig. 6.8   Definition of pressure increases across a fan with inlet and outlet ducts.

4. Refer to Fig. 6.8.
5. Find inlet total pressure p_{t1} and static pressure p_{s1}.
6. Find outlet total pressure p_{t2} and static pressure p_{s2}
7. Calculate FTP, FVP and FSP.

EXAMPLE 6.5

A centrifugal fan delivers 5 m³/s into an outlet duct of 800 mm × 800 mm. The inlet duct to the fan is 700 mm diameter. The static pressure at the fan inlet was measured as − 50 mm water gauge relative to standard atmospheric pressure. The ductwork system had a calculated resistance of 1500 Pa. Air density is 1.2 kg/m³. Calculate the pressures either side of the fan.

$$\text{air density} = 1.2 \text{ kg/m}^3$$

$$\text{inlet air velocity } v_1 = \frac{4 \times 5 \text{ m}^3\text{/s}}{\pi \times 0.7^2 \text{ m}^2}$$

$$= 13 \text{ m/s}$$

$$p_{v1} = 0.5 \times 1.2 \times 13^2 \text{ Pa}$$

$$= 101 \text{ Pa}$$

$$\text{outlet air velocity } v_2 = \frac{5 \text{ m}^3/\text{s}}{0.8 \text{ m} \times 0.8 \text{ m}}$$

$$= 7.8 \text{ m/s}$$

$$p_{v2} = 0.5 \times 1.2 \times 7.8^2 \text{ Pa}$$

$$= 37 \text{ Pa}$$

Given that

$$p_{s1} = -50 \text{ mmH}_2\text{O}$$

then

$$p_{s1} = -50 \times 9.807 \text{ Pa}$$

$$= -490 \text{ Pa}$$

and

$$\text{FTP} = \text{duct pressure drop } 1500 \text{ Pa}.$$

On the inlet side of the fan:

$$p_{t1} = p_{s1} + p_{v1}$$

$$= -490 + 101 = -389 \text{ Pa}$$

$$\text{FTP} = p_{t2} - p_{t1}$$

so

$$p_{t2} = \text{FTP} + p_{t1} = 1500 - 389 = 1111 \text{ Pa}$$

$$p_{s2} = p_{t2} - p_{v2} = 1111 - 37 = 1074 \text{ Pa}$$

To summarize:

$$\text{FTP} = 1500 \text{ Pa}$$

$$\text{FVP} = p_{v2} = 37 \text{ Pa}$$

$$\text{FSP} = \text{FTP} - \text{FVP} = 1500 - 37 = 1463 \text{ Pa}$$

FLOW MEASUREMENT IN A DUCT

A ductwork installation may be fitted with a permanent flow meter such as a venturi nozzle (Fig. 6.9), orifice plate (Fig. 6.10) conical inlet (Fig. 6.11), or flow grid (Fig. 6.12). Each of these devices introduces a permanently installed frictional resistance. The airflow entering the meter must have minimum swirl due to upstream bends, branches or fan as unstable flow produces pulsations in the measured air pressure and hence unsteady flow or velocity readings. Honeycomb flow straighteners can be used.

The airflow grid comprises one or more drilled pipes across the duct diameter. It measures the average velocity pressure. This single pressure is passed to a transducer that provides a 0–10 V electrical signal to a computerized control or energy-management system.

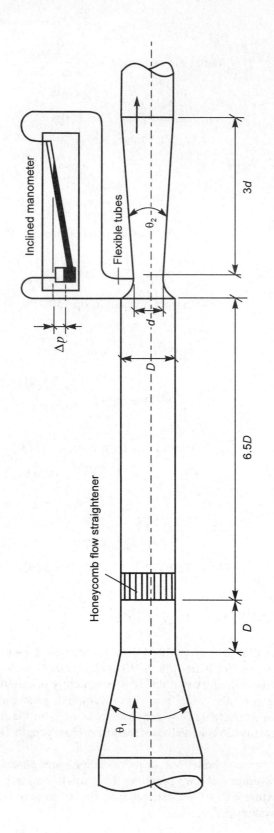

Fig. 6.9 Venturi nozzle in-duct airflow meter. $D > 1.5d$; $\theta_1 = 7°$; $\theta_2 = 15°$

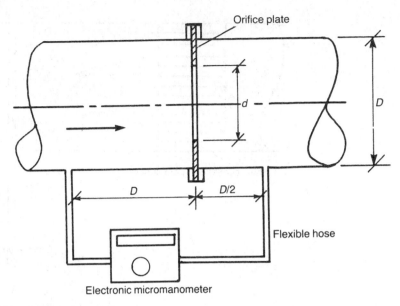

Fig. 6.10 Orifice plate airflow meter.

Portable airflow instruments such as the pitot-static tube, rotating vane anemo-meter, thermistor and hot-wire anemometer are used in commissioning tests. A rotating vane anemometer combined with a venturi hood is used to measure the airflow from a supply grille.

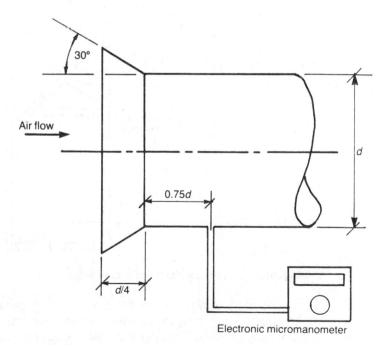

Fig. 6.11 Conical inlet airflow meter.

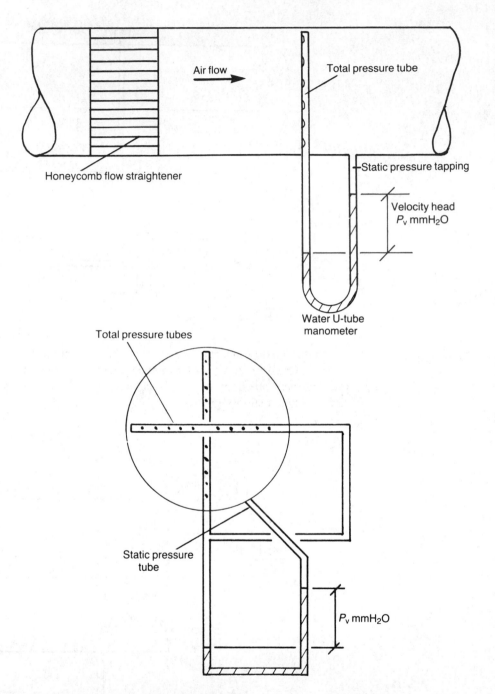

Fig. 6.12 Airflow measurement grid in a duct.

Laboratory airflow meters for research or frequent tests by a manufacturer of air-conditioning components have several diameters of upstream straight duct for flow straightening. The pitot-static tube is used for accurate flow measurements to allow calibration of a venturi or orifice plate meter.

The venturi meter causes relatively low frictional pressure loss but needs to have smooth internal surfaces and a shallow-angle expansion duct to avoid flow disturbance. The orifice plate produces a higher pressure drop and the conical inlet is appropriate to the laboratory testing of fans.

The general expression for the volume flow rate of air through the venturi, orifice plate and conical inlet meters is

$$Q = \text{coefficient} \times \pi d^2 \, (2\rho\Delta p)^{0.5}/4$$

where Q = airflow (m³/s); coefficient = around 1 for a venturi, 0.65 for an orifice plate, and 0.95 for a conical inlet; d = throat diameter (m); ρ = upstream air density kg/m³; Δp = pressure difference across meter (Pa).

EXAMPLE 6.6

A venturi meter in an air duct shows a pressure difference of 50 mm water gauge across its connections when passing air of density 1.2 kg/m³. The throat of the venturi is 150 mm diameter. The venturi flow coefficient is 1.

$$\text{venturi pressure drop } \Delta p = 9.807 \times 50 \text{ Pa}$$

$$= 490 \text{ Pa}$$

$$Q = \text{coefficient} \times \pi d^2 \, (2\rho\Delta p)^{0.5}/4 \text{ m}^3/\text{s}$$

$$= 1 \times \pi \times 0.15^2 \times (2 \times 1.2 \times 490)^{0.5}/4 = 0.606 \text{ m}^3/\text{s}$$

The pitot-static tube and inclined or electronic micromanometer is used as a reference for other methods of flow measurement. A National Physical Laboratory modified ellipsoidal-nosed pitot-static tube having an outside diameter not exceeding 1/48 of the airway diameter can be employed in airflows of up to 70 m/s and greater than 1 m/s. A traverse across three diameters in order to measure the velocity pressure at 24 locations is made for circular ducts. A total of 48 points can be used for flow measurement in rectangular ducts. Figures 6.13 and 6.14 show the measurement positions. The average air velocity at the test section is

$$v = (2p_d/\rho)^{0.5} \text{ m/s}$$

Where p_d is the SMR value of the dynamic pressures from the pitot-static traverse, that is, the square of the mean of the square roots of the j individual dynamic pressures, from

$$p_d = \left(\frac{p_{v1}^{0.5} + p_{v2}^{0.5} + \ldots + p_{vj}^{0.5}}{j} \right)^2 \text{ Pa}$$

EXAMPLE 6.7

Calculate the air volume flow rate in a 350 mm internal diameter air-conditioning duct when the pitot-static tube traverse revealed the following dynamic pressures in millimetres water gauge: 1.55, 1.58, 1.87, 1.96, 2.06, 1.23, 1.34, 0.85, 1.56, 1.54, 1.8, 1.9, 2, 1.2, 1.3, 0.8, 1.5, 1.52, 1.75, 1.85, 1.9, 1.18, 1.28, 0.75. The density of the air on the day of test was 1.2 kg/m³.

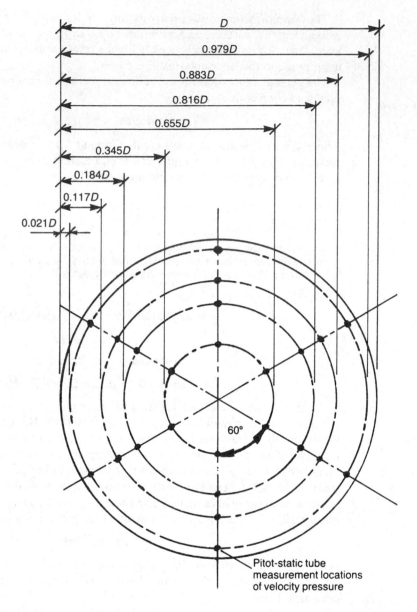

Fig. 6.13   Locations for pitot-static tube traverse across three diameters, 24 points.

Calculate the SMR value from the series of 24 dynamic pressures:

$$p_d = \left(\frac{(1.55)^{0.5} + (1.58)^{0.5} + \ldots + (0.75)^{0.5}}{24} \right)^2$$

$$= 1.219 \, \text{mmH}_2\text{O}$$

$$= 1.219 \times 9.807 \, \text{Pa}$$

$$= 12 \, \text{Pa}$$

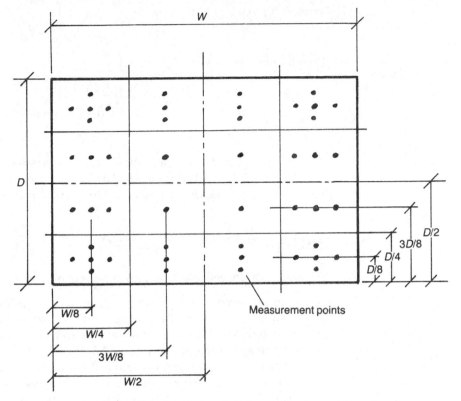

Fig. 6.14   Locations for a 48-point pitot-static tube traverse in a rectangular airway.

Average air velocity in the airway:

$$v = (2 \times 12/1.2)^{0.5} \text{m/s}$$

$$= 4.47 \text{ m/s}$$

The air volume flow rate:

$$Q = \pi d^2 \times v/4 \text{ m}^3/\text{s}$$

$$= \pi \times 0.35^2 \times 4.47/4 \text{ m}^3/\text{s}$$

$$= 0.43 \text{ m}^3/\text{s}$$

EXAMPLE 6.8

If an airflow grid is installed in the air duct used in Example 6.7, what average velocity pressure, velocity and volume flow rate would be indicated? Comment on the difference between the two calculation procedures.

The arithmetic average of the 24 dynamic pressures, $(p_{v1} + p_{v2} + \ldots + p_{vj})/j$, is 1.51 mmH$_2$O

$$p_d = 1.51 \text{ mmH}_2\text{O} = 1.51 \times 9.807 \text{ Pa} = 15 \text{ Pa}$$

Average air velocity in the airway:

$$v = (2 \times 15/1.2)^{0.5} \text{ m/s} = 5 \text{ m/s}$$

The air volume flow rate:

$$Q = \pi d^2 \times v/4 \text{ m}^3/\text{s}$$

$$= \pi \times 0.35^2 \times 5/4 \text{ m}^3/\text{s}$$

$$= 0.48 \text{ m}^3/\text{s}$$

It would appear that the flow grid overestimates the flow rate by $((0.48 - 0.43)/0.43) \times 100\% = 11.6\%$. The accuracy of the flow grid can be calibrated on site by checking it against a pitot-static traverse.

DUCTWORK SYSTEM DESIGN

The design of a ventilation or air conditioning duct system is as follows.

1. Find the supply- and extract-air quantities for each room. The sensible heat formula is used to find the supply airflow rate. The extract-air quantity may be slightly less than the supply quantity so that the room is pressurized and all air leakages become outwards. This avoids incoming draughts of unconditioned air. All the airflows into and out of the rooms must be accounted for as flow rates in l/s. It is the flow rates between rooms that create draughts under doors and the differences between room static air pressures.
2. Draw a single line layout for both the supply and extract ductwork systems on the building plans.
3. Check that the duct system can be coordinated with the other services, the architectural features of the building and the structural design.
4. Ascertain that both horizontal and vertical service shafts are available for the intended route.
5. Decide the positions for supply-air diffusers, terminal units and extract grilles. Refer to manufacturers' literature on discharge air velocity and throw of air from grille. It is the position of the supply-air grille that determines the air-movement pattern generated within the room.
6. Draw a schematic layout of the ductwork system similar to Fig. 6.15. This will be used as the working drawing for calculations and data.
7. Sketch the air-handling plant to large size. This is to identify the plant items such as filters, heater batteries and fans showing their method of connection.
8. Sketch large details of any complicated items of ductwork, such as changes of section or tortuous routes within plant rooms.
9. Mark on the drawings the locations of airflow grids or other permanently installed flow meters, terminal air filters, fire dampers and air volume flow-balancing dampers.
10. Locate the positions for air-pressure control dampers.
11. Decide the possible positions for noise attenuators. These may be either side of the fans, at terminal units or grilles, or may be acoustic linings for some ducts.

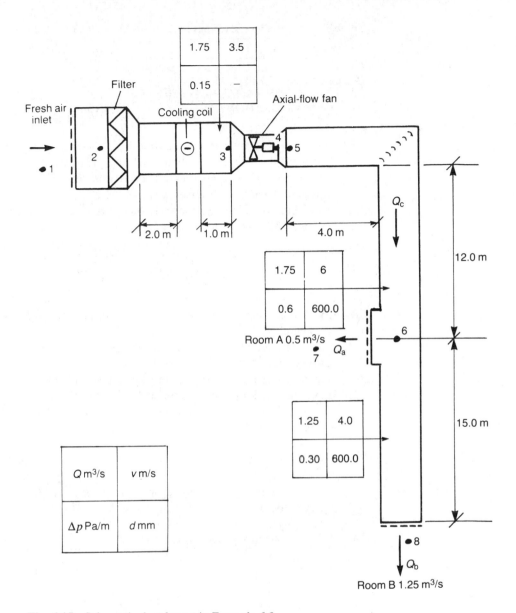

Fig. 6.15 Schematic duct layout in Example 6.9.

12. Decide the maximum air velocities for each part of the duct system from Table 6.1 depending upon the proximity of ducts to occupied rooms.

13. Summate all the airflows through the duct system. Where the air temperature remains constant, volume air flows can be added in l/s or m^3/s. Air-temperature changes due to zone or terminal heater batteries will produce changes to the air density and volume flow rate. In such cases, air mass flow rate in kg/s needs to be used.

14. Once all the duct air mass flow rates are fixed, the correct air density is used to calculate the volume flow rate in each duct.

Table 6.1 Limiting air velocities in ducts for low-velocity system design

Application	Main duct v(m/s)	Branch duct v(m/s)
Hospital, concert hall, library, sound studio	5.0	3.5
Cinema, restaurant, hall	7.5	5.0
General office, dance hall, shop, exhibition hall	9.0	6.0
Factory, workshop, canteen	12.5	7.5

15. The air mass flow rate supplied by the air-handling plant will be equal to the sum of all the air mass flow rates leaving the ductwork.
16. Decide the limiting air-pressure drop rate through the ducts. This may be chosen as 1 Pa/m for low-velocity systems.
17. Find a suitable diameter for each duct, either using Figure C4.2, from CIBSE (1986c) (the flow of air in round ducts), or calculating it from

$$Q = \pi d^2 v/4 \quad \text{so} \quad d = \left(\frac{4Q}{\pi v}\right)^{0.5} \text{m}$$

where Q is in m^3/s, v is in m/s and duct internal diameter d is in m.

18. Check that there is sufficient space within the false ceilings and service shafts for the ducts.
19. Assess the overall sizes needed for the service ducts to pass these air ducts and other services.
20. Decide whether the proposed duct system is practical.
21. If the ducts cannot be accommodated within the allocated spaces, it may be necessary to increase the air velocities. This will reduce the duct diameters but at the cost of increased frictional resistance, fan power and possible noise generation.
22. Record the design information for each section of duct on the schematic drawing.
23. Enter the data on to a spreadsheet (Table 6.2) to enable calculation of the pressure losses.

Table 6.2 Duct sizing data for Example 6.9

Sect.	Length l(m)	$\Delta p/1$ (Pa/m)	v (m/s)	p_v (Pa)	k	Fit (Pa)	Duct (Pa)	Total (Pa)	Index
1–2	0	0	2.5	4	4.5	18	0	18	*
2–3	3	0.15	3.5	7	0.04	95	1	96	*
3–4	0	0	10	60	0.04	2	0	2	*
4–5	0	0	10	60	0.12	7	0	7	*
5–6	16	0.6	6	22	0.35	8	10	18	*
6–7	0	0	6	22	1.25	78	0	78	*
6–8	15	0.3	4	10	0.09	51	5	56	

Total Δp for index route = 219

Fit = pressure drop through the duct fittings (Pa)
Duct = pressure drop through the straight duct (Pa)

24. Choose the duct fittings to be used from Table C4.35, in CIBSE (1986c) or from the extracts in Figures 6.3 and 6.4.

25. Record the velocity pressure loss factors for each duct fitting on the spreadsheet.
26. Calculate the velocity pressure in each duct.
27. Multiply the duct length by the pressure loss rate for each section of duct and enter the duct Δp Pa on the spreadsheet.
28. Multiply the duct fitting velocity pressure loss factor by the correct velocity pressure for each section and enter the fitting Δp Pa.
29. Enter all fixed pressure drops through items of plant such as filters, dampers, inlet and outlet grilles, diffusers, cooling and heating coils.
30. Identify the index circuit; this is the duct run offering the highest resistance to flow. The index route is the result of a series of resistances. Other ducts are branches from this route and are in parallel with the index system.
31. Calculate the fan total pressure rise to offset the loss along the index route.
32. Draft the ductwork system on to the scale drawings of the building to validate the design and coordinate it with all the other design features, such as structure, other services and architecture.

EXAMPLE 6.9

The duct system shown in Fig. 6.15 is to be installed in a false ceiling over an office. The air-handling plant, comprising the fresh air inlet, filter, chilled water cooling coil and an axial flow fan, is located in a plantroom. Office A is supplied with 0.5 m³/s and office B has 1.25 m³/s. Find suitable sizes for the ducts and state the performance specification for the fan.

The limiting air velocities are 2.5 m/s through the fresh air inlet grille and filter, 3.5 m/s through the cooling coil, 10 m/s through the fan and 6 m/s in the ducts. The fresh air inlet is constructed from 45° louvres having a free area of 50% and a velocity pressure loss factor of 4.5. The filter, supply grille and cooling coil have air pressure drops of 75 Pa, 50 Pa and 20 Pa respectively. The contractions in the duct are at an angle of 45° and the enlargement is at 30°. The ducts are to be sized for an air temperature of 20 °C d.b. The data is to be entered on to the schematic drawing (Fig. 6.15) and the spreadsheet (Tables 6.2 and 6.3)

The air density $= 1.2$ kg/m³, so $p_v = 0.5 \times 1.2 \times v^2$ Pa, $p_{v2} = 0.6 \times 2.5^2 = 4$ Pa. Pressure drop through the fresh air inlet grille, $\Delta p_{12} = k \times p_{v2} = 4.5 \times 4$ Pa $= 18$ Pa.

The 45° contraction in duct 2–3 after the filter has a velocity pressure loss factor of 0.04 and this is to be multiplied by the smaller duct velocity pressure, p_{v3}. $p_{v3} = 0.6 \times 3.5^2 = 7$ Pa, Δp contraction $= 0.04 \times 7 = 0.28$ Pa and fractions of pascals are to be ignored. The pressure drop through the fittings from node 2 to node 3 is 75 Pa for the filter and 20 Pa for the coil, 95 Pa.

Figure 6.16 is a simplified version of Figure C4.2 from CIBSE (1986c), and is adequate for exercises in this book. To use the duct-sizing chart, identify the air volume flow rate to be carried, Q m³/s, noticing that both horizontal and vertical scales are logarithmic, and move horizontally to the right until the second coordinate limit is reached. This second limit may be either pressure drop rate Pa/m, air velocity v m/s or the duct diameter d m. If it is decided to choose duct diameters on the basis of a maximum pressure loss rate of, say, 1 Pa/m, then mark the point of this

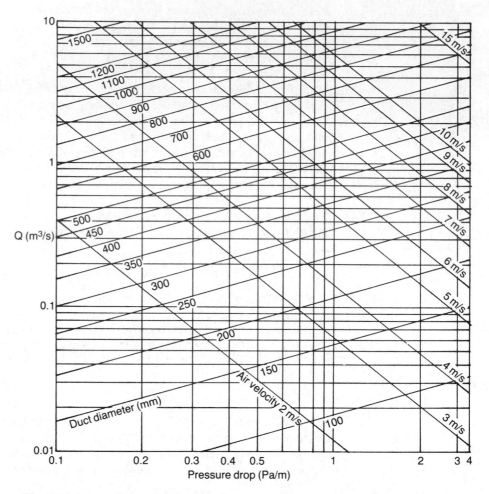

Fig. 6.16   Flow of air at 20 °C d.b. in ducts.

intersection. The diameter and velocity are estimated by interpolation between the lines: for example, at 0.3 m³/s and 1 Pa/m, a 280 mm diameter duct has an air velocity of around 4.7 m/s, which is 4.87 m/s by calculation, demonstrating the danger of casually reading a logarithmic graph. Now find the real duct size to be employed by converting the 280 mm diameter into a standard circular, flat oval or rectangular equivalent and then finding the accurate diameter and velocity.

The airflow through sections 1–6 amounts to

$$1.25 + 0.5 \text{ m}^3/\text{s} = 1.75 \text{ m}^3/\text{s}$$

For duct 2–3 at a maximum air velocity of 3.5 m/s, the pressure-drop rate in the duct will be 0.15 Pa/m, so

$$\Delta p = 3 \text{ m} \times 0.15 \text{ Pa/m}$$

$$= 0.45 \text{ Pa, say 1 Pa}$$

The air velocity through the fan is 10 m/s, so,

$$p_{v4} = 0.6 \times 10^2$$

$$= 60\,\text{Pa}$$

$$\Delta p_{34} = 0.04 \times 60$$

$$= 2\,\text{Pa}$$

Before the k factor for the enlargement 4–5 can be found, the size of duct 5–6 is read from Fig. 6.16. A 600 mm diameter duct carrying 1.75 m³/s has a pressure loss rate of 0.6 Pa/m at a velocity of 6 m/s.

$$\text{duct } \Delta p_{56} = 16\,\text{m} \times 0.6\,\text{Pa/m} = 10\,\text{Pa}$$

A mitre bend with turning vanes, $k = 0.35$, is in the duct 5–6.

$$\text{fitting } \Delta p_{56} = 0.35 \times 22\,\text{Pa} = 8\,\text{Pa}$$

Enlarger 4–5 has an area ratio A_4/A_5.

$$A_4 = \frac{1.75\,\text{m}^3/\text{s}}{10\,\text{m/s}} = 0.175\,\text{m}^2$$

$$A_5 = \frac{\pi \times 0.6^2}{4}\,\text{m}^2 = 0.283\,\text{m}^2$$

$$\frac{A_4}{A_5} = 0.175/0.283 = 0.62$$

so the sudden enlargement k factor is approximately 0.15. The concentric enlargement is at 30°, so the overall k factor is $0.15 \times 0.8 = 0.12$, $\Delta p_{45} = 0.12 \times 60 = 7$ Pa. It can be seen that great numerical accuracy is unnecessary.

The pressure loss from 6 to 7 is due to a rectangular branch. The air velocity in each duct is expected to be 6 m/s so the velocity ratios for the branches 6–7 and 6–8 are unity when compared with the velocity in the main duct 5–6. Take the branch bend to be a rectangular mitre bend: $k = 1.25$, $k_{67} = 1 \times 1.25 = 1.25$, and $k_{68} = 0.09$ for the straight-through part of the branch fitting.

$$\Delta p_{67} = 1.25 p_{v6} + \text{grille} = 1.25 \times 22 + 50 = 78\,\text{Pa}$$

Duct 6–8 carries 1.25 m³/s at a maximum of 6 m/s and this requires continuation of the 600 mm diameter duct at a pressure loss rate of 0.3 Pa/m and $v = 4$ m/s.

$$p_{v8} = 0.6 \times 4^2 = 10\,\text{Pa}$$

$$\text{duct } \Delta p_{68} = 15\,\text{m} \times 0.3\,\text{Pa/m} = 5\,\text{Pa}$$

$$\text{branch } \Delta p_{68} = 0.09 \times 10 = 1\,\text{Pa}$$

$$\text{total } \Delta p_{68} = 1 + 5 + 50 = 56\,\text{Pa including the grille}$$

The fan has to provide a rise of total pressure equal to the greatest frictional resistance. This is provided by the route 1–2–3–4–5–6–7. The branch 6–8 runs in parallel with 6–7 and has a lower resistance. When sufficient static pressure is provided at 6 to overcome the resistance of the branch 6–7, there will be more than enough to overcome the resistance to node 8. Each grille will be provided with a damper to regulate the airflow during commissioning to absorb the excess pressure in the non-index branches. It may be necessary to install balancing dampers in branches to absorb excess pressure that cannot be dropped across grille dampers.

The fan total pressure rise needed:

$$FTP = 18 + 96 + 2 + 7 + 18 + 78 = 219 \, Pa$$

The excess pressure at node $6 = 78 - 56 = 22$ Pa. The grille damper should adequately balance the two branches and equalize them so that they both appear to the fan to be index routes.

DUCT-SIZING PROGRAM

The repetitive calculations needed can be written into a spreadsheet program and adapted for each new design. Dedicated programs are intended to handle large numbers of nodes and ducts but they require that each part of the system be fully specified with the same information that the user could enter into a spreadsheet. Table 6.3 shows a spreadsheet design that can be entered into the reader's computer and spreadsheet program. It is used as follows.

Table 6.3   Spreadsheet for air-duct sizing

FILE: DUCT7079 Example 6.9   Date: 31/05/92
Duct sizing for rectangular sheet metal air ducts.
Air density = 1.2 kg/m^3 Key to columns: enter data under =====, calculated data under ****

col.	A Section	B Length l (m)	C Q (m^3/s)	D Max v(m/s)	E Max Δp (Pa/m)	F Dia. (m)
row	=====	=====	=====	=====	=====	***
14	1–2	0	1.75	2.5	0.07	0.94
15	2–3	3	1.75	3.5	0.16	0.8
16	3–4	0	1.75	10	2.16	0.47
17	fan	0	1.75	10	0	0.47
18	4–5	0	1.75	6	0.6	0.61
19	5–6	16	1.75	6	0.6	0.61
20	6–7	0	0.5	6	1.35	0.33
21	6–8	15	1.25	6	0.75	0.52
27	fan velocity pressure FVP =				49 Pa	
28	fan total pressure FTP =				194 Pa	
29	fan static pressure FSP =				145 Pa	
30	fan air volume flow Q=				1.75m^3/s	

col.	G Max Q (m^3/s)	H Q error (%)	I Likely dia m	J Likely v(m/s)	K Rectangular size Width (mm) × depth	L
row	*****	******	====	*****	=======	=====
14	1.76	–0.5	0.95	2.5	900	850
15	1.76	–0.5	0.8	3.5	700	750
16	1.75	0.1	0.5	8.9	450	450
17			0.5	8.9	450	450
18	1.75	0.2	0.6	6.2	400	800

Continued

col.	G	H	I	J	K	L
	Max Q (m^3/s)	Q error (%)	Likely dia m	Likely v(m/s)	Rectangular size Width (mm) × depth	
row	*****	******	====	*****	===============	
19	1.75	0.2	0.6	6.2	400	800
20	0.51	−2.3	0.6	1.8	400	400
21	1.26	−0.8	0.6	4.4	400	550

col.	M	N	O	P	Q	R
	Equiv. dia (mm)	Actual v(m/s)	p_v (Pa)	Fitting k	Duct (Δp) (Pa)	Fitting kp_v (Pa)
row	*****	*****	***	=====	******	*****
14	963	2.4	3	4.5	0	16
15	798	3.5	7	0.04	0	0
16	496	9.1	49	0.04	0	2
17	496	9.1	49	0	0	0
18	616	5.9	21	0.12	0	6
19	616	5.9	21	0.35	10	7
20	440	3.3	6	1.25	0	8
21	515	6.0	22	0.09	11	2

col.	S	T	U	V	W
	Plant Δp (Pa)	Total Δp (Pa)	Start p_t (Pa)	End p_t (Pa)	Index
row	======	*******	******	******	====
14	0	16	0	−16	*
15	95	96	−16	−111	*
16	0	2	−111	−113	*
17	0	0	−113	86	*
18	0	6	86	80	*
19	0	17	80	63	*
20	50	58	63	5	
21	50	63	63	0	*
27	Index = 199				

The formulae for the cells are:

cell	formula
C6	1.2 this is the air density kg/m^3
F14	= SQRT (C14*4/D14/PI( ))
G14	= − 2.0278*SQRT(E14)*(F14)^2.5*LOG(((4.05*10^ − 5)/F14) + (2.933*10^ − 5)/SQRT(E14)/F14^1.5)
H14	= (C14 − G14)* 100/C14
J14	= C14*4/PI( )/(I14)^2
M14	= 1.265*((K14*L14)^3/(K14 + L14))^0.2
N14	= C14*4/PI( )/(M14/1000)^2
O14	= 0.5*C6*N14*N14
Q14	= B14*E14
R14	= P14*O14
T14	= Q14 + R14 + S14
V14	= U14 − T14
E27	= O17
E28	= T26
E29	= E28 − E27
E30	= C17
T27	= T14 + T15 + T16 + T17 + T18 + T19 + T21

1. Number all the system nodes and enter the section numbers into the first column.

2. For each section, enter the duct length L m, air flow Q m³/s, maximum air velocity v m/s and the maximum allowable pressure loss rate Δp Pa/m.
3. The program then calculates the duct diameter required for each section and the maximum carrying capacity of each duct.
4. The difference between the maximum carrying capacity and the required Q m³/s is calculated and expressed as Q error %. If this is more than, say, 5%, enter a smaller value for maximum Δp Pa/m and repeat until satisfied.
5. Enter the likely duct diameter using increments of 50 mm.
6. Enter the dimensions of the rectangular duct to be used in mm. The circular equivalent diameter and actual air velocity are automatically calculated. Change the duct size until a satisfactory result is obtained.
7. The velocity pressure p_v Pa is calculated.
8. Enter the sum of the velocity pressure loss factors k for all the fittings in the duct section.
9. Enter the pressure drop through the air-handling plant such as filters and heater batteries in the section, plant Δp Pa.
10. Enter the starting total pressure p_t Pa for node 1; this will normally be atmospheric pressure, which is 0 Pa gauge pressure.
11. Each line is self-contained and calculations take place whenever new data is entered.
12. The total pressure at each node is displayed.
13. Enter a line of data for each section of duct.
14. Clear surplus rows of cells where sections do not exist.
15. Save the spreadsheet into a unique file name such as DUCT7079. This preserves the original larger spreadsheet for future use.
16. If the original spreadsheet does not have enough sections, copy rows down the screen to make more. Edit the new lines to ensure that the correct cell reference for air density is read into the formula.
17. The column total Δp Pa summates the pressure losses in that column. Ensure that this only contains the pressure losses through the index circuit by changing the formula.
18. The fan duty specification is shown.
19. Edit cell R18 to calculate correctly the expander Δp.
20. There are differences between the data used for the manual and spreadsheet calculations and these produce different fan total pressures.
21. The format of the formulae may vary with different spreadsheet programs.

Questions

1. State the three measurements that are made of air pressure within a duct, which direction they act, what each is used for and the scientist's name that is used to connect them. State the formulae connecting the three pressures.
2. Explain, with the aid of sketches, how the three airway pressures are measured. List all the equipment that would be needed. State how each item would be used.
3. A commissioning engineer needs to know the volume flow rate and mass flow rate of air through a 600 mm diameter duct. State all the measurements that are necessary and how they are to be acquired. Write all the formulae that would be needed and show the units of measurement used.
4. Sketch graphs of the three airway pressures changing along a duct of length L m and diameter d mm, showing the following cases.
    (a) Total and static pressures are above atmospheric.
    (b) Total pressure is above atmospheric but static pressure is below atmospheric.
    (c) The duct is on the suction side to a fan and

air total pressure is below atmospheric pressure.

(d) The duct tapers from 1 m diameter to 500 mm diameter along length L m. Pressures remain above atmospheric pressure.

(e) The duct enlarges from 300 mm diameter to 600 mm diameter while the total pressure remains below atmospheric pressure.

(f) A 500 mm diameter duct is above atmospheric pressure. The commissioning engineer omitted to seal a test hole halfway along the length of the duct.

(g) Room air returns to the air-handling plant through ducts that have inadequate joint sealing along their entire length, causing significant leakage. Total pressure at the commencement of the duct is above atmospheric. Static pressure within the duct starts at below atmospheric pressure.

5. Calculate the density of air for a temperature of 25°C d.b. when the atmospheric pressure is 101 600 Pa.

6. Calculate the temperature of air that corresponds to a density of 1.1 kg/m^3 at standard atmospheric pressure.

7. Convert the following air pressures into pascals: . 25 mmH$_2$O, 50 mb, 125 mmH$_2$O, 0.3 mH$_2$O, 0.25 b. Note that 1 b = 1 bar = 10^5 Pa and 1 mb = 10^{-3} b, so 1 mb = 100 Pa, also 1 kPa = 1000 Pa.

8. 3 m^3/s flows through a 1 m diameter duct. Calculate the air velocity.

9. Calculate the carrying capacities of air ducts of 400 mm, 600 mm, 1 m and 2 m diameters when the maximum allowable air velocity is 8 m/s.

10. The temperature of air in a 400 mm diameter duct was 32°C d.b. on a day when the atmospheric pressure was 101 105 Pa. The static pressure of the air in the duct was 45 mm water gauge below the atmosphere. The average air velocity was measured as 7 m/s. Calculate the air density, velocity pressure and total pressure.

11. 2 m^3/s are to flow through a 500 mm diameter duct at a static pressure of 300 Pa above the atmospheric pressure of 101 500 Pa and at a temperature of 26°C d.b. Calculate the air density, velocity and total pressures.

12. Calculate the pressures that occur when m^3/s flow through a 20 m long, 1300 mm diameter duct that then reduces to 1000 mm diameter and remains at 1000 mm for 20 m. The air total pressure at the commencement of the 1300 mm duct is 600 Pa above atmospheric. The reducer is the 60° concentric type shown in Fig. 6.3. The air density is 1.2 kg/m^3. The frictional pressure loss rates are 0.23 Pa/m in the 1300 mm diameter and 0.9 Pa/m in the 1000 mm diameter ducts.

13. Calculate the pressures that occur when 2 m^3/s flow through a 12 m long, 600 mm diameter duct that then reduces to 400 mm diameter and remains at 400 mm for 30 m. The air total pressure at the commencement of the 600 mm duct is 250 Pa above atmospheric. The reducer is the 30° concentric type shown in Fig. 6.3. The air density is 1.15 kg/m^3. The frictional pressure loss rates are 0.85 Pa/m in the 600 mm diameter and 6.5 Pa/m in the 400 mm diameter ducts.

14. Calculate the static regain and pressure changes that occur when 3 m^3/s flow through a 10 m long, 450 mm diameter duct that then enlarges to 700 mm diameter and remains at 700 mm for 20 m. The air total pressure at the commencement of the 450 mm duct is 400 Pa above atmospheric. The enlarger is the 30° concentric type shown in Fig. 6.3. The air density is 1.24 kg/m^3. The frictional pressure loss rates are 8 Pa/m in the 450 mm diameter and 0.86 Pa/m in the 700 mm diameter ducts.

15. A 500 mm diameter supply-air duct suddenly enlarges into a 1 m diameter plenum chamber containing filters. Calculate the static regain and pressure changes that occur when 1.5 m^3/s flows through the 35 m long, 500 mm diameter duct, enlarges, and then flows through the 1 m diameter plenum for 5 m. The air total pressure at the commencement of the 500 mm duct is 200 Pa above atmospheric. Refer to Fig. 6.3 for the enlarger pressure loss factor. The air density is 1.22 kg/m^3. The frictional pressure loss rate is 1 Pa/m in the 500 diameter and can be assumed to be 0.05 Pa/m in the 1 m diameter ducts.

16. An air-duct branch is similar to that shown in Fig. 6.7. Use the data provided to calculate all the duct pressures at the nodes.
   800 mm diameter duct 1–2 is 22 m long and carries 3 m^3/s.
   600 mm diameter duct 3–4 is 12 m long and carries 2 m^3/s.
   450 mm diameter duct 5–6 is 10 m long and carries 1 m^3/s.
   The branch offtake is a short radius bend.
   The straight-through contraction has a velocity pressure loss factor of 0.05. Air density is 1.2 kg/m^3. Pressure drop rates are: duct 1–2, 0.43 Pa/m; duct 3–4, 0.85 Pa/m; duct 5–6, 0.95 Pa/m.
   The total pressure at node 1 is 400 Pa.

17. An air-duct branch is similar to that shown in Fig. 6.7. Use the data provided to calculate all the duct pressures at the nodes.
   700 mm diameter duct 1–2 is 20 m long and carries 3 m^3/s.
   700 mm diameter duct 3–4 is 20 m long and carries 2 m^3/s.
   400 mm diameter duct 5–6 is 2 m long and carries 1 m^3/s.
   The branch offtake is a right-angled bend having several turning vanes. Use circular duct data.

Ignore the fact that turning vanes are not fitted to a circular branch. The vanes are in the branch duct. Air density is 1.2 kg/m^3.

Pressure drop rates are: duct 1–2, 0.85 Pa/m; duct 3–4, 0.4 Pa/m; duct 5–6, 1.8 Pa/m.

The total pressure at node 1 is 200 Pa.

18. A centrifugal fan delivers 3 m^3/s into an outlet duct of 600 mm × 400 mm. The inlet duct to the fan is 500 mm diameter. The static pressure at the fan inlet was measured as – 90 mm water gauge relative to standard atmospheric pressure. The ductwork system had a calculated resistance of 2000 Pa. Air density is 1.16 kg/m^3 Calculate the pressures either side of the fan.

19. The duct system shown in Fig. 6.15 is to be installed in a false ceiling over offices and corridors. The air-handling plant, comprising the fresh air inlet, filter, chilled-water cooling coil and an axial flow fan, is located in a plantroom. Office A is supplied with 1.5 m^3/s and office B has 2.5 m^3/s. Find suitable sizes for the ducts and state the performance specification for the fan.

The limiting air velocities are 2.5 m/s through the fresh air inlet grille and filter, 3 m/s through the cooling coil, 12 m/s through the fan and 5 m/s in the ducts. The fresh air inlet is constructed from 45° louvers having a free area of 60%. The filter, supply grille and cooling coil have air-pressure drops of 65 Pa, 30 Pa and 40 Pa respectively. The contractions in the duct are at an angle of 60° and the enlargement is at 40°. The ducts are to be sized for an air temperature of 20°C d.b.

(*There is more than one correct solution to this question. Duct sizes can be: section 1–2, 1200 mm × 1500 mm; section 3–4, 600 mm × 600 mm; section 5–6, 1200 mm × 700 mm; section 6–7, 800 mm × 400 mm; section 6–8, 1200 mm × 500 mm.*)

20. The ducts shown in Fig. 6.15 are an extract system removing air from two workshops where low-velocity air and heat reclaim are employed. The airflow direction arrows are to be reversed on the figure. All the duct lengths are three times those shown. The air-handling plant comprises a fan, heat-reclaim cooling coil, a noise attenuator in place of the filter shown and an exhaust grille to outdoors. Workshop A has an extract rate of 3.5 m^3/s and area B has extraction of 2.5 m^3/s. Find suitable sizes for the ducts and state the performance specification for the fan.

The limiting air velocities are 2 m/s through the exhaust grille and attenuator, 3 m/s through the cooling coil, 15 m/s through the fan and 8 m/s in the ducts. The exhaust grille is constructed from 45° louvres having a free area of 70%. The attenuator, extract grilles and cooling coil have air pressure drops of 135 Pa, 70 Pa and 90 Pa respectively. The contractions in the duct are at an angle of 45° and the enlargement is at 30°. The ducts are to be sized for an air temperature of 20°C d.b.

7 Automatic control systems

INTRODUCTION

The automatic control of air conditioning aims to satisfy the temperature and humidity specification throughout the year. This can be for the benefit of the thermal comfort of the occupants or a plant or process environment. Each application is likely to be unique. The principles of operation and main features of commonly used systems are explained. The stages in control schemes are presented in logical order. Examples of practical control strategies are given. Data such as 20% minimum fresh air intake and a 5 V control signal corresponding to a set point for air temperature are meant as examples, not as universal statements. Combinations such as fan-starting methods, detector operation, enthalpy control of dampers and temperature control require the reader to bring together different sections of the book to formulate a complete system. The reader is challenged to propose suitable designs for control systems, operating diagrams, control signals, controller configuration and wiring diagrams. The principles are explained within the text. The aim of questions that do not have a model solution is to build the reader's competence in preparation for tests and practical design work. The reader's answers to questions are for discussion with colleagues and tutor.

LEARNING OBJECTIVES

Study of this chapter will enable the user to:
1. know the component parts of control systems for air conditioning;
2. understand some terminology;
3. know the types and values of control signals;
4. understand the meaning and use of analogue signals;
5. know the voltages used for controls and actuators;
6. understand the use of thermistor temperature detectors;
7. know how digital communication takes place between computer and control equipment;
8. know how humidity is detected;
9. understand the use of enthalpy control;
10. know how pressure is detected;
11. know where and how airflow sensors are used;
12. understand weather compensators;

13. know the methods used for water-temperature detection;
14. know the meaning and use of actuators;
15. know the principles of pneumatic actuators;
16. understand the operation of solenoids, relays and contactors;
17. understand the different modes of control operation;
18. know symbols for control diagrams;
19. realize the importance of the control system design to the air-conditioning designer;
20. understand the operation of enthalpy control over fresh air ventilation;
21. draw plant-operation graphs for all-year control;
22. relate control signals to air conditions and plant status;
23. understand the use of optimum start control;
24. state the control equipment necessary for heating, ventilating and air-conditioning applications;
25. list logical operating sequences;
26. know the starting sequence for heating plant;
27. know fan-start procedures;
28. understand the control of single-duct and variable air volume air-conditioning systems;
29. identify the components in a controller;
30. understand direct digital control;
31. know how fan performance is regulated and matched to the air-conditioning system requirement;
32. understand how chilled-water refrigeration plant is controlled;
33. understand the principles of electrical wiring diagrams.

Key terms and concepts

COMPONENTS OF CONTROL SYSTEMS

Systems of automatic control consist of the principal components:

1.   air humidity sensor;
2.   air-pressure modulating or balancing dampers;
3.   air-pressure or pressure-difference sensor;
4.   air-temperature sensor;
5.   air volume flow rate sensor;
6.   analogue-digital module;
7.   central processing unit, computer;
8.   controller;
9.   dedicated programmer;
10.   disc data storage;
11.   distributed processing unit, outstation;
12.   electric or pneumatic valve and damper actuator motor;
13.   electrical power cable;
14.   electropneumatic and pneumatic-electric relay;
15.   electropneumatic transducer
16.   flow-modulating valve and damper;
17.   fluid-pressure-modulating or balancing damper;
18.   indicator instrument, analogue or digital, panel-mounted;
19.   instrument rack;
20.   isolating valve and damper;
21.   liquid crystal display;
22.   low-voltage control wiring, 0–10–24 V;
23.   mimic diagram on visual display unit;
24.   motorized potentiometer;
25.   motorized valve, 2-, 3- or 4-way
26.   pressure switch;
27.   printed circuit card, plug-in module;
28.   printer;
29.   programmable logic controller;
30.   rectifier, alternating to direct current;
31.   relay;
32.   standby electrical supply on mains failure;
33.   switches, on/off, manual, selector, or electronic;
34.   thermostat;
35.   time clock, digital or motorized;

36. transducer, detector signal to 0–10 V signal
37. transformer;
38. visual display unit (VDU);
39. water-temperature sensor;
40. weather compensator.

The standard electrical signal used between detector, controller and computer is 0–10 V stabilized direct current from a rectifier. The current is kept constant at a value between 5 and 20 mA. The variable voltage is an analogue of the condition being measured. Thus a 5 V signal would represent 20 °C if the range of temperature being measured was 10–30 °C. The electrical power to drive valve and damper motors is operated at 24 V or 240 V a.c. Each controller has an alternating-current-to-direct-current rectifier to provide the signal voltage.

A temperature sensor is usually a thermistor. A fixed current of around 5 mA passes through the thermistor. Changes in detected temperature alter the electrical resistance of the thermistor. Ohm's law states that applied voltage equals current multiplied by resistance. The voltage drop through the thermistor changes with the detected temperature. The detector output signal varies from zero to ten volts (0–10 V). This analogue output signal is either used directly by the controller or converted into a digital signal in binary code. Detected temperatures can be averaged by connecting two or more thermistor temperature detectors in series and parallel so that the whole circuit has the same resistance as only one detector. Figure 7.1 shows how this can be achieved. Space temperature t_1 °C is measured with two thermistor detectors that are connected in series. The t_2 °C and t_3 °C temperatures are sensed by separate thermistors that are connected in series. The combined resistance of the four thermistors is equal to that of one thermistor.

Each thermistor produces outputs in the range 0.1–10 V at a current of 5 mA. The variation of thermistor resistance R Ω with voltage is, from Ohm's law,

$$\text{current } I = \frac{V \text{ volts}}{R \text{ ohm}}, \text{ so } R = \frac{V}{I}$$

$$\text{at 10 V, } R = \frac{10\ V}{5\ mA} = \frac{10}{5 \times 10^{-3}}\ \Omega = 2000\ \Omega$$

$$\text{at 0.1 V, } R = \frac{0.1\ V}{5\ mA} = \frac{0.1}{5 \times 10^{-3}}\ \Omega = 20\ \Omega$$

The electrical resistance circuit that is equivalent to the thermistor connections is shown in Fig. 7.1. Series resistances are added, $R_1 + R_2$ and $R_3 + R_4$. The whole circuit resistance R is found by adding the reciprocals of parallel resistances:

$$\frac{1}{R} = \frac{1}{R_1 + R_2} + \frac{1}{R_3 + R_4}$$

When each thermistor has a resistance of 20 Ω,

$$\frac{1}{R} = \frac{1}{20 + 20} + \frac{1}{20 + 20}$$

$$= \frac{1}{40} + \frac{1}{40}$$

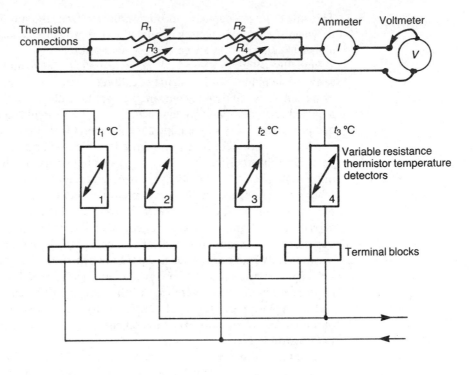

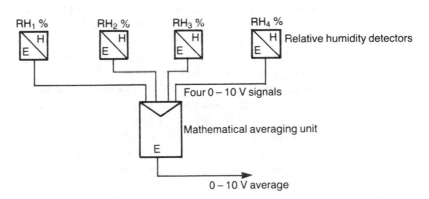

Fig. 7.1 Averaging temperature and humidity conditions.

$$= \frac{2}{40}$$

$R = 20\,\Omega$, the same as one thermistor

Temperature averaging is useful where there are significant variations within a room, zone or group of rooms or spaces. An averaging analogue unit contains amplifiers to process the mathematical calculation of the voltages. Figure 7.1 shows the use of a mathematical averaging unit with four humidity-detector inputs. Digital controllers employ software programs for the mathematical processing.

Humidity sensors detect a change in the electrical resistance between two conductors. The conductors are fused on to plastic and coated with lithium chloride salt.

The salt is hygroscopic, absorbing moisture from the air. The electrical resistance between the two conductors is related to the moisture absorbed. The 0–10 V output signal is calibrated to air percentage saturation.

Enthalpy detectors sense both air temperature and humidity, often in a combined sensor. The output 0–10 V signal is calibrated into a measure of the specific enthalpy of the air. They are used to compare the specific enthalpy of the incoming fresh air into a ducted air-conditioning system with that of the recirculated room air. The air-mixing dampers or heat-recovery equipment is operated from these enthalpy signals.

Air or gas pressure is sensed with a flexible diaphragm that separates two chambers. One chamber is connected to the air duct or space where the pressure is to be measured. The other chamber is connected either to atmospheric pressure or to the other pressure space. A change in pressure between the two compartments causes the diaphragm to flex. The physical movement of the diaphragm operates a variable capacitor in a signal processor to generate a 0–10 V output. The sensor is a pressure transmitter that sends its output to a digital display, dial gauge or computer.

An airflow rate transmitter is a pressure sensor that is connected to an averaging velocity pressure probe or flow grid as in Figs 6.1 and 6.12. The airflow measured is proportional to the square root of the pressure difference. The output signal has had the square root processed electronically. Pressure can be sensed with bellows or a bourdon tube that is expanded on rise of pressure and these are frequently used with instruments.

An external air-temperature sensor is a thermistor within a weatherproof rigid plastic or cast alloy box. The temperature within the box is affected by solar radiation, wind and precipitation. The signal processor can be programmed to compensate the indoor plant operation for these weather conditions.

A water-temperature sensor is a thermistor in a brass casing that is inserted into the waterway. An output signal of 0–10 V is proportional to the water temperature. Temperature-sensing methods utilize a bimetal strip, bimetal rod and tube, liquid and vapour-filled bellows, a thermocouple and an electrical resistance coil. They are for local indicators or are used with automatic control. Temperature detectors take anything from 30 s to 5 min to change their output signal in response to a change in the measured variable (CIBSE, 1986, Section B11).

An actuator is a device that converts the control signal into an action. Such actions are to operate valves and dampers that control fluid flow or to stop and start flows. Electric motor actuators can be operated on the line voltage of 240 V or extra-low voltage of 24 V a.c. The motor is either bidirectional, reversing, to drive the valve or damper in both directions, or spring return. The device may be normally closed or normally open depending upon the safe state when electric power is removed. The motor output shaft is driven through reduction gears or a linear screw thread. The motor run time can be 2 min from fully open to closed condition.

Pneumatic actuators use only air pressure for their motive power. They are spark-free, require only a transducer to convert an electrical or electronic control signal into an air pressure and a 1–3 bar air-pressure supply. It can become cheaper to install pneumatic actuators when many valves are needed. A control that combines electronic signals and pneumatic actuators is called a hybrid system. A pneumatic actuator comprises a diaphragm that is pushed outwards by increasing air pressure against a return spring to open the valve or damper controlling the fluid flow. On removal of the controlling pneumatic pressure, the spring returns the final control

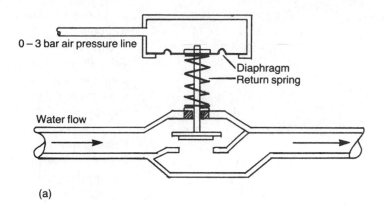

(a)

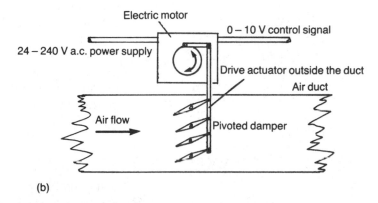

(b)

Fig. 7.2 Operating principles of pneumatic and electric actuators. (a) Pneumatic activation on water flow. Valve is normally open; increase of air pressure closes valve. (b) Electric motor drive on flow-control clampers. Motor rotation is reversible to open and close clamper heads.

element, valve or damper, to its original position. This can be normally open or normally closed. A double-acting actuator requires pneumatic pressure to move the output shaft in both directions and there may not be an automatic fail-safe feature. Figure 7.2 shows the operating principles of electric and pneumatic actuators.

Electric solenoids are used to open and shut valves on fluid pipelines and to open and close electrical switches. They are usually employed in fail-safe situations where power is required to keep them open.

The terms used to describe control systems include the following:

Actuator: valve or damper motor.
Analogue: a variable electrical voltage or current is transmitted between elements of the control system to initiate actions. 0–10 V d.c. of 0–5 mA are used.

Boost:	temporary operation of the heating or cooling plant at full output.
Cascade:	a controller that receives two or more input 0–10 volt signals. The first signal is processed and then combined with the second signal for further processing to generate the output signal.
Controlled condition:	the physical state of the controlled variable.
Controlled variable:	the temperature, pressure, speed or humidity that is being controlled.
Controller:	the equipment that receives input data and produces an output signal that is used for control action.
Controller action:	the mathematical relationship between the input and output signals created by the controller.
Dead time or zone:	the time between a change of input signal to a controller and the commencement of change in the controlled variable.
Desired value:	the condition of the controlled variable that is to be maintained.
Differential:	a band of values of the controlled variable that is needed to initiate a response from the controller.
Direct digital control, ddc:	binary bytes are communicated between elements of the control system.
Error:	the difference between the actual value and the desired value of the controlled condition.
Final control element:	the valve, damper or switch that operates upon the controlled variable.
Inertia:	the resistance to change of the whole system. This includes the time taken to detect an error, send a signal to the controller, operate the final control element and change the state of the controlled variable. The thermal storage capacity of the building's surfaces may be a part of the overall inertia.
Input signal:	the signal received by the controller. This is normally the 0–10 V signal from the detector.
Lag:	the time between a control signal being sent to a final control element and that element starting to respond. This may be less than dead time.
Offset:	the controlled variable is maintained at a deviation from the set point or desired value.
Output signal:	the mathematically processed output signal from a controller to the final control element.
Proportional:	the output signal from the controller is proportional to the deviation of the controlled variable from the set point.
Set point:	the setting of the controller that is necessary to achieve the desired value of the controlled variable.

Controllers operate in a variety of methods, as follows.

On/off:
The controlled medium is switched on or off. A differential exists between the controlled conditions at the on and off switching points. The swing in the value of the controlled condition will be larger than the differential. This is due to the inertia of the whole system and the building.

Incremental:
There are several fixed steps between on and off. Multiple boiler and refrigeration multi-cylinder compressor systems are controlled this way. The effect of each step is smoothed by the inertia of the fluid distribution system.

Proportional, P:
The output signal from the controller is proportional to the difference between the set point for the controlled variable and the sensed condition. The response is in proportion to the problem.

Floating:
The valve or damper actuator is moved at a fixed speed. The controller gives pulses of movement to bring the controlled condition within a neutral band where no pulses are made. The controlled condition varies continuously and the actuator is used frequently.

Integral, I:
The output signal from the controller is time-dependent. The valve or damper actuator is moved at a speed that is in proportion to the error signal from the controlled condition. The detection of a large error causes the maximum speed response of the actuator. This is analogous to an emergency stop when driving a vehicle.

Derivative, D:
The valve or damper actuator is moved at a speed that is in proportion to the rate of change in the controlled variable. This is to match fast load changes. It is usually employed with another mode.

Proportional plus integral, P + I:
The combination provides stable proportional control and the ability of the integral action to minimize offset.

Proportional plus integral plus derivative, P + I + D:
The ability to match fast load changes is added to P + I.

Direct acting:
The controller causes the valve or damper to close on increase of detected temperature, pressure or humidity. This is the normal action for heating system control.

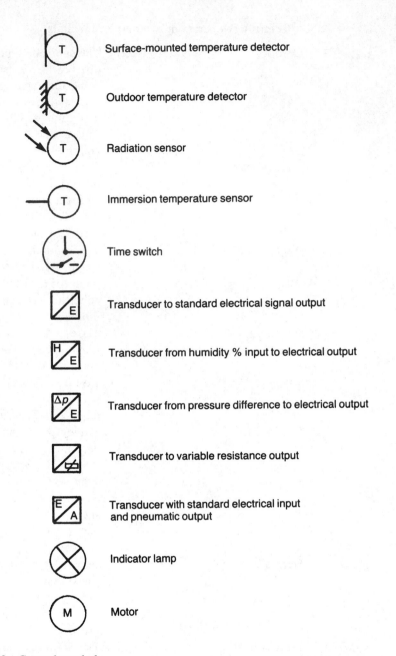

Surface-mounted temperature detector

Outdoor temperature detector

Radiation sensor

Immersion temperature sensor

Time switch

Transducer to standard electrical signal output

Transducer from humidity % input to electrical output

Transducer from pressure difference to electrical output

Transducer to variable resistance output

Transducer with standard electrical input
and pneumatic output

Indicator lamp

Motor

Fig. 7.3   Control symbols.

Reverse acting: The controller causes the valve or damper to
                                 open on increase of detected temperture,
                                 pressure or humidity. This is the normal action
                                 for cooling system control.

    The DIN, German Industrial Norms, symbols used for control diagrams are shown
in Figs 7.3, 7.4 and 7.5.

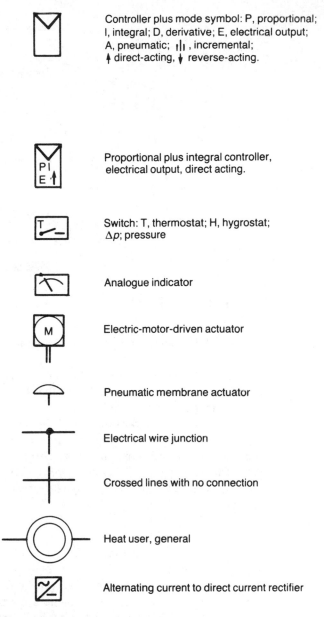

Controller plus mode symbol: P, proportional; I, integral; D, derivative; E, electrical output; A, pneumatic; ₁|₁ , incremental; ↑ direct-acting, ↓ reverse-acting.

Proportional plus integral controller, electrical output, direct acting.

Switch: T, thermostat; H, hygrostat; Δp; pressure

Analogue indicator

Electric-motor-driven actuator

Pneumatic membrane actuator

Electrical wire junction

Crossed lines with no connection

Heat user, general

Alternating current to direct current rectifier

Fig. 7.4 Control symbols.

CONTROL SYSTEM DIAGRAM

The air-conditioning design engineer makes the initial analysis of the requirements and likely systems of control. These are discussed with other interested parties to ensure integration with the other services. A full description of the system logic is made. Schematic diagrams are used to demonstrate the location of all the components. Diagrams of the electrical wiring are needed. It is particularly important to

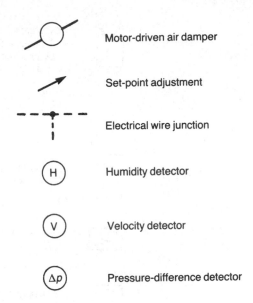

Motor-driven air damper

Set-point adjustment

Electrical wire junction

(H)   Humidity detector

(V)   Velocity detector

(Δp)   Pressure-difference detector

Fig. 7.5   Control symbols.

specify all the connection points between the wiring and the elements of the control system.

Figure 7.6 shows the automatic control schematic diagram for the variation of outdoor air during the year. When the specific enthalpy of the outdoor air, h_o, exceeds that of the extracted room air, h_r, in summer, it is reduced in volume flow. This is to avoid overheating the room or to minimize the cooling plant load. Air-temperature and humidity detectors are located in the return-air duct from the room and in the fresh air inlet duct. Display devices allow manual reading of the data. The enthalpy controller compares the two sets of data. The output signal from the controller is 0–10 V. A 5 V signal corresponds to equal specific enthalpy of the two air streams, h_o/h_r of 1. In winter, when h_o is much less than h_r and h_o/h_r is 0.25, the controller output signal is 10 V. In summer, when h_o is much greater than h_r and h_o/h_r is 2, the controller output signal is 0 V.

An output of 0 V corresponds to the minimum fresh air inlet quantity during the summer. This is typically 12 l/s per person occupying the building. An output signal of 10 V corresponds to the minimum fresh air inlet quantity in winter. 5 V occurs when the maximum quantity of outdoor air can be passed through the building, perhaps 100%. The heating and cooling plant are switched off when h_o/h_r is 1. All the winter outdoor-air conditions lie within the range 0–5 V. Between these limits, the outdoor-air intake is gradually reduced. 5 V may correspond to 100% fresh air and 0 V to 20% fresh air. A controller output of 2.5 V will always produce 60% fresh air: that is, halfway between 20% and 100%. All the summer conditions are in the 5–10 V range. A 10 V signal will produce the minimum summer fresh air inlet quantity, say 20%. A controller output of 7.5 V will always produce 60% fresh air, that is, halfway between 100% and 20% on the summer programme. The controller will always know the difference between summer and winter by the voltage signal.

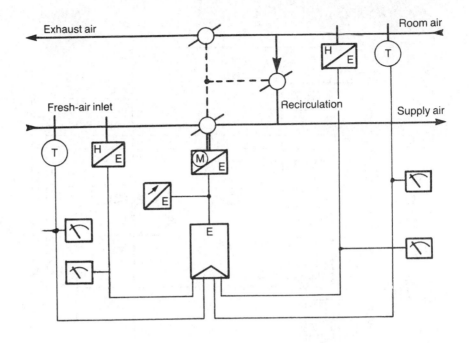

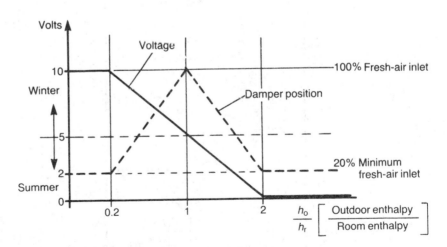

Fig. 7.6 Enthalpy control of fresh air inlet quantity.

HEATING AND VENTILATING CONTROL

A typical heating and ventilating control schematic is shown in Fig. 7.7. Cooling is not provided. The boiler plant has additional controls for frost protection, ignition cycle, flue-induced draught fan and other connected services such as hot-water storage. The ventilation plant has an optimum start and stop time-switch controller

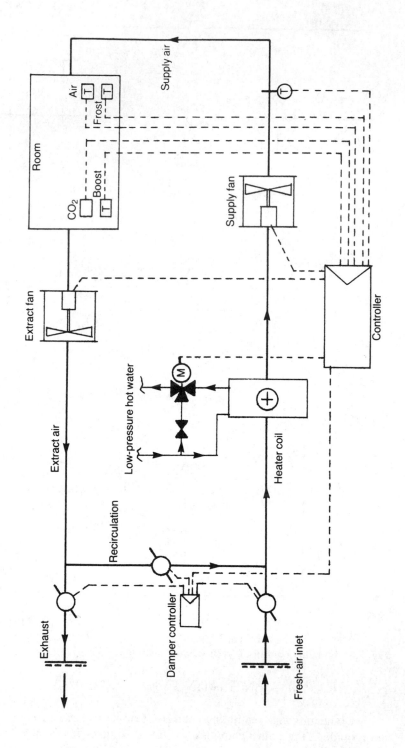

Fig. 7.7 Heating and ventilating system control.

that decides when to activate the plant. The user inputs the desired time of occupancy. An algorithm within the controller calculates when it is necessary to activate the heating to raise the occupied space to the desired value at the time of occupancy. The calculation includes the current space air temperature, outdoor temperature and the length of the off period. A frost temperature detector within the building is used to override the overnight and weekend off instruction. A minimum indoor air temperature of 10 °C will prevent internal freezing and serious condensation. The optimum-start software can be adapted by the user to allow variation of the system performance.

The starting sequence may be as follows.

1. Overnight frost protection.
2. Fresh-air inlet and exhaust-air dampers remain shut overnight.
3. Optimum start time.
4. Start signal sent to boiler plant.
5. Boiler plant start sequence is initiated:
    (a) combustion air and flue fans start;
    (b) airflow detectors confirm fan operation;
    (c) time delay of 1 min to purge combustion passage;
    (d) water-circulation pumps started;
    (e) water-flow sensors confirm circulation;
    (f) ignition transformer and spark activated;
    (g) time delay to confirm ignition device;
    (h) gas valve or oil pump and valve opened;
    (i) fuel is ignited; ignition remains active;
    (j) photocell registers presence of flame;
    (k) ignition spark maintained for 1 min or longer;
    (l) heat output controlled from flow water temperature detector;
    (m) multiple boilers start and are controlled by a multistep controller.
6. The two- or three-port flow-control valve on the low-pressure hot-water flow or return at the air-heater coil is held in its normally open position.
7. The heater coil reaches working temperature.
8. A water-temperature detector in the return pipe from the coil can be used to signal correct performance.
9. The frost detector is switched out of circuit.
10. The boost temperature detector is activated.
11. The supply air fan is started.
12. A time delay of 30 s prior to starting the extract fan.
13. Flow switches in the supply and extract ducts confirm air flow.
14. The boost limit temperature detector signals the end of the boost period.
15. Room-air temperature detector is activated and assumes its normal control function over the water-flow control valve.
16. The time of commencement of occupancy is used to open the fresh air inlet and exhaust dampers to their minimum settings.
17. The three damper motors are interlocked to operate in synchronization.
18. The external air-temperature sensor, or weather compensator, resets the room air temperature set point or the supply-air duct temperature to minimize energy use.

19. An occupancy detector measuring room CO_2 may be used to increase the fresh air inlet quantity.
20. At the end of the occupancy period, the fresh air and exhaust dampers are shut and the heating plant is switched off.
21. The recirculation damper and heater-coil flow control valve are run to their open positions.
22. The heating water pump continues to run for 10 min to distribute remaining heat into the building before being switched off.
23. The supply and extract fans run for 15 min to distribute heat from the heater coil before being switched off.
24. The plant reverts to control from the frost detector.

SINGLE-DUCT VARIABLE AIR TEMPERATURE CONTROL

A single-duct air-conditioning system is shown in Fig. 7.8. It may incorporate features previously described that are not repeated here. The main aspects of its operation are as follows.

1. Temperature of the air supplied to the conditioned space and that of the extracted air are both detected.
2. Supply-air temperature has minimum and maximum limits.
3. The cascade controller resets the set point of the supply-air temperature according to the return-air duct temperature. The reset schedules and sequence graphs are shown in Fig. 7.9.
4. The 0–10 V output from the cascade controller operates the heating valve, mixing dampers and cooling-coil valve in sequence.
5. The three mixing dampers have a controller that provides free cooling from outdoor air and extract room air temperature detectors.
6. When the outdoor air temperature is higher than the air temperature extracted from the room, the damper controller closes the fresh- and exhaust-air dampers to their minimum opening. The recirculation damper is opened.
7. The minimum fresh- and extract-air damper setting is adjustable.
8. The outdoor air temperature is used by a controller to reset the room extract air temperature set point. This allows increases in the room temperature as outdoor temperature rises. This can be permissible for comfort cooling. It reduces refrigeration plant load and running cost.

A simplified block diagram of the internal components of the controller for the three dampers is shown in Fig. 7.10. The thermistor circuit electrical resistance signals from the outdoor-air and room-air temperature detectors enter the controller. Each signal is converted into 0–10 V by a transducer. The voltage is processed and its set point and proportional band can be adjusted. The two signals are then compared and the higher voltage prevails. When the outdoor air-temperature signal is higher than that from the room air temperature, the higher value is used to close the damper to its minimum opening. A damper position indicator signal is compared with the chosen voltage. If the damper is not in its minimum opening, the signal operates a changeover switch to drive the damper motors. An output voltage signal to the damper motors causes the 24 V a.c. power circuit to close the dampers.

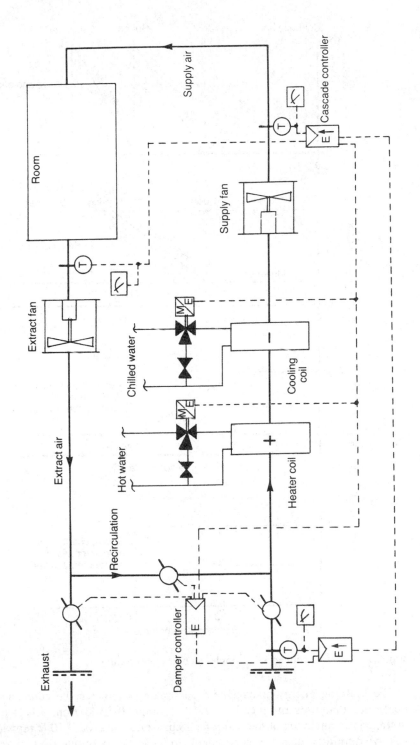

Fig. 7.8 Single-duct air-conditioning control.

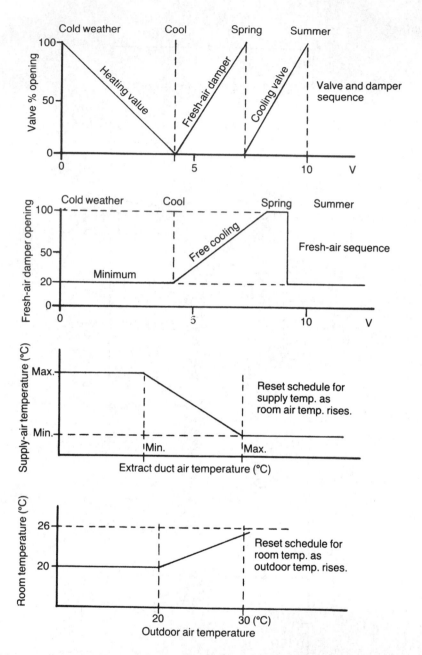

Fig. 7.9 Control sequences for single-duct air conditioning.

The building energy-management computer system can be connected into any controller that uses the standard 0–10 V range (Chadderton, 1991, p. 82). The voltage is an analogue of the control systems status, that is, if 10 V represents 100% relative humidity, then 5 V corresponds to 50%. The computer has an analogue-to-digital conversion card in its data bus. The digital bytes representing the voltage signal are processed by the software. The computer communicates in both directions

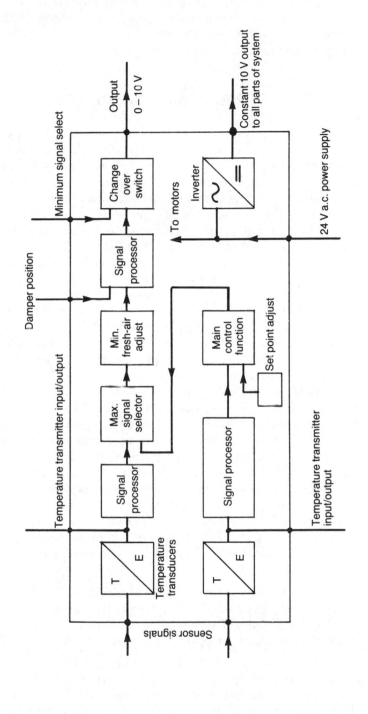

Fig. 7.10 Internal components of a mixed-air controller.

with the controller. Several controllers can be accessed, interrogated and reset from the remote computer. A local intelligent microprocessor outstation may be used between a group of controllers and the supervising computer. Each controller and outstation has a digital address. The supervisory computer scans the control and data system every few seconds or minutes as required.

SINGLE-DUCT VARIABLE AIR VOLUME CONTROL

This air-conditioning system supplies the occupied spaces with a single supply-duct air temperature. It is a cooling system. Terminal reheating can be provided by a low-pressure hot-water heating coil or electric resistance element. The supply-air duct temperature is scheduled against the outdoor air temperature. A weather compensator may be used to account for sunshine, wind and rain. Different orientations of the buildings may be grouped in different zones. Enthalpy control of the fresh-air intake may be used to minimize refrigeration use. The part-load performance needed for most of the year is met by reducing the supply air into each room. An arrangement for the room variable air volume controller is shown in Fig. 7.11. A room air-temperature detector is located in the room or within the extract-air duct or false ceiling. A 24 V a.c. power supply provides the 0–10 V stabilized and rectified control-signal circuit. A 10 V room air-temperature signal corresponds to maximum cooling. The volume-control damper is open and the heating-coil valve is shut. The controller output from 10 V down to 5 V is caused by a reducing cooling load from the room. This is found by the room air temperature falling. The volume-control damper is scheduled to close to its minimum setting at 5 V. The minimum air flow may be 20% of the full load condition. It is necessary to maintain the Coanda effect of the supply air entering the space at reduced flows. The cooling requirement within the room has now ceased. Further fall of the room air temperature means that heating is called for. At 4 V, the controller commences to open the two- or three-port heating valve. At 0 V the heating valve is at full load and is open. The controller directs 24 V a.c. power to the valve and damper motors. The reverse of the control process occurs as the space moves from a heating load to one requiring cooling.

The air-handling plant supply and extract fans interface with a duct system requiring fluctuating airflow rates. It is wasteful of electrical energy to run fans against volume-control dampers. Noise could arise from the excess pressure difference across room air terminal units. The fan delivery pressure is reduced by one of several possible methods:

1.   variable-frequency control (0–50 Hz) of the fan motor;
2.   variable-pitch inlet guide vanes on the inlet to a centrifugal fan;
3.   variable-pitch blades of an axial-flow fan;
4.   a variable-resistance damper on the fan inlet or outlet.

It is preferable to reduce the speed of the fan to lower its discharge air pressure (Sheldrake, 1991). This results in considerable savings in electrical energy consumption, lower duct pressure, reduced air leakage from the ducts, lower air velocity, less noise and less wear on the electric motor, drive belts and bearings. This is achieved with a variable-frequency controller that takes 0–10 V signals. The cost of this approach is up to a 10% increase in electrical power consumption to run the

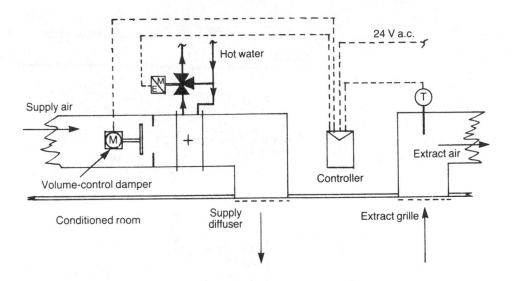

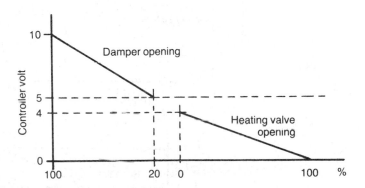

Fig. 7.11 Room variable air volume control.

frequency inverter. This 10% electrical loss produces an equal heat gain into the plant room. The use of variable-pitch inlet guide vanes and axial fan blades reduces the efficiency of the fan to lower its performance. The fan operates at a constant speed and motor power consumption is reduced. Hydraulic actuators are used to move the blades and vanes. Inserting a damper to reduce the air volume flow rate through the duct system while running the fan at full load is not good engineering practice. It is analogous to driving a car on full throttle and using only the brake to moderate the speed.

Figure 7.12 shows how the air-handling plant supply and extract fans are control-led in a variable air volume system. A variable frequency control (VFC) is designed to maintain a constant duct air pressure. Partial closure of the room terminal unit VAV damper increases the resistance of the ductwork system. The fan output is constrained to the shape of the curve shown. An increase in the duct system resist-ance causes the fan to supply less air at the elevated pressure. The duct air-pressure

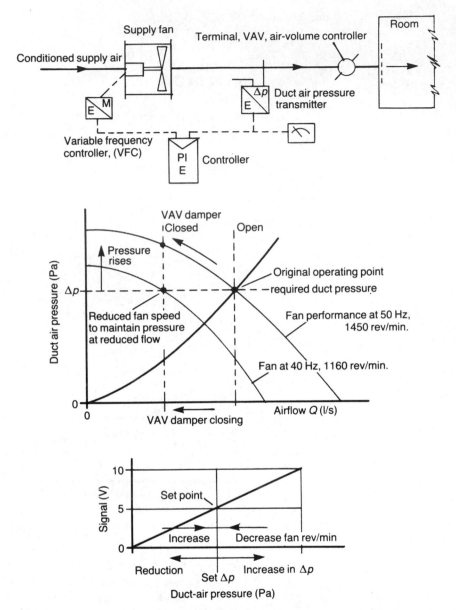

Fig. 7.12   Fan output control for a variable air volume system.

transmitter detects the rise by comparing the duct air with the atmosphere. The air-pressure set point corresponds to 5 V. An increase in duct pressure is scheduled to the 5–10 V signal band. A signal above 5 V causes the fan-speed controller to reduce speed until 5 V is established. A reduction in duct air pressure occurs when the variable volume terminal dampers open in response to rising room-air temperature. Duct air pressure that falls below the set point is scheduled to the 0–5 V range. A signal below 5 V causes the fan-speed controller to increase speed until 5 V is established. A duplicate pressure transmitter, pressure controller and variable-frequency controller are fitted to the extract fan and duct. The supply and extract fans will be different in size and power consumption. The controlled pressure in the extract system is not identical to that in the supply system. An alternative to duct-pressure control for the extract fan is to detect the volume flow rate in the supply and extract ducts. Reduction in the supply airflow triggers a corresponding lowering of the extract flow with a fan-performance controller.

A variable volume air-conditioning plantroom control system is shown in Fig. 7.13. Enthalpy control of the fresh air and exhaust dampers provides low-cost cooling. The supply-air fan speed is controlled from duct air pressure. The extract fan is controlled from air-velocity detectors in the supply and extract air ducts to balance the flows. The plant only needs a chilled-water cooling coil. Terminal hot-water coils accomplish the heating demand. A heating coil in the plantroom may be needed for frost protection and boost heating on cold starts during the winter. The building is protected from cold weather by closure of the external air dampers and recirculating the room air. The fresh and exhaust air dampers remain at their 20%, 2 V, minimum opening during the cooler weather until a room-cooling need is identified. Increasing amounts of fresh air are admitted until the fresh air enthalpy equals the extracted room-air enthalpy. The dampers are moved to their minimum fresh air intake positions. The controller signal is reset to 0 V. The controller commences opening the chilled-water valve on the cooling coil. The controller signal varies from 0 to 10 V over the 0–100% chilled-water flow through the cooling coil. The supply-air temperature is scheduled to the external air temperature to avoid excessive cooling and to minimize refrigeration plant load.

CHILLED-WATER PLANT

Chilled water is provided from a variable-output refrigeration plant. A chilled-water flow temperature of 6 °C and return of 10 °C is often suitable for air conditioning. The plant serves several cooling coils. The coils are in the zone air-handling units and room terminal fan-coil or induction units. Controlling the cooling output from the refrigeration machine and matching it to the current load is a similar problem to that with a boiler and a heating system. The heating or cooling load varies with the outdoor air temperature while the heat supply is generated in steps.

The cooling output of rotary compressors and absorption refrigeration machines can be smoothly variable. Reciprocating-compressor refrigeration machines are single- or multicylinder. Each cylinder has a reciprocating piston and spring steel valves. To reduce the compressor performance, the suction valve is held open with hydraulic pressure. The valve actuator is operated by the oil pressure from the lubrication pump. All the pistons reciprocate whether the valves are open or not. Cylinders which are unloaded in this way do not pump refrigerant. This reduces the mass flow rate of refrigerant vapour through the evaporator. The evaporator lowers the temperature of the water circulating through the chilled-water circuit. The cooling performance of the plant is reduced in steps corresponding to the number of cylinders that are unloaded.

To achieve close control over the chilled-water temperature, two multicylinder compressors are used. One compressor leads the other. The lag compressor becomes activated when the lead machine is fully operational. The lag machine is the first to

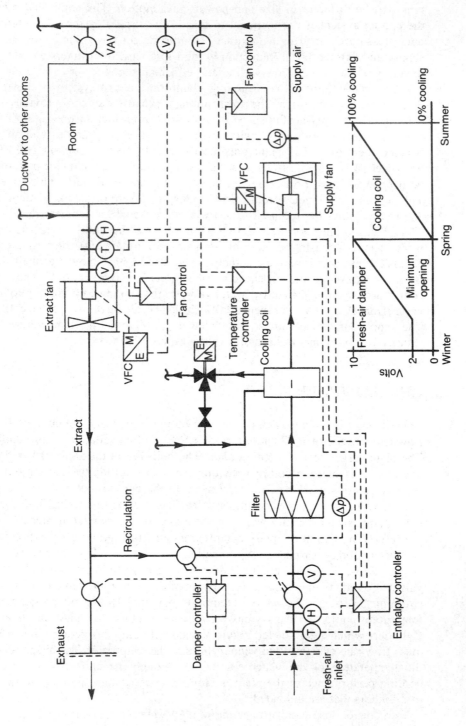

Fig. 7.13 Plantroom control scheme for variable air volume.

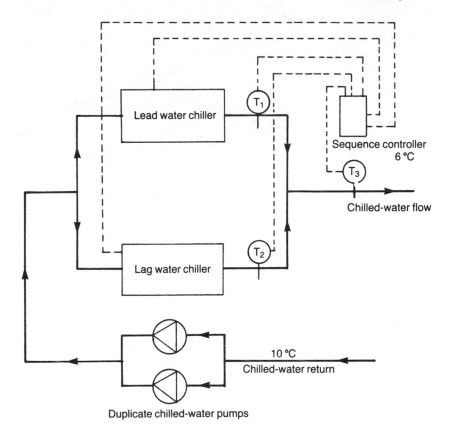

Fig. 7.14 Sequence control of refrigeration chilled-water plant.

be switched off. The result of the capacity-control steps is to form a smoothed control of the chilled-water temperature. Figure 7.14 shows the arrangement of a chilled-water refrigeration plant. Figure 7.15 shows the resulting mixed-flow water temperature leaving the compressors. The design flow and return water temperatures are 6 °C and 10 °C at full cooling load. The compressor control maintains a leaving-water temperature of 6 °C plus or minus a tolerance of 0.5 °C. A cooling-load dead band of 1.5 °C around the set point is the fluctuation in the flow water that corresponds to one capacity step. 6 °C is produced for all cooling load conditions within the building. A reducing demand for cooling within the building causes the return-water temperature to fall to 6 °C when the cooling load is removed.

While one refrigeration machine is unloaded, return water is still circulated through its evaporator. The combined-flow water leaving the plant is the mixture of two streams, one chilled, the other not. The two compressor machines are of equal size and chilled-water flow rate. To achieve a leaving mixed-flow temperature of 6 °C from a chilling machine and an unloaded machine, the chiller set point is 4 °C. The refrigerant R22, chlorodifluoromethane $CHClF_2$, is evaporated at 3 °C. The chilled water leaving the evaporator will approach 3 °C on occasions. Freezing of the water must be avoided. Lower-temperature systems for industrial process cooling utilize brine at subzero temperatures. The mixing of water as low as 3 °C with

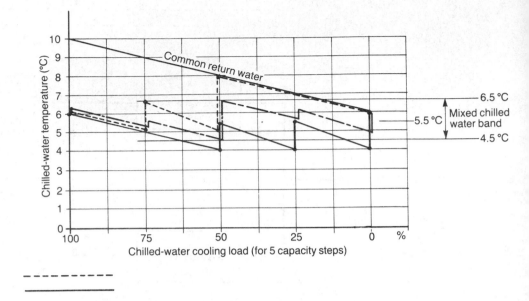

Fig. 7.15 Temperature schedule for a chilled-water plant: —-, lag machine leaving-water temperature; ___, lead machine leaving-water temperature; -.-.-, mixed chilled water leaving temperature.

uncooled water at 10 °C produces the desired value of 6 ± 0.5 °C. When it does not, another capacity step is loaded or unloaded as appropriate.

ELECTRICAL WIRING DIAGRAM

A diagram of the electrical wiring shows all the components of the power equipment and control system, how they are connected and the logic of their operation. A simple wiring diagram for part of a ventilation system control is shown in Fig. 7.16. A single-phase 240 V a.c. circuit is indicated. The protective conductor, earth, has not been shown. The making of switches completes the circuit between the line and neutral conductors and current flows. When a solenoid coil is energized in a circuit, a second switch contact is made. The metal actuator is attracted towards the solenoid coil by magnetic attraction. The movement of the actuator pulls the switch into the closed position. This energizes the controlled circuit that may be at a higher voltage and current. A 10 V control signal can switch on a 415 V three-phase circuit to a fan motor. The actuator does not conduct electricity between the low- and higher-voltage circuits. This arrangement of solenoid, actuator and switch is called a relay or contactor. A relay is shown in Fig. 7.17. Energizing the solenoid coil with an external electrical current causes the metal core to become magnetized. The magnetic field attracts the metal actuator towards the core. Movement of the actuator closes the switch contacts in the controlled circuit. Removal of the control-circuit energizing current switches off the magnetic field. The return spring pulls the actuator back to its off position.

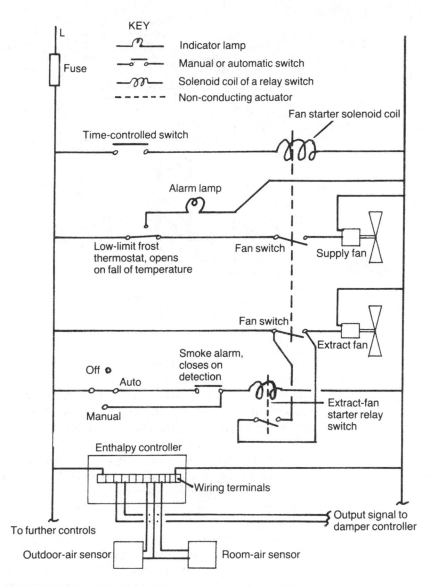

KEY

Indicator lamp

Manual or automatic switch

Solenoid coil of a relay switch

Non-conducting actuator

L

Fuse

Fan starter solenoid coil

Time-controlled switch

Alarm lamp

Low-limit frost
thermostat, opens
on fall of temperature

Fan switch

Supply fan

Fan switch

Extract fan

Off

Auto

Smoke alarm,
closes on
detection

Manual

Extract-fan
starter relay
switch

Enthalpy controller

Wiring terminals

To further controls

Output signal to
damper controller

Outdoor-air sensor

Room-air sensor

Fig. 7.16 Simplified electrical wiring diagram for air-conditioning equipment.

Figure 7.16 shows a 240 V line conductor serving a time-controlled switch. The line conductor has a fuse or overload micro circuitbreaker. The time switch closes at the designated time and energizes the fan-starter solenoid coil. The fan-starter relay closes the power switch to the supply-air fan. The extract-fan relay switch is closed simultaneously. If the low-limit frost-thermostat switch is closed, electrical power flows to the fan starters or direct on-line motors. Each fan may have a starter control such as star-delta, time delay, soft start or variable-frequency speed control. If the low-limit thermostat detects frost, the thermostat switch contact opens and the alarm circuit is illuminated. The fans cannot be started. Additional lamps may be in the fan wiring to indicate their operation. When the fans are switched off, the smoke

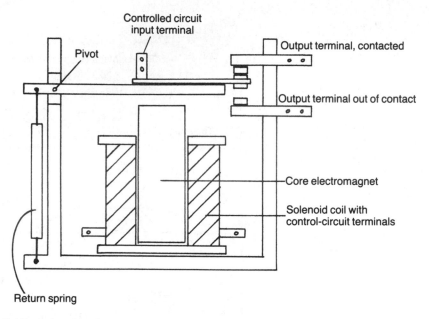

Fig. 7.17 Solenoid relay.

alarm closes the automatic detection circuit. This energizes the relay that provides power to the extract fan. This bypasses the time switch and low-limit thermostat. The operation of the smoke-alarm circuit can be manually tested or switched off. The enthalpy controller for the fresh air damper is shown. A row of numbered terminals is provided for installation work. Temperature and humidity detectors have a two-wire cable for a common input voltage and an output signal. The enthalpy controller processes the signals and sends its output to the damper controller.

The wiring diagram of a control system is shown in Fig. 7.18. Each controller has a numbered terminal block. A common 24 V a.c. and neutral cable is connected to each controller and power-operated actuator. The controller has an electronic rectifier to produce stabilized 10 V d.c. The 10 V is used as the control signal by the detectors and signal processors. The analogue outputs from the controller to actuators are 0–10 V. An analogue-to-digital circuit board may be within the controller. This allows direct digital control (DDC) from the supervisor computer. Status and qualitative digital output data from the controller are transferred through the building energy management (BEMS) computer network cable. These data can be transferred through the 240 V a.c. cables, (mains-borne signals) to the supervisor. Each package of data is transmitted by its controller at a discrete frequency in the MHz range. Each controller has its own digital address. Signal interference with the 50 Hz system or other possible sources of electronic noise is unlikely. The MHz harmonic frequencies produced by variable-frequency controllers on electric motors need to be checked for possible interference.

A switch is used for control. For direct current and single phase it is single-pole and only opens the line conductor. An isolating switch on single phase is double-pole. Both the line and neutral conductors are switched open. Three-phase wiring has three line conductors and a neutral. Each phase is a different colour. Yellow, blue and red are used. The phase currents are equal. Each phase is switched

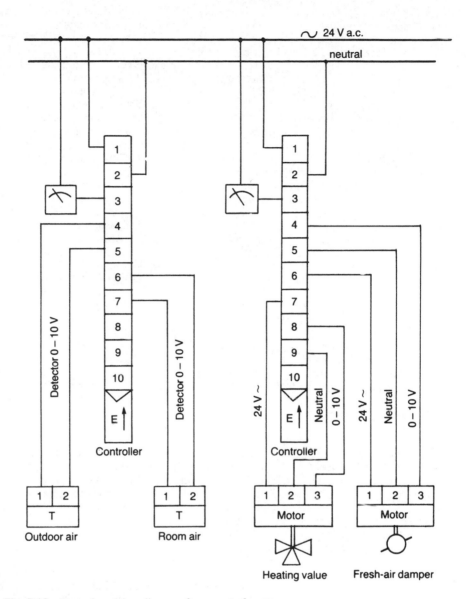

Fig. 7.18 Part of a wiring diagram for a control system.

open or connected simultaneously. A three-phase wiring diagram shows the three line and neutral conductors. Isolating switches are triple-pole and neutral (TP&N) Electric motors of 1 kW and above power consumption are normally on three phase (Chadderton, 1995, Chapter 13).

Questions

Discussion questions require the use of the text and may benefit from additional information. It is expected that references, manufacturers' literature, colleagues and tutor will be utilized to increase the reader's breadth of knowledge. It may be necessary to assume source data to answer some questions.

1. List the components of an automatic control system for:
   (a) domestic gas fired central heating;
   (b) ducted-air heating and ventilation system in

a single-story factory used for assembly of electronic components;

(c) single-duct variable air temperature air-conditioning system for a lecture theatre for 120 people;

(d) single-duct fan-coil air-conditioning system serving a 12-story office building. Chilled-water refrigeration compressors and oil-fired boilers are used.

2. Explain what is meant by analogue and digital values in control.

3. Explain how an analogue signal can be created and used to represent room air temperature.

4. Draw a graph of the output signal from a thermistor temperature sensor that has an operating range of 0–40°C and a constant current of 5 mA. A linear voltage output of 0–10 Volt is produced by the thermistor. 10 V corresponds to 40 °C air temperature. Calculate the voltage that corresponds to an air temperature of 25°C and the resistance of the thermistor at this value.

5. Explain the sensing and operating principle of:
   (a) thermistor temperature detector;
   (b) humidity detector;
   (c) enthalpy detector;
   (d) air pressure transmitter;
   (e) a detector to measure the air volume flow rate in a duct and use its value for automatic monitoring and control;
   (f) weather compensator;
   (g) water-temperature detector.

6. An open-plan occupied space is served by a single-duct variable-temperature air-conditioning system. Explain how the average air temperature and humidity of the space can be sensed and used for the control of the hot- and chilled-water diverting control valves. Sketch the arrangement of the detectors, controllers and actuators on a plant schematic.

7. A thermistor produces a 5 V d.c. output when passing a control current of 10 mA. Four thermistors are used to find the average of four air temperatures in a lecture theatre. Draw a suitable wiring circuit that would produce an average voltage for use by the controller. Calculate the resistance of the circuit. Validate your connection design by calculating the circuit resistance from first principles.

8. Explain the use of enthalpy control of the fresh air intake to a ducted ventilation or air-conditioning system. State why it is used and what limitations may arise.

9. State the types of actuator used to control heating, ventilating and air-conditioning equipment. Sketch and describe their operating principles.

10. Discuss the use of pneumatic actuators. Include in your discussion their operating principles, the plant necessary, their interaction with electrical, electronic and digital signals, their advantages and limitations.

11. Explain, with the aid of sketches, what an electric solenoid, relay and contactor is. State what each device is used for.

12. Explain how a personal computer is used to monitor and control an air-conditioning system. State how 24 V, 240 V and 415 V a.c. electric-powered actuators and plant are interfaced with the binary code used by the computer and network cables. Give examples of the software used by the supervising computer and the engineering management personnel in dealing with the data accessed.

13. Discuss, with examples, how lags occur in the detection and control of air conditioning within an occupied building.

14. Explain the following terms: proportional, integral, derivative, proportional plus integral, proportional plus integral plus derivative, offset, differential, boost, controlled variable, controller, dead time, set point.

15. State the ten controller operation methods used. Briefly describe the principle of each method.

16. A public entertainment theatre and conference centre seats 500 people. The basic layout of the air-conditioning system is shown in Fig. 7.7. Add a chilled-water cooling coil and three-port diverting valve on the chilled-water circulation. The room condition is to be maintained at 20°C d.b. ± 2°C and 50% percentage saturation ± 10%. The outdoor air temperature varies from – 10°C d.b. to 32°C d.b. during the year. Design an automatic control system that will maintain thermal comfort throughout the year. Specify the types of detection and control equipment necessary. Draw a schematic diagram of the air-handling plant and control system to describe its components, locations and modes of control. Draw operating graphs for the controls to demonstrate the voltage signals that correspond to plant status. Describe the logical operation of the control sequence. Chilled water is available at a flow temperature of 6°C and hot water is at 82°C. Ignore the boiler and refrigeration plant operations. Frost protection of the building and air-handling plant is needed owing to the low external air temperature. Make any assumptions that may be considered necessary. The design may be discussed with colleagues, the tutor and suppliers of control systems.

17. Explain the logical operation of the control of a variable air volume air-conditioning system. Draw the plant and control equipment schematic diagram.

18. Discuss the four methods used to control the output performance of centrifugal and axial flow fans. Illustrate the methods to show their application. Sketch the effect of each method on a fan-performance graph. Each graph is to identify the duct system resistance, the control effect and the combined performance point. State the energy savings, advantages and limitations of each method.

19. Explain the methods used to control the chilled-water output performance of refrigeration plant. Include the various types of refrigeration system and compressor.

20. Two three-cylinder R22 refrigeration reciprocating-compressor and water-chilling evaporator sets are to produce a flow temperature of 5 °C when the return water is at 11 °C. The minimum chilled-water temperature is 3 °C. The tolerance of the chilled water temperature control is ±0.5 °C. Each cylinder corresponds to a cooling-load dead band of 1 °C around the set point. Design a control schematic for the refrigeration plant. Draw a capacity-control graph to show the control steps and the mean chilled-water flow temperature that will be produced.

21. Draw an electric wiring diagram, similar to Fig. 7.16, for the heating and ventilating system in a shop. The supply and extract fan motors are single-phase. The fresh air inlet fan has an elec-

tric resistance heater that raises the supply air to 20 °C. A low-temperature limit thermostat in the supply duct switches the supply fan off after a 2 min time delay. The extract fan is started from a 30 s time-delay unit after the supply fan has started. A time-controlled switch activates the fans and two fan-powered single-phase electric resistance heaters in the shop. An air thermostat for each heater switches them on and off. Each fan and heater has an on-status indicator lamp. A frost thermostat is set at 8 °C in the shop and it overrides the time switch. Three smoke detectors switch on a smoke-extract fan. The smoke-extract system is always operational and it has a ready-status indicator lamp, a smoke-alarm indicator lamp and an audible smoke alarm.

22. Draw the wiring diagram for a three-phase 415 V power supply to two air-conditioning fans, a refrigeration compressor and a steam humidifier. Each item has a triple-pole and neutral isolating switch. The humidifier is controlled from a 10 V control signal from a humidity detector within the air duct. The supply-duct air temperature controls a hot-water valve. There is an isolating switch for all the air-conditioning circuits. An overload circuit-breaker protects the whole system. The plantroom has a three-phase distribution board. Single phase is used by the temperature and humidity controller.

8 Commissioning and maintenance

INTRODUCTION

The chapter introduces the meaning of commissioning and the scope of work that is normally carried out. An air-conditioning system may be allied to other mechanical services systems such as heating plant. Maintenance of air conditioning has energy-consumption implications, and also health and safety at work legislative obligations. The main elements of commissioning and maintenance are included but the topics and lists are not exhaustive. The commissioning and maintenance of the air-handling system are prominent. The information provided for water systems, automatic control, refrigeration and heat generation is general rather than specific in detail. Items are listed alphabetically and not in their order of importance or timing. The reader's answers to descriptive questions should be discussed with colleagues or tutors. Questions require the use of the information within this chapter, references, manufacturers' recommendations and details of specific applications.

LEARNING OBJECTIVES

Study of this chapter will enable the reader to:
1. Understand the purposes and importance of commissioning;
2. know the range of information needed to commission an air-conditioning system;
3. recognize the plant and systems that are connected with the commissioning of air conditioning;
4. identify the visual data checks needed;
5. know the safety criteria for the safe activation of electrical systems;
6. know when to commission plant and systems;
7. understand how fluid pressure tests are conducted;
8. know the cleanliness and hygiene criteria for systems;
9. understand fan and pump starting and speed control methods;
10. recognize how the stages of commissioning relate to the other members of the construction team;
11. know the checks and data needed when commissioning fans;

12. understand the meaning of variable-frequency control of fans;
13. know how to conduct a duct air-leakage test;
14. use practical duct air-leakage test data;
15. devise suitable test, commissioning and maintenance data record forms and logs;
16. know how to set to work and regulate an air-duct system;
17. understand what is meant by proportional balancing;
18. know the information needed by the commissioning engineer;
19. identify the instruments used during commissioning;
20. know how airflow rates are measured;
21. know the principles of operation of gas detectors;
22. know the application of gas detectors;
23. recognize why gas detectors are used;
24. understand the use of tolerance in measurement;
25. know the methods used in the measurement of ventilation rates;
26. know what is needed to commission automatic control systems;
27. recognize the need for a commissioning sequence document;
28. know the items requiring maintenance;
29. identify maintenance schedules;
30. understand the use of standby plant;
31. know how and why systems are cleaned and disinfected;
32. know how systems are maintained in clean and disinfected condition;
33. recognize the possible sources of airborne health hazards;
34. know the principal maintenance items.

Key terms and concepts

aerosols	252	cooling tower	252
air duct	239	damper	242
anemometer	244	data	235
automatic controls	249	descaled	238
biocide	252	detergent	252
catalyst	246	disinfectant	252
chlorine	252	documentation	235
cleaning	238	drive belt	236
client	234	electrical	236
codes	234	electrochemical	247
commissioning	234	engineers	234
contamination	238	epoxy resin	252

COMMISSIONING

The purpose of commissioning an air-conditioning system is to set it into operation, regulate flows and verify that it performs according to the specified design. The completed installation is visually inspected and has certain measurements taken. The measured data will be unique to the installation rather than the test figures supplied by manufacturers under standard conditions. The accurate and logical recording of data is essential. Commissioning work can extend over months or years in large applications. Several groups of engineers and clients may have access to the data. The client needs records of performance for future use during maintenance, repair, refurbishment and for continual energy and plant management.

Commissioning codes are in use for air distribution (CIBSE, 1971), boiler plant (CIBSE, 1975), automatic control (CIBSE, 1973), refrigerating (CIBSE, 1991) and water-distribution systems (CIBSE, 1989). These represent a standard of works and detail that competent engineers will hand over to the client upon completion of the commissioning process. Commissioning procedures extend over a wide range of specialist areas such as refrigeration, computers, electrical power and electronics. It is recommended that a single authority should have overall responsibility for commissioning. This may be one contractor or consultant, with subcontracted specialists.

Some data reading will be witnessed by the client to ensure that satisfactory performance has been agreed. Output numeric and graphical data from supervisory computer systems will form part of the information. Such graphical representation of system execution is termed a 'trend log'. Plant status is recorded as timed events such as the starting of a fan or the opening of a motorized valve. Each event and time is recorded and often printed at line printers as it happens. Computer records of trends and status are stored on hard disc or magnetic tape for a predetermined interval upwards from 24 h.

Commissioning documentation may be incomplete. The owner can recommission the plant to prepare records of the operational parameters of the entire system.

INFORMATION REQUIREMENT

Commissioning information includes:

1. description of the plant and its function;
2. design data on temperature, pressure, humidity and flows;
3. description of how to start and stop the systems;
4. location of all plant and distribution items;
5. fault diagnosis data, alarm limits and corrective action;
6. manufacturers' instructions and literature;
7. spare parts and sources of supply, stock list;
8. plant operational instructions;
9. maintenance schedule with frequency;
10. record drawings of all services;
11. plant schematic logic diagrams, framed and displayed;
12. lubrication charts, lubricants list, frequency;
13. valve and damper list, number, location and setting;
14. logbook of work carried out, data, date and signature.

The plant and systems that are directly connected with air conditioning are:

1. air-distribution ductwork, fans and terminal units;
2. artificial lighting;
3. building energy management computer-based control;
4. dedicated automatic controls;
5. electrical power;
6. fire and smoke detection plus active smoke control;
7. fire-fighting systems;
8. heat generation;
9. hot- and chilled-water distribution;
10. interaction with the manufacturing process;
11. refrigeration, heat reclaim and heat rejection;
12. security and alarm systems;
13. telecommunications;
14. transportation systems.

VISUAL DATA

The system is visually inspected prior to being started to ensure that it is operational and safe. Some items that are checked are:

1. access hatches and test holes in air ducts closed;
2. air- and water-pressure sensors;

3. air- and water-temperature sensors;
4. air-duct joints sealed;
5. air filters, access panels and driving motors;
6. air intake and exhaust louvres;
7. air volume control dampers, electric motors and linkages;
8. all component bolts, fixings and supports secure;
9. anti-vibration mountings;
10. boiler and burner equipment;
11. builder's works ducts clear of debris and dust;
12. cooling-coil condensate tray and drain;
13. correct polarity of electrical connections to motors;
14. drain water seals filled and drains functional;
15. drive guards in place and secure;
16. electrical switches, circuit breakers and fuses;
17. electrical wiring, earth conductor and insulation;
18. external cleanliness of all plant and distribution;
19. fan blades, bearings, drive belts and mounting bolts;
20. fan drive-belt tension;
21. fire dampers in air ducts;
22. flue system;
23. fresh and correct grade of lubricant to motor and fan bearings;
24. refrigeration compressor lubrication;
25. fuel supply;
26. grilles and diffusers in rooms fully open;
27. heater and cooler batteries;
28. hot- and chilled-water pumps and valves;
29. interior of air ducts for cleanliness;
30. noise attenuators in ducts;
31. steam or water-spray humidifier;
32. switch and circuit-breaker operation;
33. thermal insulation securely in place;
34. water quality;
35. water-storage tanks and float valves;
36. water supplies, tanks and float valves.

ELECTRICAL ITEMS

Before switching on the electrical power to any circuit, the following checks are made:

1. access equipment, ladders, available;
2. all wire terminal connections tightly connected;
3. cables electrically insulated;
4. carbon dioxide fire extinguisher available;
5. clean and dry equipment and floors;
6. competent and qualified electrical engineer;
7. correct current rating of fuses and circuit breakers;

8. correct isolating switches in place and operative;
9. correct voltage available for each item;
10. cover plates and cubicle doors closed;
11. earth cables and bonds to utilities in place;
12. electrical insulation tests satisfactory;
13. equipment undamaged;
14. lighting equipment operational;
15. line and neutral wires correctly connected;
16. mechanical equipment, such as fans, ready for use;
17. no unshrouded wires within control panels;
18. packing removed from switches, contactors and motors;
19. services electrically bonded to the earth terminal;
20. spare fuses and lamps available;
21. test certificates ready;
22. test equipment available;
23. thermal cut outs operational;
24. wiring of control equipment complete.

SETTING TO WORK

Parts of the installation are tested during the construction phase. Tests are to ensure that the components are safe and will retain their containment and distribution function when fully operational. Large plant such as boilers, refrigeration machines and fans cannot be tested without having their connecting distribution services filled and usable. Pipework, air-duct systems and cables are tested in sections as each is completed. The test fluid and electrical current may not be the same as will be finally used, but is compatible with the system materials. Such tests are as follows.

1. *Pressurization of air ducts.* The leakage of air from ductwork is measured by sealing the ends of a section and blowing air into it with a test fan. The static air pressure in the duct is raised to between 200 Pa and 2000 Pa depending upon the pressure class of the duct system.
2. *Pressurization of pipework.* Prior to filling with water, gas or refrigerant, an air compressor pressurizes the network up to 5 b depending upon the design working pressure. The test pressure is held for an hour. Loss of pressure is investigated with a soap solution being applied to joints.
3. *Pressurization of drainage systems.* The ends of above- and below-ground drain pipework are temporarily sealed and water seals are filled. A static air pressure of 100 mm water gauge is applied with a hand pump. For the system to pass the test, the test pressure is to remain above 75 mm water gauge during a 5 min period without further pumping.
4. *Water-fill test.* The completed hot- or chilled-water pipework system is filled with water. It is visually inspected for leaks. The water pressure is increased either by connecting the water main or by connecting an air compressor to increase the test pressure to 5 b or more, depending upon the design working pressure. Leak inspection is conducted for an hour.

When the water-distribution system is proven to be reliable and pressure-tested, it is drained of test water, flushed with clean water to remove debris, metal swarf and surplus jointing material, and filled. Large plant is descaled and cleaned internally by the circulation of chemicals in water under cold and hot conditions. The filled system is ready for boiler or refrigeration plant operation.

The air-distribution system is ready for use when all the components are installed, grilles and dampers are fully open and the building is sufficiently clean to allow air circulation. This will usually be when the internal decorations, floor and ceiling finishes are in place. The initial air circulation will contain dust and debris from within the air ducts, air-handling plant, rooms and false ceilings that avoided the cleaning operation. High-efficiency air filters, terminal mixing boxes, induction units, duct attenuators, finned coils, velocity, pressure and humidity sensors and variable air volume boxes are susceptible to dirt contamination. Air filters used for commissioning may be replaced prior to handover to client. The discharge of dust from supply grilles into conditioned spaces can cause secondary damage or contamination. The initial starting of fans is made when appropriate temporary protective measures are taken.

Tests and commissioning that are conducted during cold weather require attention to possible frost damage. Hot-water air-heating coils are to be either drained or passing the correct hot-water flow rate prior to starting the fresh-air inlet flow. Automatic control systems and automatic fabric air filters are set to manual control. Boiler and refrigeration plant remain off load.

The supply-air fan is started. Switching methods are a direct on-line switch starter, timed soft start that provides a reduced initial voltage, or a variable-frequency inverter control (VFC). A VFC start allows the fan to be accelerated from zero to maximum revolutions per minute during an adjustable time. Run the fan up to speed as gradually as available from the method installed. The extract or recirculation fan is started 30 s or longer after switching on the supply-air fan. This is to avoid overloading the electrical supply with the simultaneous start of two high-current motors. It may be unsafe to run only one fan owing to the creation of high or low duct or room-air static pressures. Damage might be caused to ductwork, room doors or partition walls. The building contractor is kept informed of test operations and advised if structural faults could be caused. Where excessive pressure variations could be generated, the supply and extract air fans are interlocked with a short time delay. The second fan is automatically started after the first fan. Fans should always be started at light load with reduced supply voltage or frequency to reduce wear on the drive belts and bearings, minimize starting current and duct-air pressure surge. Closure of the main supply-air duct damper can be used with a direct on-line starter as this minimizes the electrical power demand.

Activation of the fans is accompanied by checks;

1. anti-vibration mountings and duct air connectors secure;
2. belt drive security;
3. dynamic balance of the rotating items acceptable;
4. electric motor speed and current correct;
5. fan and motor shaft rotation direction;
6. fan rotation speed correct;
7. full-load start;
8. harmonic frequencies from VFC do not cause interference;
9. light-load start;
10. lubricant seals on bearings not leaking;

11. motor and fan bearing temperature correct;
12. running-in period may be several days;
13. spark-free motor operation;
14. starting and stopping controls operational;
15. three-phase circuits have equal current;
16. vibration and noise acceptable.

Frequent starting of electric motors on fans, pumps and refrigeration compressors is to be avoided. The high initial current required, over double the running current, will cause overheating of the motor, belt drive, switch, fuse and circuit breaker, leading to premature failure. Up to six starts per hour may be allowable, if cooling intervals are long enough.

Variable-frequency control of fan motor speed works by digitally reproducing alternating current from the input 50 Hz supply. The output from a VFC is controllable from 0 Hz up to more than 50 Hz. The rotation speed of an a. c. motor is directly proportional to the applied frequency. 50 Hz is 3000 rev/min, but an induction motor has slip, so a maximum of 2900 rev/min is obtainable. 25 Hz produces 50% full speed, 1450 rev/min. The VFC can accelerate and decelerate the fan during a variable time and hold the speed at any value. In doing this, it generates a wide range of frequencies into the kHz and MHz ranges. These harmonic frequencies are injected into the neutral electrical cable and distributed to other parts of the building. These high-frequency electromagnetic emissions can cause radio and electrical interference with other systems. The commissioning process should include their measurement and analysis for secondary effects.

Acceleration meters can be attached to fan and motor bases to measure the vibration frequency and movement. Anti-vibration mountings are designed to absorb a particular frequency and limit the deflection of the fan. The correct functioning of the mountings is checked. Excess vibration that is transmitted to the building structure may produce unacceptable low-frequency noise at some distance from the source. Fans have a critical, resonant, rotation frequency that may be as low as 100 rev/min. Constant rotation at this speed will equalize the forcing and natural vibration frequencies of the rotating mass. The amplitude of vibration can become infinite at resonance, leading to rapid failure of the mountings. It is essential to accelerate the fan through its resonant frequency without pausing. Vibration measurements are particularly important near critical rev/min values.

When the running-in period has been satisfactorily completed, the fan is started at full-load condition with dampers fully open and the regulation activity is commenced.

DUCT AIR-LEAKAGE TEST

Low-pressure class ducts have an internal static pressure limit of ± 500 Pa and a maximum design air velocity of 10 m/s (HVCA, 1982). The test arrangement is shown in fig. 8.1. A test fan supplies air into a completed duct section and maintains the test pressure. The ductwork is inspected for leaks. The leakage rate is equal to the rate of air flowing into the duct through the test fan. An airflow meter (Fig 6.9, 6.10 and 6.11), records the leakage rate. The allowable leakage rate for low pressure category ducts, Q l/s per m^2 of duct surface area, is found from:

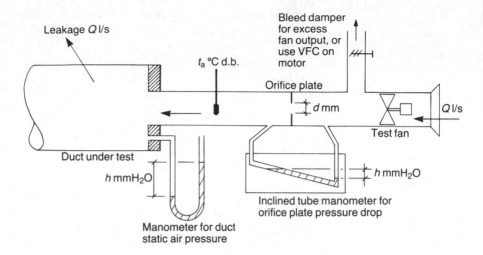

Fig. 8.1   Air-leakage testing of ductwork.

$$Q = 0.027 \, p_s^{0.65} \text{ l/s}$$

where p_s = static air pressure maintained during the test (Pa).

The leakage airflow can be measured by a venturi, orifice or conical inlet flow meter from the equation

$$Q = \frac{C \pi d^2}{4} (2\rho \Delta p)^{0.5} \text{ m}^3/\text{s}$$

where C = flow coefficient, (approximate values are 1 for a venturi, 0.65 for an orifice and 0.95 for a conical inlet), d = orifice or venturi throat internal diameter (m), ρ = air density (kg/m^3), and Δp = pressure drop through meter (Pa).

$$\rho = 1.205 \times \frac{273 + 20}{273 + t} \times \frac{101\,325 + p_s}{101\,325} \text{ kg/m}^3$$

where t = temperature of the air in the test duct (°C).

The test procedure is as follows.

1. Seal the air duct to be tested with inflatable bags, polythene sheets or blank plates inserted at flanged joints.
2. Allow the joint sealant to cure.
3. Air-handling plant such as filters, heater and cooler batteries are not included in leakage tests.
4. Tests may be witnessed by the client's representative.
5. Record all test data on a standard form.
6. Install test equipment and run until stable conditions are found.
7. The leakage rate is to be stable for 15 min.
8. Switch the test fan off.
9. Wait until duct static pressure drops to zero.
10. Immediately switch on the test fan.
11. Verify that the same test results are found.

12. Record all the data, sign, date and witness the forms. A sample test data sheet is shown in Table 8.1.

Table 8.1 Data sheet for sample duct air-leakage test

Air leakage test data	
Test number	1001
Date	8 February 1997
Client	Watt Air plc
Job	South wing VAV, Chilworth office
Contract number	TC/150144
Part 1, installation	
(a) Duct section	nodes 34–88
(b) Ducts shown on drawings	HV/4007
(c) Surface area of duct	65 m^2
(d) Test static air pressure	300 Pa
(e) Leakage equation	$Q = 0.027 p_\mathrm{s}^{0.65}$ l/s
Part 2, test	
(a) Duct air static pressure reading	280 Pa
(b) Duct air temperature	22 °C
(c) Maximum allowable leakage	68 l/s
(d) Airflow measuring device	venturi
(e) Meter flow coefficient C	0.98
(f) Flowmeter range l/s	10–200
(g) Flowmeter serial number	D342
(h) Flowmeter throat diameter	50 mm
(i) Flowmeter calibration certificate	attached
(j) Pressure drop at flowmeter	25 mm H$_2$O
(k) Interpreted duct leakage l/s	47
(l) Switch fan off and repeat	done
(m) Test duration	46 min
Part 3, conclusion	
(a) Result of test	satisfactory
(b) Test engineer	A. N. Smith, Airsure Commissioning Ltd
(c) Witness	I. Care, Watt Air plc

EXAMPLE 8.1

Use the airflow test data shown in Table 8.1 and calculate the results.

Allowable duct leakage is found from

$$Q = 0.027 p_\mathrm{s}^{0.65} \text{ l/s per m}^2 \text{ duct surface area}$$

$$p_s = 280 \text{ Pa}$$

$$\text{duct area} = 65 \text{ m}^2$$

$$Q = 0.027 \times 280^{0.65} \text{ l/m}^2 \text{ s} \times 65 \text{ m}^2$$

$$= 68.38 \text{ l/s}$$

$$\rho = 1.205 \times \frac{273 + 20}{273 + t} \times \frac{101\,325 + p_s}{101\,325} \text{ kg/m}^3$$

$$= 1.205 \times \frac{273 + 20}{273 + 22} \times \frac{101\,325 + 280}{101\,325} \text{ kg/m}^3$$

$$= 1.2 \text{ kg/m}^3$$

$$C = 0.98, d = 0.05 \text{ m}, \Delta p = 25 \text{ mmH}_2\text{O}$$

$$\text{pressure } \Delta p \text{ Pa} = 9.807 \times h \text{ mmH}_2\text{O}$$

$$\Delta p = 25 \times 9.807 \text{ Pa}$$

$$= 245 \text{ Pa}$$

$$Q = \frac{C\pi d^2}{4}(2\rho\Delta p)^{0.5} \text{ m}^3/\text{s}$$

$$= \frac{0.98 \times \pi \times 0.05^2}{4} \times (2 \times 1.2 \times 245)^{0.5} \text{ m}^3/\text{s}$$

$$= 0.047 \text{ m}^3/\text{s}$$

$$= 47 \text{ l/s}$$

AIRFLOW REGULATION

Figure 8.2 illustrates a typical air-duct system with the information needed by the commissioning engineer. The system commences with the balancing dampers and grille deflectors and dampers fully open. The design calls for a specific airflow of 1 m³/s through duct B and 2 m³/s through duct C. The balancing dampers at B and C need to be set to the position that ensures that the correct proportion of the total air supply from duct A passes into each branch. When this is done for each branch, the duct system will be proportionally balanced.

The proportions are

$$Q_A = 3 \text{ m}^3/\text{s, or } 100\%$$

$$Q_B = 1/3 \text{ m}^3/\text{s, or } 33\%$$

$$Q_C = 2/3 \text{ m}^3/\text{s, or } 67\%$$

Whatever airflow is found in duct A, 67% of it must be directed into duct C. Throttling the airflow with dampers B and C will increase the overall system resistance and reduce the airflow at A. Repeated flow measurements and damper adjustment are made until the desired proportions are stabilized. The same operation is carried out at successive locations in an upstream direction.

The correct system airflow can then be achieved by throttling the main damper or changing the fan speed, rev/min. Fan and motor combinations are normally selected to deliver slightly greater airflow than called for at the design operating point. Fan speed is changed by using different pulley diameters on the fan or motor shaft, the slack in the belt drive being taken up with a sliding motor support. The electric motor runs at a constant 2900 rev/min. If the fan pulley is the same diameter as the motor pulley, the fan also runs at 2900 rev/min. This is generally

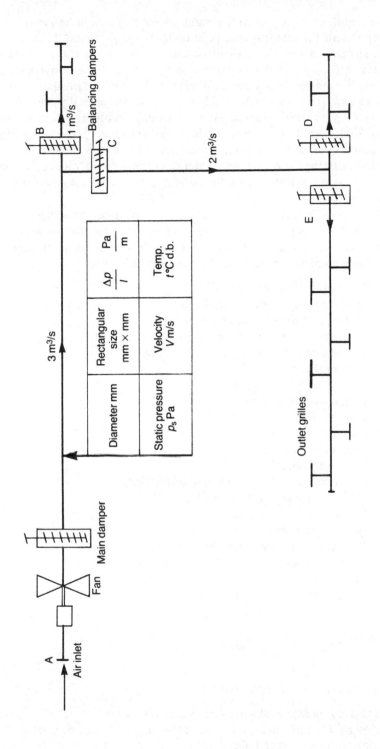

Fig. 8.2 Commissioning information for an air-duct distribution system.

undesirable owing to the aerodynamic noise that would be generated. Fan speeds of up to half the motor speed, 1450 rev/min, are preferred. If the correct duct-system airflow cannot be provided by the fan as installed, then a smaller fan-shaft pulley will increase the fan speed and allow a higher flow rate. If the fan is delivering too much airflow, as is expected, a larger fan-shaft pulley will reduce fan rev/min and the airflow delivered. This is generally the most economical solution. A variable frequency control would simply be set to a frequency lower than 50 Hz until the correct airflow was achieved. Either method of fan-speed reduction saves operational energy consumption and is preferable to throttling the airflow with the main damper. A damper introduces additional frictional resistance into the system that the fan motor has to overcome. This is paid for at the electricity kWh meter.

Air ductwork varies in size from 100 mm diameter thin-sheet galvanized metal up to walk-in room dimensions constructed from builder's-work concrete blocks. Specialist applications may have thick stainless steel ducts for toxic substances in air. The information needed by the commissioning engineer includes:

1. air-duct internal dimensions;
2. air ductwork and plant locations;
3. air static pressure at controlled locations;
4. air temperature in ducts under operational conditions;
5. air volume flow grid locations and type;
6. damper design flow rate and pressure drop;
7. damper locations and type;
8. design airflow rate in each duct;
9. design air velocity in each duct;
10. drive motor speed;
11. duct material thickness;
12. duct test pressure;
13. ductwork materials and method of jointing;
14. electrical supply voltage and frequency;
15. fan design rotation speed;
16. fan type and characteristic curve;
17. filter pressure drop when clean and dirty;
18. location of access hatches into the ductwork;
19. location of test holes;
20. pressure drop through heating and cooling coils.

INSTRUMENTS NEEDED

Rotating-vane anemometer

The rotating vane anemometer (Fig. 8.3) gives a stable mean velocity due to its large diameter (100 mm) and the inertia of the blades. It can be used at the throat of a venturi, or hood, which collects all the air issuing from a grille. This enables one reading of air velocity to be multiplied by the cross-sectional area of the throat to find the total air volume flow rate. Calibration is achieved by placing the anemo-

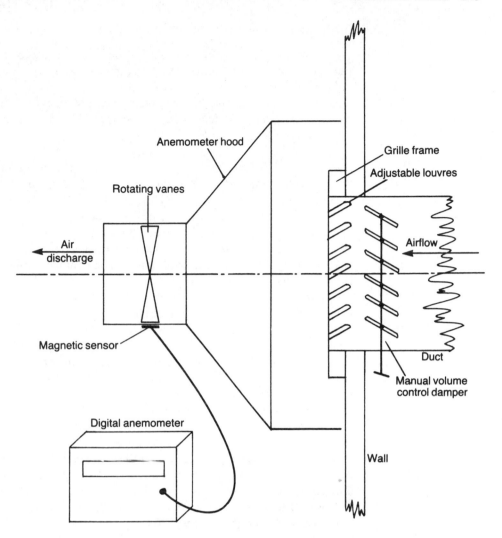

Fig. 8.3   Rotating-vane anemometer and hood for measuring air discharge from a grille.

meter in a jet of air from a test duct. The mean air velocity of the jet is know from pitot-static tube readings within the duct.

Pitot-static tube and manometer

This is used for measuring the air-velocity pressure and static pressure in ducts (Figure 6.1). An inclined-tube liquid manometer can be used and this requires no external power supply or calibration. An electronic micromanometer can provide a wider range of pressure readings, it can have greater accuracy, its output can be sent to a data logger, chart recorder or remote computer and it can be left in place unattended. Static pressure measurements are needed at strategic locations within ductwork, for example either side of fans (Fig. 6.8), at controlled pressure locations and at air inlets to induction and variable air volume units and dual-duct mixing boxes.

Gas detectors

These are used in the commissioning of boilers to measure the constituents of the flue gas. Adjustments are made to the combustion air supply to optimize the boiler efficiency. They also may be employed in leak detection to trace the presence of methane or other combustible or explosive gas. Manual inspection of the insides of large oil or water tanks and underground service tunnels or drains is accompanied by gas detection.

It is unreasonable to expect that airflow can be measured with absolute accuracy by the commissioning team, or that it is necessary. The equipment has tolerance limits. Airflows rarely produce steady velocity readings. The internal surfaces of the duct-work will be far from smooth. The airflows are turbulent and fluctuating. A skilled operator will produce carefully obtained data, but the time necessary to balance out all supposed errors is likely to be uneconomic. It is important that the design airflow is achieved within the conditioned space at the commissioning stage. The minimum is − 0% of the design airflow.

Commissioning is conducted with the air filters in their clean state. Soiling of the filters will provide increased resistance and air-pressure drop. The filters are changed at a predetermined pressure drop. The ductwork system has a greater airflow when the filters are clean. This initial flow gradually reduces as the filter clogs. Some air-conditioning systems have a motor-driven damper to introduce a variable resistance that is compatible with the changes in the filter. This equalizes the filter resistance at an externally constant pressure drop.

The accumulation of errors that is inherent while acquiring airflow data leads to an overall tolerance on the desired flow. It is reasonable to be satisfied with increased airflows in steps of 5%, 10%, 15% and 20% depending upon the designer's criteria and the need to provide an overall satisfactory installation at reasonable cost. Excessive airflow may cause undesirable noise or draught that is noticeable by the users of the conditioned space. A balance needs to be made between these factors and agreed at an early stage (CIBSE, 1971). The designer specifies the air flow tolerance as design Q l/s + 20% − 0%.

GAS DETECTORS

The oxygen content of air is 20.9% by volume, most of the remainder being nitrogen. The presence of unwanted gas may displace or consume oxygen to a dangerously low level. An alarm condition is usually 19% oxygen in air. Leaking oxygen cylinders used for welding in confined spaces can increase the oxygen content of air. This increases the flammability of materials and consequent danger. The presence of toxic gas needs to be detected (Chapter 5). Three types of detector are used, as follows.

Pellistor sensors

These detect flammable gas and vapours. Two coils of platinum wire are each embedded in a bead of alumina. One bead is impregnated with a catalyst that promotes oxidation. An electrical current passes through each coil that raises the bead temperature. When the target gas is present, the heated catalyst causes the gas to be oxidized, that is, combustion takes place. The local combustion further increases the temperature of the coil. Increasing the wire temperature increases its resistance. The voltage applied to the coil circuit remains constant, so the current flowing through the coil reduces. The coil without the catalyst remains unaffected and has a constant current flow. The imbalance between the two electrical circuits is detected in a Wheatstone bridge circuit and an output reading of gas percentage is calibrated on the output display. Pellistor sensors can last in service for years. They can be poisoned by halogens, lead or silicone substances.

Thermal conductivity sensors

Two thermistor beads have a current passing through them. A thermistor is a semi-conducting material made from oxides of manganese and nickel. Their resistance varies with temperature (Encyclopaedia Britannica, 1980). One thermistor is exposed to the gas–air atmosphere while the other is sealed within an air container. They are connected in a Wheatstone bridge circuit. The rate of heat loss from the detector depends upon the thermal conductivity of the gas–air mixture. The imbalance between the two thermistor circuits is a measure of the gas in the atmosphere. Carbon dioxide, helium, hydrogen and methane can be detected by this method.

Electrochemical sensors

These are used to measure toxic gas and oxygen. An electrochemical sensor is a fuel cell having two electrodes and a liquid electrolyte. The sensing head has a permeable membrane covering the electrolyte. Gas diffuses from the atmosphere through the membrane and reacts on the surface of the electrodes to produce a flow of ions and electrons. The ions flow through the electrolyte to the other electrode. The reaction produces a flow of electrical current through the external wiring circuit. The thermal efficiency of fuel cells is around 90% (Encyclopaedia Britannica, 1980). The electrical current produced is directly proportional to the gas concentration being measured and is displayed as percentage gas in air. Oxygen detectors are replaced annually but hydrogen sulphide and other sensors last for up to three years.

Figure 8.4 shows the sensors described and Fig. 8.5 shows a portable gas-detecting instrument.

VENTILATION-RATE MEASUREMENT

The rate of natural ventilation in a building can be measured with the technique described for the leak testing of air ducts. Figure 8.1 shows how a space can be pressurized and the airflow through it measured. A typical test kit for a complete house would include a replacement door with the test fan and airflow meter attached. The building is pressurized to about 25 Pa above the atmosphere and the leakage

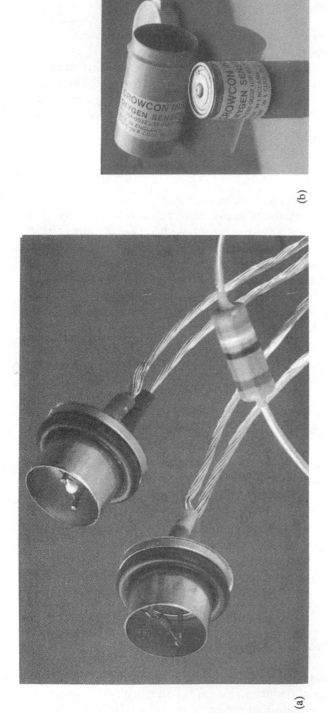

(a)

(b)

(c)

(d)

Fig. 8.4 Sensors for gas detectors: (a) Pellistor sensors; (b) oxygen sensor; (c) hydrogen sulphide sensor; (d) thermal conductivity sensor (courtesy of Crowcon Detection Instruments Limited).

Fig. 8.5 Portable gas detector for oxygen, methane and hydrogen sulphide (courtesy of Crowcon Detection Instruments Limited).

rate measured. This flow represents the natural ventilation rate under normal conditions.

An alternative is to release a tracer gas such as nitrous oxide or helium into the room to achieve a measurable percentage concentration and then switch on the mechanical ventilation system. The concentration of the tracer gas is measured at suitable intervals. The rate of decay of the gas is used to calculate the ventilation rate in air changes per hour or air flow rate (Chadderton, 1991, p. 112).

COMMISSIONING AUTOMATIC CONTROL SYSTEMS

Activation of the air- and water-distribution and electrical systems allows the automatic control equipment to be commissioned. The work necessary includes checking the following.

1. Actuating motors for dampers and valves have full movement.
2. Air and water temperature and pressures are correct.
3. Airflow dampers are lubricated and move fully.
4. Airflow meters are ready for use.
5. All detectors, valves and dampers are numbered and listed.
6. Automatic sequence of switching is correct.
7. Calibration of detectors and flow meters is completed.
8. Controls respond in intended direction and scale.
9. Client is informed of commissioning work.
10. Commissioning drawings and specifications are provided.
11. Commissioning timetable or sequence is published.
12. Control valves and dampers have full movement.
13. Control valves are correctly located.
14. Control valves have the correct pipe connections.
15. Controllers operate over their correct range of values.
16. Correct control is maintained at low flow rates.
17. Data display and storage meet design criteria.
18. Desired states of measured variables can be stabilized
19. Detector locations are marked on drawings.
20. Detectors are not subject to external influences.
21. Electrical interlock, circuit breaker and switches are in place.
22. Electrical items are operational and safe.
23. Electrical systems are operational and safe.
24. Fan-pressure rise is within design tolerance.
25. Full airflow is within design tolerance.
26. Gas, water, electricity and drainage services are functional.
27. Isolating-gate valves are open.
28. Main contractor is informed of commissioning activity.
29. Major building operations are completed.
30. Manufacturers' instructions are complied with.
31. Measuring instruments have calibration certificates.
32. Occupants are informed of commissioning work.
33. Pipework systems are tested and operational.
34. Plant switches on and off at correct settings.
35. Pressure-balancing valves are ready for use.
36. Pressure sensors are operative.
37. Pump-pressure rise is within design tolerance.
38. Schematic system logic drawings are current edition.
39. Sensing elements are sampling representative conditions.
40. Signal transmitters and outstations are functioning.
41. Specification of equipment is correct.
42. Steady conditions within the building can exist for tests.
43. Supervisory computer system is fully functioning.
44. Temperature and humidity detectors are correct range.
45. Temperature, pressure and flow settings are listed.
46. Water flow is within design tolerance.

Where pneumatic control systems are in use, the additional commissioning operations include the following.

47. Air compressors are tested and operational.
48. Air compressors are clean and free of leakage.
49. Air-drying systems are functional.
50. Air-pressure reducing set is clean and operational.
51. Air volume flow rate is within design tolerance.
52. Branch-pipe air pressures are within design tolerance.
53. Compressed-air receiver tank is pressure-tested.
54. Dual-compressor changeover is operational.
55. Electrical power to air compressor is complete and tested.
56. Pneumatic control actuators are functional.
57. Pneumatic pipelines are clear and pressure-tested.
58. Pneumatic pressure is maintained within design tolerance.
59. Pressure switch operates compressor correctly.
60. Safety pressure-relief valves are set and functioning.

COMMISSIONING SEQUENCE

The commissioning work can take from days to months depending upon the system size, complexity and detail needed. A sequence of activity should be published by the design team. This will itemize each step to be undertaken. The designers need to initiate the work as only they are in a knowledgeable position at the time. The formalization of a plan of action will ensure that all interested parties are included.

MAINTENANCE SCHEDULE

Maintenance is organized for regular work at specified intervals and at the expiry of the running hours of specific plant. Scheduled maintenance is the planned execution of replacement and servicing operations. This takes place prior to plant breakdown and is the culmination of knowledge and experience. Lamps, drive belts, electric motors and lubrication fluids are replaced as their service period finishes although they remain operational. Their efficiency will have deteriorated and replacement is desirable or essential. Unscheduled maintenance is due to breakdown. When the building has essential functions such as 24 h computing equipment, manufacturing or health care, standby plant is switched in automatically. In crucial applications, a third plant item is at standby. The first standby is a duplicate fan, pump, boiler or refrigeration compressor that is fitted in parallel with the running item. Changeover between the running and standby plant is made at frequent intervals as part of the maintenance schedule. The third standby requires some manual intervention to bring it into service. It might be an item kept in storage on the site or permanently connected into the circulation system.

 Air-conditioning systems are susceptible to the accumulation of dust, dirt, bacteria and rust within air ducts, air-handling plant and cooling towers. Inspection of the internal surfaces is necessary to ascertain the need for cleaning, repair and disinfection. Contamination of the air passing through plant and ducts can lead to illness, outbreaks of legionnaire's disease and cases of sick building syndrome (Chadderton, 1991, p. 116; Sykes, 1988a, 1988b).

Water aerosols cause the dispersal of micro-organisms and they come from:

1. dehumidifying cooling coils;
2. drainage systems;
3. evaporative humidifiers;
4. fountains and garden sprinklers;
5. piped hot- and cold-water systems;
6. showers;
7. spas, whirlpool baths and therapy pools;
8. spray-washing equipment such as for vehicles and processes;
9. sprayed chilled-water cooling coils;
10. water-spray humidifiers.

Cleaning and disinfection of equipment is carried out:

1. prior to commissioning;
2. after a shut down of five days or more;
3. after system alterations;
4. when the cleanliness of the system is in doubt.

The internal surfaces of air ducts, cooling towers and water storage tanks are cleaned with sprayed detergent solution. Slime, rust and debris are loosened by brushing or grit-blasting and then vacuumed. Cross-contamination from other items is avoided by cleaning all parts of the system simultaneously. Pipework is cleaned with chemical dispersant and then disinfected by circulating chlorine solution for 6 h. Water-storage tanks can have their internal surfaces coated with epoxy resin that is compatible with potable water supplies.

Chlorine is used as the biocide (disinfectant), but it has no detergent-cleansing property. Chlorine reacts with organic matter, ferrous salts and hydrogen sulphide that may be present in water or on wetted surfaces. Its disinfecting ability can be neutralized by them and other biocides Residual chlorine must be present in the disinfecting solution after such reactions have taken place. Sodium hypochlorite solution is used and it contains up to 15% free chlorine. The disinfectant solution is made from sodium hypochlorite to obtain a minimum free residual chlorine concentration of 5 mg/l (ppm).

The cleaned and disinfected system is drained, flushed and filled with treated water for continued use. The water authority is informed before the discharge of chlorinated water into the public sewer. Cooling-tower water is dosed with 30 mg/l of sodium hypochlorite. The water condition is regularly monitored so that it remains unfavourable to the proliferation of micro-organisms (HMSO, 1989).

The maintenance of air-conditioning systems includes the following:

1. air-filter material cleaning or replacement;
2. air-filter pressure drop monitoring;
3. air-intake louvres clear;
4. bearing wear assessed on rotary plant;
5. building energy management computer system diagnostic check;

6. chemical dosing of water measured;
7. clean and disinfect cooling towers;
8. clean and disinfect water-storage tanks;
9. cleaning surfaces of air- and water-distribution systems;
10. clean surfaces of electric motors and fans;
11. combustion efficiency measurements on boiler plant;
12. compressor pressure and temperature controls working;
13. cooling-tower packing in good condition;
14. corrosion check made on all metalwork;
15. disinfection of air- and water-distribution systems;
16. electric motor carbon brush replacement;
17. electric motor full-load current measured;
18. electrical insulation resistance inspected and tested;
19. electrical switches and circuit breakers operational;
20. fan-belt drive tension adjustment;
21. float valves operational;
22. flowmeters calibrated;
23. hours-run meter readings taken;
24. investigate vibration and noise sources;
25. leakage from pipes or plant assessed for repair;
26. lubricate and test air duct dampers;
27. lubrication of electric motor bearings;
28. lubrication of fan bearings;
29. manufacturers' maintenance instructions followed;
30. no loose bolts, supports or plant connections;
31. operation and maintenance documents completed;
32. outstation telecommunications functioning;
33. overflow and drain systems clear;
34. paintwork and corrosion protective finishes in place;
35. pipeline strainer cleaning;
36. pressure sensors calibrated;
37. pressures of air, water and refrigerant measured;
38. records of maintenance and alterations up to date;
39. refrigerant pressures and temperatures correct;
40. room-terminal grilles clean and quiet;
41. standby pump and fan changeover to running operation;
42. standby refrigeration, boiler, fan and pump tested;
43. steam humidifier descaled;
44. storage of spare parts and chemicals checked;
45. temperature and humidity sensors calibrated;
46. temperature, pressure and humidity transmitters operational;
47. valve operation verified, motor-driven and manual;
48. valve-packing gland leakage checked;
49. visual inspection of thermal insulation.

Questions

1. A leakage test on completed ductwork revealed the following data: duct static pressure 45 mmH$_2$O, duct surface area 120 m^2, duct air temperature 12 °C, orifice-plate flow coefficient 0.67, orifice throat diameter 70 mm, orifice pressure drop 160 mmH$_2$O. Calculate whether the

duct system meets the maximum leakage criteria when it is calculated from $0.027\,p_s^{0.65}$ l/s per m^2 duct surface area.

2. A commissioning engineer is to carry out a leakage test on a section of completed ductwork. On the day of test, the duct surface area is measured as 93 m^2, duct air temperature 14 °C, orifice-plate flow coefficient 0.68, orifice throat diameter 50 mm and orifice pressure drop 230 mmH$_2$O. Calculate the duct static pressure, h mmH$_2$O, that must be maintained by the test fan in order for the duct system to meet the maximum leakage criteria when it is calculated from $0.027\,p_s^{0.65}$ l/s per m^2 duct surface area.

3. A leakage test on a section of completed ductwork with a surface area of 260 m^2, duct air temperature 18 °C, venturi flow coefficient 0.97, throat diameter 60 mm and venturi pressure drop is 190 mmH$_2$O. The duct static pressure is held at 62 mmH$_2$O by the test fan. Calculate whether the duct system meets the maximum leakage criteria when it is calculated from $0.009\,p_s^{0.65}$ l/s per m^2 duct surface area.

4. Write a complete schedule for the commissioning work necessary on the air-handling equipment only for a single-duct air-conditioning system serving a lecture theatre. The system is similar to that shown in Fig. 1.1. Heat is provided from a low-pressure hot-water two-pipe system from a boilerhouse. Cooling is provided by a chilled-water two-pipe system.

5. Explain the difference between scheduled and unscheduled maintenance work on an air-conditioning system. State the items that will require replacement, their likely length of normal service and whether they should be held in storage on the site.

6. Describe how failure of plant during normal use is overcome to maintain the air-conditioning service.

7. Discuss the approach to a suitable maintenance programme for the air conditioning in the following applications:
   (a) a large general hospital;
   (b) a university;
   (c) residential buildings in the UK;
   (d) office accommodation in Brisbane;
   (e) a manufacturing building including the containment of biological and radioactive materials;
   (f) comfort control for the workplace in the UK.

8. A refrigeration condenser is cooled with water that is circulated to an evaporative cooling tower on the roof of a hospital in a city. State the actions that are taken to ensure that the tower operation will not cause health hazards to the patients and the public.

9. Design a maintenance log for a city office building

that has low-pressure hot-water heating and a single-duct heating and ventilation system. There is no refrigeration plant. Maintenance work is to be carried out by contractors. The office users will allow some interruption of the heating and ventilating systems during 0900–1700 h for repairs.

10. Discuss the reasons for using standby plant in an air-conditioned building. Include the implications on plantroom size, storage commitment, stock control, capital and recurring costs for the user.

11. List the frequency of visual checks, physical measurement, changeover of running plant and planned replacement of the items and systems in an air-conditioned building of your choice from the types given in question 7.

12. Find the maintenance records for a building that is accessible to you. This may be a residence, office, factory, warehouse, college or university. Request the help and cooperation of the professionally employed maintenance engineering staff where this is applicable. Write a report on the comprehensiveness of the records that are kept. Make appropriate recommendations as to improvements that should be made.

13. Compare the quality of maintenance work and its recording when it is conducted by employees of the organization that uses the site, and contract companies. This may involve the acquisition of information from several sources that are dissimilar. Discuss the advantages and the budgetary control implications of each method.

14. Draw a schematic diagram of part of an air-conditioning system. Show the location of all the necessary valves, dampers and controls. Number all the controls and test points and produce a schedule of their data. Write on the diagram all the data that will be needed by the commissioning and maintenance engineers.

15. Design a maintenance log for the refrigeration plant of an air-conditioning system that serves a 10 000 m^2 floor area complex of offices and computer-manufacturing facilities. Evaporative cooling towers are located on the roof. Refrigeration plant is in a ground-floor plantroom. Each room has a chilled-water cooling coil and local temperature control. Make any assumptions about the systems that are necessary. State the intervals between servicing and replacement work.

16. Acquire manufacturers' literature that demonstrates the use of computer-screen system-logic diagrams for air-conditioning systems. Note how the air circulation, detectors and control systems are represented. Choose a different type of air conditioning system form Chapter 1 and create an equivalent diagram. List all the data points to be used, with sample data.

17. Use the solution to question 4, 14 or 16 and list all the points that are to be connected into the automatic control system. A point is where a detector, switch, control panel, outstation, modem or other item is wired into the automatic control system. The control, commissioning and maintenance engineers need to know this information.

18. Explain the advantages gained when the original designer of an air-conditioning system publishes the schedules for commissioning and maintenance work. State the information that is included and the form that the documentation should take. State how a computer-based system can aid good maintenance practice.

19. List the order in which each part of an air-conditioning system will be commissioned. State the condition required of the building works for each stage of commissioning.

20. Explain the following:
    (a) why frequent starting of fans and pumps is to be avoided;
    (b) why the internal surfaces of air ducts, cooling towers, water storage tanks and water circulation pipework systems need to be cleaned and disinfected;
    (c) how cleaning and disinfection work is conducted;
    (d) the methods available for the starting of the electrical motor drive on a fan;
    (e) why the first installed air filters will be temporary;
    (f) proportional balancng;
    (g) the need for vibration measurements;
    (h) harmonic interference from electrical equipment;
    (i) fan-speed regulation;
    (j) full-load start;
    (k) why and where gas detectors are used;
    (l) tolerance limits;
    (m) how cross-contamination between services is avoided;
    (n) where water aerosols originate;
    (o) what reduces the effectiveness of chlorine dosing and disinfection.

21. List the instruments that may be needed to commission an air-conditioning system. Describe, with the aid of sketches, the operating principle and method of use of each instrument.

9 Fans and systems

INTRODUCTION

The types of fans and their applications are introduced. Fan and system characteristic curves are demonstrated through the use of manual and spreadsheet calculations. Spreadsheet calculation files are listed. Fan performance calculations are explained with worked examples. Fan starting and the methods of controlling the air delivery from fan and ductwork systems are discussed. Energy consumption is calculated for the different methods of control. Commissioning and maintenance procedures for fan installations are listed.

LEARNING OBJECTIVES

Study and use of this chapter will enable the user to:

1. know the types of fans used in building services;

2. apply fan types correctly;

3. understand the use of fan plenums;

4. calculate the opening force on a plenum door;

5. understand the use of fan characteristic performance curves;

6. calculate fan pressure and power curves manually and with a spreadsheet;

7. generate fan characteristic curves manually and on a spreadsheet;

8. understand and calculate the fan motor power over a range of air flows;

9. know why the control of fan pressure may be needed;

10. know the methods that are used to control the performance of a fan;

11. use the fan laws to predict the performance of fans under different conditions;

12. calculate the characteristic resistance of a ductwork system and plot it on a fan performance graph;

13. find the fan and system operating point;

14. know the fixed motor speeds that are used;

15. calculate and plot fan performance data for different fan speeds;

16. know the results of using fans that are connected in series;

17. know the components of the electrical power used by a fan;

18. know the methods of starting and controlling fans;

19. analyse the energy costs for the different methods of controlling the air flow in a duct system, manually and with a spreadsheet;

20. know the methods used to protect electric motors;

21. know the advantage of running electric motors at varying speeds;

22. know the commissioning and maintenance procedures for fan and air-ductwork installations.

Key terms and concepts

Key terms and concepts

FAN TYPES

Fans are the prime movers for air-conditioning and ventilation systems. They are used to exhaust vitiated air and flue gases, as well as to generate the required air movement within the occupied spaces of the building. They are a source of acoustic energy and the possible creation of noise for the occupants of the treated building and for those outside it. Fans are rotodynamic machines that are used to generate low to medium increases in the pressure of air or other gases. The passage of each blade of the fan impeller imparts a pulse of energy to the air flow. The number and geometry of the blades, and the rotational speed of the impeller, determine the frequency of the pulsations in the air stream and the noise that is produced. The linear speed of the outer tip of the fan impeller is a factor in the material design of the rotating components.

The impeller blade shape, complexity, dimensions, constructional material, rotational speed and the shape of the enclosure all depend upon the performance that is required and the application of the fan. Low-cost fans are formed from cut sheets of flat steel. Their blades may be flat or slightly curved. These appear in domestic appliances and in vehicle cooling and ventilation. High-cost fans have aerofoil cross-sectional blades, a high-pressure volute casing, they can be run at high speeds and they are used in high air-pressure and air-flow applications. Blade types are shown on Fig. 9.1.

The fan types and their applications are as follows.

1. Propeller – where there is a low frictional resistance and high air volume flow rate system requirement, such as exhaust fans and outdoor air cooled heat exchangers. Fig. 9.2.

2. Axial: ducted air and hot gas exhaust systems – where the fan has to be in line with the ductwork owing to space restriction, and the pressure rise is limited. Fig. 9.3.
3. Forward-curved centrifugal: packaged fan coil, air-handling and air-conditioning units; roof-mounted exhausts; low-cost ducted supply and exhaust ventilation, air conditioning. Fig. 9.4.
4. Backward-curved centrifugal: large duty ventilation and air conditioning with high frictional resistance and air flow.
5. Mixed flow – where the fan casing is to be in line with the ductwork but a higher fan pressure than can be produced by an axial fan is required. Fig. 9.5.

Centrifugal fans are installed as part of packaged ventilation and air-conditioning equipment such as room air-conditioners, heating and cooling fan-coil units, prefabricated air-handling units, and outdoor dry and evaporative heat exchangers. Where a high pressure and air-flow duty is needed in extensive ducted systems, the centrifugal fan is usually free-standing installed within a plant room. In large air-flow applications the fan plant room also serves as a path for the air flow. In this case, the fan plant room becomes a plenum chamber that is maintained at a different static pressure than that of the atmosphere. A fan plenum chamber may be constructed from concrete and brickwork as part of the building structure. The building materials that are used, the quality of the surface finishes and the air tightness of the plenum are carefully controlled to ensure reliable long-term performance and low mainten-

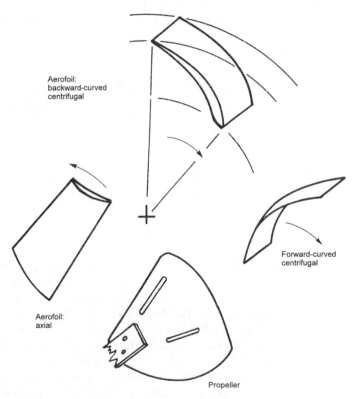

Fig. 9.1 Types of fan blade.

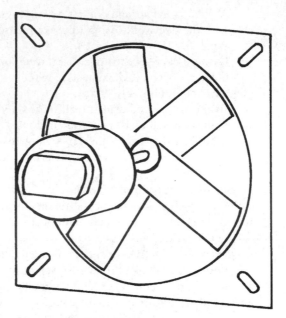

Fig. 9.2   Propellor fan.

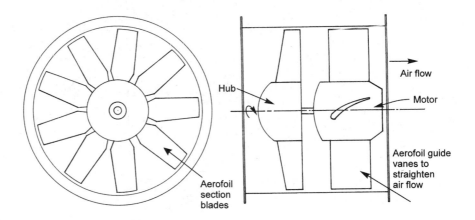

Hub

Air flow

Motor

Aerofoil
section
blades

Aerofoil guide
vanes to
straighten
air flow

Fig. 9.3   Axial-flow fan.

ance requirements. Large builders work plenums and air ducts for filters and
coils have been used for television studios and swimming-pool ventilation plants.
The centrifugal fan, motor and belt drive for metal ducted systems may be installed
within an acoustic plenum. The inlet and outlet air ducts in the plenum are fitted
with noise attenuators. Manual access into a fan plenum is usually barred while the
fan is in operation. The plant control system will include the facility of switching the
fan, or fans, off while the plant operator and the maintenance engineer are still
outside the plenum. An air-tight access door is provided into a fan plenum. The
door is designed to maintain the desired attenuation, air sealing and a safe means
of access.

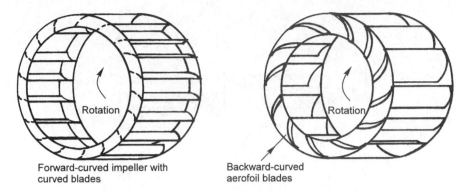

Forward-curved impeller with curved blades

Backward-curved aerofoil blades

Fig. 9.4 Forward- and backward-curved centrifugal fan impeller.

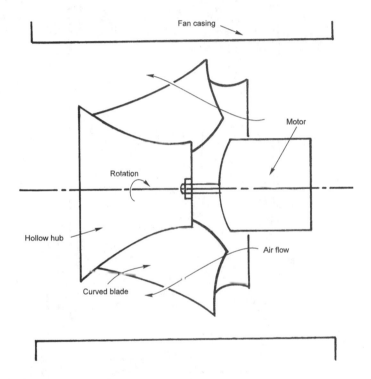

Fig. 9.5 Mixed-flow fan.

When manual access is allowed into a fan plenum while the fan is running, safeguards are required. These include interior illumination, interior light switch, emergency and exit lighting, warning notices, ear defenders, eye protection, protective overalls, covers to all rotating components, local fan-motor isolating switch, fan-motor interlock contactor to switch the fan off if the plenum door remains open for, say, 30 seconds, and a limit on the manual force that is required to open, or close, the door against the air-pressure difference. A plenum door that is being sucked shut by a negative air static pressure within the enclosure can require a significant opening force, will not remain open naturally while the maintenance person is carrying tools, replacement drive belts and

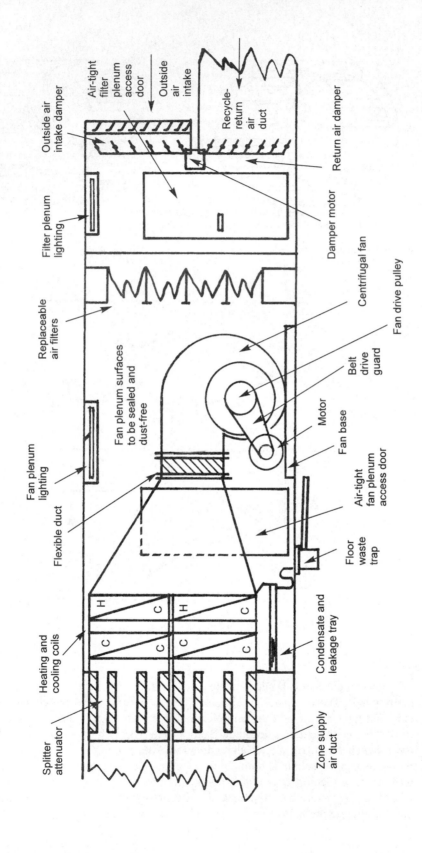

Fig. 9.6 Components of a supply air fan plenum.

air filter panels through the doorway, and can slam shut and entrap the person, or parts of the person, as well as being difficult to open from the inside. There is a danger that loose items will be sucked into the fan, drive belt, air filter or coil during maintenance work or inspections. Good, safe practice, is to allow access to pressurised plant rooms and plenums only when the plant can be switched off. Figure 9.6 shows the components of a fan plenum.

EXAMPLE 9.1

A backward-curved centrifugal fan maintains a static air pressure of 100 Pa below that of the atmosphere, within the fan acoustic plenum. An access door of 750 mm width and 2000 mm height allows entry for regular maintenance of the air filters and drive belt. The door is hinged along one vertical side and has a handle in the opposite side. Calculate the manual force that is necessary to open the door.

The air-pressure difference creates a force on the door

$$\text{force } F_1 \text{ N} = \text{pressure } p \text{ Pa} \times A \text{ m}^2$$

$$A = 0.75 \text{ m} \times 2 \text{ m}$$

$$= 1.5 \text{ m}^2$$

$$p = 100 \text{ Pa}$$

$$1 \text{ Pa} = 1 \text{ N/m}^2$$

$$F_1 = p \text{ N/m}^2 \times A \text{ m}^2$$

$$= 100 \times 1.5 \text{ N}$$

$$= 150 \text{ N}$$

The force, F_1 N, on the door that is created by the difference of air pressure acts equally over the surface area of the door. F_1 will be considered to act at the centre of the door. When an opening force, F_2 N, is applied to the door handle, the person is applying an opening torque, T_O N m, that acts on the door hinge. This opening torque is

$$\text{Opening torque } T_O \text{ N m} = \text{manual force } F_2 \text{ N} \times \text{distance } L_2 \text{ m}$$

$$L_2 = 0.75 \text{ m}$$

The opening torque must equal the closing torque, overcome the frictional resistance of the hinge, move the door away from its perimeter air-tight seals and move the door against the wind-induced drag force. The simple closing torque represents the minimum torque that is needed to open the door.

$$\text{Closing torque } T_c \text{ N m} = \text{closing force } F_1 \text{ N} \times \text{distance } L_1 \text{ m}$$

$$T_o = T_c$$

$$L_1 = 0.375 \text{ m}$$

$$T_c = 150 \text{ N} \times 0.375 \text{ m}$$

$$= 56.25 \text{ N m}$$

$$T_o = F_2 \text{ N} \times 0.75 \text{ m}$$

$$56.25\,\text{N m} = F_2\,\text{N} \times 0.75\,\text{m}$$

$$F_2 = \frac{56.25\,\text{N m}}{0.75\,\text{m}}$$

$$= 75.0\,\text{N}$$

$$= 75.0\,\text{N} \times \frac{1\,\text{kg}}{9.81\,\text{N}}$$

$$= 7.65\,\text{kg}$$

This is equivalent to the force required to lift an 8 litre bucket that is full of water, but pulling horizontally, while the inward rush of air is sucking the door shut, the engineer is attempting to hold the door open, switch on the lights, carry equipment into the plenum and, once inside, hold the door ajar to stop it slamming shut and trapping a hand, foot or filter frame. The designer is obligated to avoid such unsafe practice.

FAN CHARACTERISTIC CURVES

Characteristic performance curves are used in graphical form as the result of tests on geometrically similar fans. The performance of a fan is specified by the pressure rise that it generates at the volume flow rate of air that is passing through it. Typical curves of fan total, static and velocity pressures, fan motor power and a duct-system resistance for propeller and axial-flow fans are shown in Figs 9.7 and 9.8. Figure 9.9 is the listing of the spreadsheet cell formulae that were used to generate Fig. 9.8 for an axial-flow fan. The formulae and data can be entered into a spreadsheet program and saved as file named AXIAL.WKS. Format the column widths to ten (10)

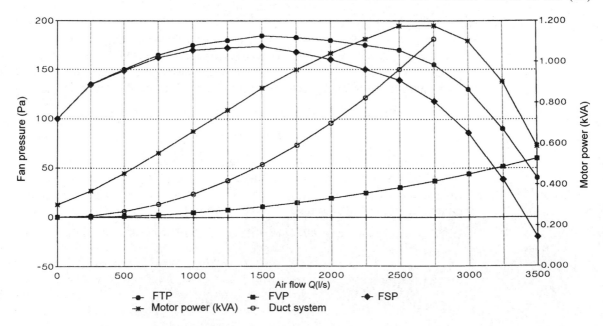

Fig. 9.7 Propellor fan, fan speed 24 Hz.

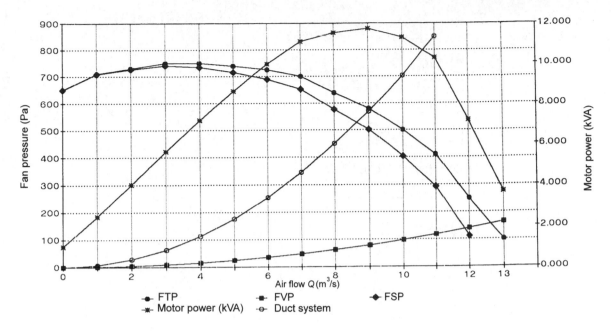

Fig. 9.8 Axial fan, fan speed 24 Hz.

characters for the whole worksheet. Copy the repeated lines as ranges of cells, rather than typing each new formula, and then check the cells against the published list. A chart is produced by selecting the correct ranges of data for each axis; Q m^3/s for the X axis, FTP for the $Y$1 axis, FVP for the $Y$2 axis, FSP for the $Y$3 axis, duct-system resistance for the $Y$4 axis and motor power for the right-hand $Y$5 axis. The user selects the chart titles, legends, line-graph format, font formats, border, grid lines, line colours and markers, depending upon the software being used.

Figures 9.10 to 9.12 show typical fan and system performance curves for forward-curved centrifugal, backward-curved centrifugal and mixed-flow fans. The spreadsheet list that is provided for an axial fan can be saved into different file names such as PROP.WKS for the propellor fan, FORCENT.WKS for the forward-curved centrifugal fan, BACKCENT.WKS for the backward-curved centrifugal fan and MIXED.WKS for the mixed-flow fan. Each of these files can then be edited with the data for new fan types for examples and questions within this book, assignments or design office work.

Notice on the charts that when the fan total pressure coincides with the fan velocity pressure, the fan static pressure is zero. This point defines the end of the fan performance curve. All the fan pressure development is being used to generate kinetic energy and there is no pressure available to overcome duct resistance. The commencement of the fan total pressure curve is not so easy to define. When a fan is operating against a closed and air-tight dampered duct, there is no air flow delivered by the fan. Air only recirculates within the fan casing. No power is being delivered to the air system as the product of Q l/s and FTP Pa is zero. The driving motor is consuming power to overcome the overhead losses, as described later. Heat that is generated by the electric motor when it is within the air stream in a direct-drive arrangement, the air compression by the impeller and the friction within the shaft

```
A1:   (G)  'Fan Curve
B1:   (G)  'for:
C1:   (G)  'Axial
D1:   (G)  'Fan
A2:   (G)  '**********
B2:   (G)  '**********
C2:   (G)  '**********
D2:   (G)  '***
A3:   (G)  'Date:
B3:   (G)  '15/5/96
C4:   (G)  'INPUT DATA
E4:   (G)  'CALCULATED
C5:   (G)  '—————
E5:   (G)  '—————
A6:   (G)  'Fan speed
C6:   (F0) 1440
D6:   (G)  'RPM
E6:   (F0) +C6/60
F6:   (G)  'Hz
A7:   (G)  'Outlet vel
B7:   (G)  'ocity V =
C7:   (F1) 10
D7:   (G)  'm/s
A8:   (G)  'Outlet dia
B8:   (G)  'meter D =
C8:   (F0) 1000
D8:   (G)  'mm
A9:   (G)  'Blade angl
B9:   (G)  'e     B =
C9:   (F0) 20
D9:   (G)  'degrees
A10:  (G)  'Outlet are
B10:  (G)  'a     A =
C10:  (F3) @PI*(C8/1000)^2/4
D10:  (G)  'm2
A11:  (G)  'Air Densit
B11:  (G)  'y   rho =
C11:  (F2) 1.2
D11:  (G)  'kg/m3
A12:  (G)  'System des
B12:  (G)  'ign    Q =
C12:  (F0) 8000
D12:  (G)  'l/s
A13:  (G)  'FTP requir
B13:  (G)  'ed   FTP =
C13:  (F0) 450
D13:  (G)  'Pa
A14:  (G)  'Motor
B14:  (G)  '     PF =
C14:  (F2) 0.9
D14:  (G)  '    Power
E14:  (G)  'factor
A15:  (G)  'Motor
B15:  (G)  '     El =
C15:  (F0) 60
D15:  (G)  '%   effici
```

Fig. 9.9   (pp. 266–269).

E15: (G) 'ency
A16: (G) 'Impeller
B16: (G) '   E2 =
C16: (F0) 90
D16: (P0) '%   mean v
E16: (G) 'alue in op
F16: (G) 'erating ra
G16: (G) 'nge
A17: (G) 'Minimum po
B17: (G) 'wer   P   =
C17: (F2) 1
D17: (G) 'kVA at zer
E17: (G) 'o air flow
A21: (G) '   Q m3/s
B21: (G) '   Q l/s
C21: (G) '   Q m3/s
D21: (G) 'FTP Pa
E21: (G) 'V m/s
F21: (G) 'FVP Pa
G21: (G) 'FSP Pa
H21: (G) ' System Pa
I21: (G) ' Motor kVA
A22: (G) '   *****
B22: (G) '   *****
C22: (G) '   ******
D22: (G) '   ******
E22: (G) '   ****
F22: (G) '   ******
G22: (G) '   ******
H22: (G) ' *********
I22: (G) ' *********
A24: (F0) 0
B24: (F0) 0
C24: (F3) +B24/1000
D24: (G) 650
E24: (F2) +C24/$C$10
F24: (F1) 0.5*$C$11*E24^2
G24: (F0) +D24-F24
H24: (F0) +$C$13*(B24/$C$12)^2
I24: (F3) +$C$17+B24*D24/($C$14*$C$15*$C$16*100)
B25: (F0) 1000
C25: (F3) +B25/1000
D25: (G) 710
E25: (F2) +C25/$C$10
F25: (F1) 0.5*$C$11*E25^2
G25: (F0) +D25-F25
H25: (F0) +$C$13*(B25/$C$12)^2
I25: (F3) +$C$17+B25*D25/($C$14*$C$15*$C$16*100)
A26: (F0) 2
B26: (F0) 2000
C26: (F3) +B26/1000
D26: (G) 730
E26: (F2) +C26/$C$10
F26: (F1) 0.5*$C$11*E26^2
G26: (F0) +D26-F26
H26: (F0) +$C$13*(B26/$C$12)^2
I26: (F3) +$C$17+B26*D26/($C$14*$C$15*$C$16*100)

Fig. 9.9   *cont'd.*

```
B27:  (F0)  3000
C27:  (F3)  +B27/1000
D27:  (G)   750
E27:  (F2)  +C27/$C$10
F27:  (F1)  0.5*$C$11*E27^2
G27:  (F0)  +D27-F27
H27:  (F0)  +$C$13*(B27/$C$12)^2
I27:  (F3)  +$C$17+B27*D27/($C$14*$C$15*$C$16*100)
A28:  (F0)  4
B28:  (F0)  4000
C28:  (F3)  +B28/1000
D28:  (G)   750
E28:  (F2)  +C28/$C$10
F28:  (F1)  0.5*$C$11*E28^2
G28:  (F0)  +D28-F28
H28:  (F0)  +$C$13*(B28/$C$12)^2
I28:  (F3)  +$C$17+B28*D28/($C$14*$C$15*$C$16*100)
B29:  (F0)  5000
C29:  (F3)  +B29/1000
D29:  (G)   740
E29:  (F2)  +C29/$C$10
F29:  (F1)  0.5*$C$11*E29^2
G29:  (F0)  +D29-F29
H29:  (F0)  +$C$13*(B29/$C$12)^2
I29:  (F3)  +$C$17+B29*D29/($C$14*$C$15*$C$16*100)
A30:  (F0)  6
B30:  (F0)  6000
C30:  (F3)  +B30/1000
D30:  (G)   725
E30:  (F2)  +C30/$C$10
F30:  (F1)  0.5*$C$11*E30^2
G30:  (F0)  +D30-F30
H30:  (F0)  +$C$13*(B30/$C$12)^2
I30:  (F3)  +$C$17+B30*D30/($C$14*$C$15*$C$16*100)
B31:  (F0)  7000
C31:  (F3)  +B31/1000
D31:  (G)   700
E31:  (F2)  +C31/$C$10
F31:  (F1)  0.5*$C$11*E31^2
G31:  (F0)  +D31-F31
H31:  (F0)  +$C$13*(B31/$C$12)^2
I31:  (F3)  +$C$17+B31*D31/($C$14*$C$15*$C$16*100)
A32:  (F0)  8
B32:  (F0)  8000
C32:  (F3)  +B32/1000
D32:  (G)   640
E32:  (F2)  +C32/$C$10
F32:  (F1)  0.5*$C$11*E32^2
G32:  (F0)  +D32-F32
H32:  (F0)  +$C$13*(B32/$C$12)^2
I32:  (F3)  +$C$17+B32*D32/($C$14*$C$15*$C$16*100)
B33:  (F0)  9000
C33:  (F3)  +B33/1000
D33:  (G)   580
E33:  (F2)  +C33/$C$10
F33:  (F1)  0.5*$C$11*E33^2
G33:  (F0)  +D33-F33
```

```
H33:  (F0)  +$C$13*(B33/$C$12)^2
I33:  (F3)  +$C$17+B33*D33/($C$14*$C$15*$C$16*100)
A34:  (F0)  10
B34:  (F0)  10000
C34:  (F3)  +B34/1000
D34:  (G)   500
E34:  (F2)  +C34/$C$10
F34:  (F1)  0.5*$C$11*E34^2
G34:  (F0)  +D34-F34
H34:  (F0)  +$C$13*(B34/$C$12)^2
I34:  (F3)  +$C$17+B34*D34/($C$14*$C$15*$C$16*100)
B35:  (F0)  11000
C35:  (F3)  +B35/1000
D35:  (G)   410
E35:  (F2)  +C35/$C$10
F35:  (F1)  0.5*$C$11*E35^2
G35:  (F0)  +D35-F35
H35:  (F0)  +$C$13*(B35/$C$12)^2
I35:  (F3)  +$C$17+B35*D35/($C$14*$C$15*$C$16*100)
A36:  (F0)  12
B36:  (F0)  12000
C36:  (F3)  +B36/1000
D36:  (G)   250
E36:  (F2)  +C36/$C$10
F36:  (F1)  0.5*$C$11*E36^2
G36:  (F0)  +D36-F36
H36:  (F0)  +$C$13*(B36/$C$12)^2
I36:  (F3)  +$C$17+B36*D36/($C$14*$C$15*$C$16*100)
A37:  (G)   13
B37:  (F0)  13000
C37:  (F3)  +B37/1000
D37:  (G)   100
E37:  (F2)  +C37/$C$10
F37:  (F1)  0.5*$C$11*E37^2
G37:  (F0)  +D37-F37
H37:  (F0)  +$C$13*(B37/$C$12)^2
I37:  (F3)  +$C$17+B37*D37/($C$14*$C$15*$C$16*100)
```

Fig. 9.9 Spreadsheet cell contents for an axial-flow fan characteristic curve

arrangement, the air compression by the impeller and the friction within the shaft bearings, is not dissipated, and the temperature within the fan casing will rise. Operating a fan against a closed damper could be used as a means of controlling the air flow and static pressure within the duct system during start up for short time periods. It may be necessary to control static pressure within the ventilated room during the starting of high-pressure supply and exhaust fans, for example, where room air contaminants must be contained within the ventilated spaces through the room air static pressures. However, for the cost of a modulating damper, duct air-pressure detector, damper motor and automatic controller, an electrical soft starter or a variable-speed drive provides a higher quality performance and energy savings, when lower speeds are used.

At zero fan air delivery, the fan velocity pressure is zero, while the fan total and static pressures coincide. Increasing the system air flow by opening the fan outlet damper or, more likely, the system-balancing dampers, produces the general performance curves shown in Fig. 9.1. The fan supply air-flow region between zero and

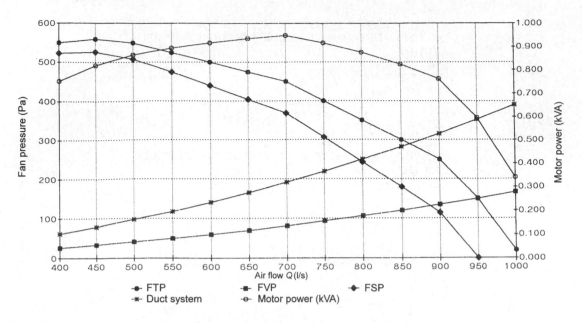

Fig. 9.10 Forward-curved centrifugal fan speed 12 Hz.

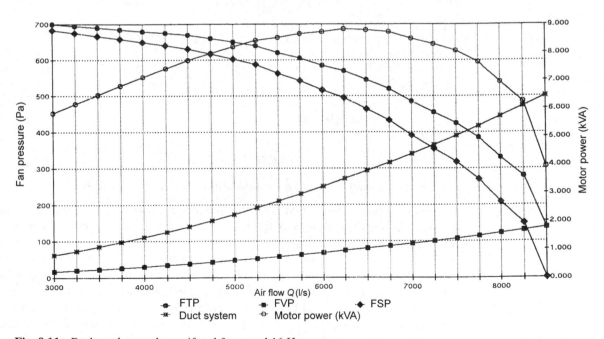

Fig. 9.11 Backward-curved centrifugal fan speed 16 Hz.

that at the peak of the fan total pressure curve is not normally used for fan selection. The impeller efficiency is low in this region and fan operation can be unstable. The central portion of the highest fan total pressure curve is used for the selection of a fan to meet the duct-system design. The fan-motor power curve is termed as being non-overloading when a peak power consumption is reached within the operational

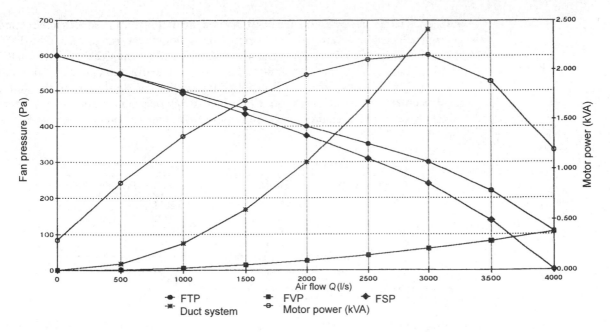

Fig. 9.12 Mixed-flow fan, fan speed 16 Hz.

range of system air flow, as shown in Fig. 9.1. An overloading fan-motor power characteristic is where the power continues to increase as air flow increases. When a fan has an overloading power curve, it may be possible to overload the driving motor and cause the excess current circuit-breakers or the motor thermal protection to operate.

Manufacturers predict the pressure output of a new size of fan from a knowledge of the performance of geometrically similar models that have been tested. Curves of pressure against volume flow are scaled upwards and downwards depending upon the rotational speed of the new fan. New fan models are subsequently tested to confirm their predicted performance.

The selection of a suitable fan is made by plotting the ductwork-system characteristic curve on to the fan performance curve. The intersection of these curves shows where the fan and system will operate together. A graphical enlargement of the cross-over region allows the designer to make any tolerance adjustments that are desired. The operating point is selected to maximize the fan-impeller efficiency.

The curves of fan total pressure, FTP Pa, against air volume flow rate, Q l/s, correspond to a smooth polynomial curve. FTP is the fan total pressure rise in Pascal and Q is the air volume flow rate delivered by the fan in l/s. The use of mathematical models of fan characteristic curves is demonstrated on spreadsheets and charts (Chadderton, *Building Services Engineering Spreadsheets*). These allow the fan speed to be changed. The system characteristic curve is used to locate the operating point of the installed fan and duct system.

The highest practical motor rotational speed on a 50 Hz electrical power supply is normally 2900 rev/min. 50 Hz corresponds to 3000 rev/min. The 60 Hz supply that is used in some countries corresponds to 3600 rev/min. There is always a slip between the alternating current frequency in the stator of the motor and the rotor shaft speed. Fans are either directly driven from the motor shaft or with a v-belt drive running on pulleys. A larger pulley on the fan shaft than on the motor shaft produces a reduction gearing effect. Fan rotational speeds are usually within the range of 350 to 1450 rev/min in order to minimize the generation of noise and mechanical stress within the rotating components. Some manufacturers show curves for fan speeds up to 4000 rev/min. These have a smaller diameter pulley on the fan shaft than on the motor shaft in order to raise the gear ratio. Up to 100 dB can be produced in the outlet duct at such speeds and attenuators may be required. The normal range of fan speeds is shown in Table 9.1.

Table 9.1 Fan speeds

No. of poles in motor	Motor speed (Hz)	Motor speed (rev/min.)
2	48	2880
4	24	1440
6	16	960
8	12	720

FAN DATA CALCULATION

The equations that are used to produce the fan and duct system characteristic curves are:

Fan outlet area A m^2,

$$A = \text{width W mm} \times \text{height } H \text{ mm} \times \frac{1 \text{ m}^2}{10^6 \text{ mm}^2}$$

$$= \frac{W \times H}{10^6} \text{ m}^2$$

Fan air outlet velocity v m/s,

$$v = \frac{\text{air flow } Q \text{ l/s}}{\text{outlet area } A \text{ m}^2} \times \frac{1 \text{ m}^3}{10^3 \text{ l}}$$

$$= \frac{Q}{A \times 10^3} \text{ m/s}$$

Fan velocity pressure FVP Pa,

$$\text{FVP} = 0.5 \times \rho \, \frac{\text{kg}}{\text{m}^3} \times \frac{Q^2 \, \text{l}^2}{\text{s}^2} \times \frac{1}{A^2 \, \text{m}^4} \times \frac{1 \text{ m}^6}{10^6 \, \text{l}^2}$$

$$= 0.5 \times \rho \, \frac{Q^2 \, \text{kg}}{A^2 \times 10^6 \, \text{m s}^2} \times \frac{1 \text{ N s}^2}{1 \text{ kg m}} \times \frac{1 \text{ Pa m}^2}{1 \text{ N}}$$

$$= 0.5 \times \rho \, \frac{Q^2}{A^2 \times 10^6} \text{ Pa}$$

Also, fan velocity pressure FVP Pa,

$$\text{FVP} = 0.5 \times \rho \, \frac{\text{kg}}{\text{m}^3} \times v^2 \, \frac{\text{m}^2}{\text{s}^2}$$

$$= 0.5 \times \rho \times v^2 \frac{\text{kg}}{\text{m s}^2} \times \frac{1 \text{ N s}^2}{1 \text{ kg m}} \times \frac{1 \text{ Pa m}^2}{1 \text{ N}}$$

$$= 0.5 \times \rho \times v^2 \text{ Pa}$$

The fan velocity pressure is calculated from the fan air velocity and the density of the air flowing through the discharge duct. The density of the air at this location depends upon the dry-bulb temperature and the barometric pressure of the air within the duct. The air is subject to frictional heating by the fan and any heat output from a close-coupled electric driving motor. Belt-drive motors are remote from the air within the duct, but will be within a fan plenum. The duct air static pressure should be added to the barometric air pressure in order ascertain the air density.

Air density ρ,

$$\rho = 1.2 \times \frac{p_a + p_s}{101\ 325} \times \frac{273 + 20}{273 + t} \frac{\text{kg}}{\text{m}^3}$$

where

$p_a =$ barometric pressure Pa

$p_s =$ static air pressure within the duct Pa

Note that p_s can be positive or negative.

Fan static pressure FSP,

FSP = fan total pressure FTP − fan velocity pressure FVP Pa

$\quad$ = FTP − FSP Pa

If the pressure is in millimetres water gauge, mm H_2O,

multiply it by the gravitational acceleration, g m/s^2, to

convert it to Pascal. That is:

pressure, $p = g \times H$ Pa

where $p =$ pressure, Pa

$\quad g =$ gravitational acceleration, 9.907 m/s^2

$\quad H =$ manometer head, mm H_2O

p Pa $= 9.907 \times H$ mm H_2O

The ductwork-system resistance varies with the square of the air-flow quantity. The general form for this is:

$$\frac{\Delta p}{Q^2} = \text{constant for the ductwork system. So,}$$

$$\frac{\Delta p_1}{(Q_1)^2} = \frac{\Delta p_2}{(Q_2)^2}$$

where suffixes 1 and 2 refer to different flow rates. Once the design flow rate and pressure drop due to ductwork resistance are known, that is at condition 1, the expected system pressure drop at any other rate of flow, condition 2, can be calculated from the ratio.

$$\Delta p_2 = (Q_2)^2 \times \frac{\Delta p_1}{(Q_1)^2}$$

$$= \Delta p_1 \times \frac{(Q_2)^2}{(Q_1)^2}$$

The fan laws used to predict the performance of a geometrically similar fan that is to be operated at a different speed, with air of a different density or with an impeller of a different diameter are:

air flow $Q = K \times \rho \times N \times D^3$

pressure rise $\Delta p = K \times \rho \times N^2 \times D^2$

fan air power $P = K \times \rho \times N^3 \times D^5$

where K = constant of proportionality (dimensionless)

N = rotational speed of the impeller Hz

D = impeller diameter m

The fan performance at the second condition is found by rearrangement of the laws of proportionality, where the known data are

$$Q_1 = K \times \rho_1 \times N_1 \times D_1^3$$

$$\text{and } K = \frac{Q_1}{\rho_1 \times N_1 \times D_1^3}$$

The required data are

$$Q_2 = K \times \rho_2 \times N_2 \times D_2^3$$

$$\text{so, } K = \frac{Q_2}{\rho_2 \times N_2 \times D_2^3}$$

$$K = \frac{Q_1}{\rho_1 \times N_1 \times D_1^3}$$

$$= \frac{Q_2}{\rho_2 \times N_2 \times D_2^3}$$

The volume flow provided by a geometrically similar fan at speed N_2, diameter D_2 and air density ρ_2 is found from

$$\frac{Q_2}{\rho_2 \times N_2 \times D_2^3} = \frac{Q_1}{\rho_1 \times N_1 \times D_1^3}$$

$$Q_2 = \frac{Q_1}{\rho_1 \times N_1 \times D_1^3} \times \rho_2 \times N_2 \times D_2^3$$

$$Q_2 = Q_1 \times \frac{\rho_2}{\rho_1} \times \frac{N_2}{N_1} \times \frac{D_2^3}{D_1^3}$$

Similarly, the fan pressure rise at a geometrically similar fan for a different air density, speed and impeller diameter is

$$\Delta p_2 = \Delta p_1 \times \frac{\rho_2}{\rho_1} \times \frac{N_2^2}{N_1^2} \times \frac{D_2^2}{D_1^2}$$

and the fan air power required is

$$P_2 = P_1 \times \frac{\rho_2}{\rho_1} \times \frac{N_2^3}{N_1^3} \times \frac{D_2^5}{D_1^5}$$

EXAMPLE 9.2

The Clune fan Company has designed a mixed-flow fan to exhaust humid air at atmospheric pressure and a temperature of 30°C d.b. from swimming-pool buildings. The prototype fan passed 500 l/s, had an impeller of 300 mm diameter and developed a fan static pressure of 100 Pa when running at 1440 revolutions per minute with a two pole motor. The input electrical power to the motor of the fan tested was 110 W when connected to a single-phase 240 V alternating-current supply. Predict the performance of a 1000 mm diameter fan that will be run at 12 Hz when it is fitted with an 8 pole 415 V motor, for the manufacturer.

The larger fan is to be used at the same atmospheric pressure and air temperature as the prototype. The air densities,

$$\rho_1 = \rho_2$$

so

$$\frac{\rho_2}{\rho_1} = 1$$

$$N_1 = 1440 \frac{\text{rev}}{\text{min}} \times \frac{1 \text{ min}}{60 \text{ s}} \times \frac{1 \text{ Hz s}}{1 \text{ rev}}$$

$$= 24 \text{ Hz}$$

$$N_2 = 12 \text{ Hz}$$

$$Q_2 = Q_1 \times \frac{N_2}{N_1} \times \frac{D_2^3}{D_1^3}$$

$$= 500 \times \frac{12}{24} \times \frac{1000^3}{300^3} \text{ l/s}$$

$$= 9260 \text{ l/s}$$

$$\Delta p_2 = \Delta p_1 \times \frac{N_2^2}{N_1^2} \times \frac{D_2^2}{D_1^2}$$

$$= 100 \times \frac{12^2}{24^2} \times \frac{1000^2}{300^2} \text{ Pa}$$

$$= 278 \text{ Pa}$$

$$P_2 = P_1 \times \frac{N_2^3}{N_1^3} \times \frac{D_2^5}{D_1^5}$$

$$= 110 \times \frac{12^3}{24^3} \times \frac{1000^5}{300^5} \text{ W}$$

$$= 5658 \text{ W}$$

FAN TESTING

Fans are tested in accordance with British Standard 848: Fans for General Purposes. The performance of a fan is specified from:

1. air volume flow rate
2. inlet air static pressure
3. outlet air static pressure
4. electrical power consumed at the fan shaft
5. fan rotational speed
6. air temperature handled by the fan
7. density of the air handled by the fan.

DUCT SYSTEM CHARACTERISTICS

The frictional resistance of the air-duct system is found from the D'Arcy equation,

$$H = \frac{4 \times f \times L \times V^2}{2 \times d \times g} \text{ m air}$$

where H = frictional resistance of air duct m air

f = duct surface friction factor (dimensionless)

L = duct length m

v = air velocity m/s

d = duct internal diameter m

The air pressure rise at the fan is

$$\Delta p = \rho \times g \times H \text{ Pa}$$

$$= \rho \times g \times \frac{4 \times f \times L \times v^2}{2 \times d \times g} \text{ Pa}$$

where Δp = air pressure drop in duct Pa

ρ = air density kg/m^3

The velocity of air that flows through a circular duct is found from

$$Q = \frac{\pi \times d^2}{4} \text{ m}^2 \times v \, \frac{\text{m}}{\text{s}}$$

where Q = air volume flow rate m^3/s

The velocity is

$$v = \frac{4 \times Q}{\pi \times d^2} \frac{\text{m}}{\text{s}}$$

Thus the system pressure drop is found from

$$\Delta p = \rho \times g \times \frac{4 \times f \times L}{2 \times d \times g} \times \left(\frac{4 \times Q}{\pi \times d^2} \right)^2 \text{Pa}$$

$$= \rho \times g \times \frac{4 \times f \times L}{2 \times d \times g} \times \frac{16 \times Q^2}{\pi^2 \times d^4} \text{Pa}$$

$$= \rho \times \frac{4 \times f \times L}{2 \times d} \times \frac{16 \times Q^2}{\pi^2 \times d^4} \text{Pa}$$

$$= \frac{\rho \times 32 \times f \times L}{\pi^2 \times d^5} \times Q^2 \text{Pa}$$

For a particular air-ductwork system, the friction factor, f, is a fixed quantity as all the ducts, their material and shape and fittings are built. The length, L m, and the diameter, d m, of the constructed duct system are fixed. The density of the air flowing through the duct system will vary only as a result of the actions of the automatic control system and outdoor air variations. The basic design for the air density is fixed. The air-flow rate through the duct system will vary in response to the action of volume-control dampers, temperature controls and the gradual increase in frictional resistance of the air filter. For a specific installation, the pressure drop through the duct system is characterized by

$$\Delta p = K \times Q^2 \text{Pa}$$

where the constant $K = \dfrac{\rho \times 32 \times f \times L}{\pi^2 \times d^5}$

FANS IN SERIES

The installation of a supply-air fan and an extract-air fan for a space is an example of fans that are connected in series. For comfort air-conditioning and ventilation systems, the occupied room is almost always maintained at, or very close to, the ambient atmospheric pressure so as not cause noticeable air movement at doorways and at openable windows. Buildings can be maintained at a small positive pressure above that of the atmosphere so that there will not be an ingress of unconditioned outdoor air. Kitchens, chemical, biological and nuclear radiation areas and toilet accommodation are maintained at a static air pressure that is slightly below the surrounding rooms so that vitiated air flows outward only through the exhaust system that is dedicated to that area. Normally constructed buildings are not air tight and outward leakages through the porosity of building materials and cracks around window and door frames, as well as the designed air flow through door openings, are used to exhaust some of the outdoor air intake to positively pressurized areas. The static air pressure maintained within the room or building is a balance between the residual supply-air fan pressure, air leakage from the space and the suction pressure that is generated by the extract-air fan. The residual static pressure within the space, or room, is calculated as if it were a section of the duct system.

Two or more axial-flow fans may be connected in series to increase the fan static pressure available and to provide step control of the fan static pressure. This method of connection does not increase the air-flow rate as both fans pass the same air

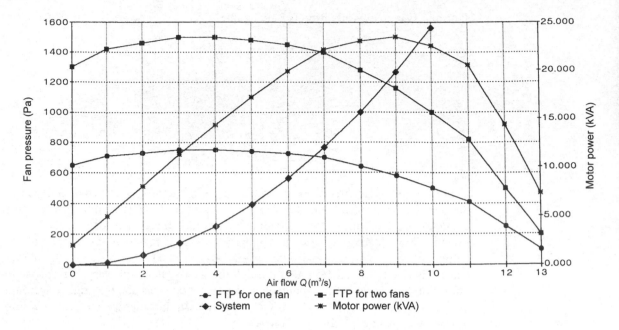

Fig. 9.13   Two axial fans connected in series.

quantity. Guide vanes are fitted between the fan impellers to remove the swirl that is imparted to the air by each axial impeller. The overall performance characteristic for two axial-flow fans that are connected in series is shown in Fig. 9.13. The cumulative curve is found by doubling the fan total pressure at each air-flow rate. The fan velocity pressure curve remains the same as for one fan. The fan static pressure curve is found by subtracting the total and velocity pressures. The motor power consumed is twice that for the single fan. The sound power level of the combined identical fans is an increase of 3 dB over that for one fan.

FAN POWER

The input electrical power to be supplied to the motor is found from:

1. useful fluid power consumed;
2. hydraulic power loss at the impeller;
3. hydraulic power loss within the impeller casing;
4. power loss due to air leakage from the fan;
5. power loss in the fan-shaft bearings;
6. power loss in the belt drive and pulleys;
7. loss of power within the electrical driving motor.

These losses of available power are classified into three groups: hydraulic, mechanical and electrical. Losses are usually expressed in percentage terms. The overall efficiency of the fan and drive system is found from the product of the hydraulic, mechanical and electrical efficiencies. The term for electrical efficiency is power factor.

$$\text{Power factor} = \frac{\text{motor useful output power kW}}{\text{apparent electrical input power kVA}}$$

The term kVA, kilovolt-ampere, refers to the electrical input power to the motor. When the power factor is 1.0, kVA are equal to kW, but as with most motors, the useful output power of a fan motor, in watts, is less than the electrical input power, in kVA, that has to be paid for at the meter. Electrical motors typically have a power factor in the range 0.5 to 0.7. Such low power factors are corrected with capacitors that are connected in parallel with the motor. Capacitor power-factor correctors store and release electrical charge during alternating currents and make it possible to achieve a power factor of 0.9 to 0.95. The original equipment designer and the subsequent energy auditor work to achieve the highest practical power factor, or overall electrical efficiency, at an affordable cost.

The electrical current per phase taken by the motor is

$$\text{current} = \frac{\text{volt ampere}}{\sqrt{3} \times \text{volt}} \text{ ampere}$$

The electrical services designer needs to know the full-load phase running and starting currents. The input-fan total power requirement of a fan and motor installation is found from

$$\text{Input power } P = Q \frac{1}{s} \times \text{FTP Pa} \times \frac{1}{\text{PF}} \times \frac{1}{\text{imp \%}} \times \frac{1}{\text{motor \%}} \times R$$

where imp % = fan impeller efficiency %

motor% = overall electrical motor efficiency %

PF = motor power factor

R = overall motor drive power ratio

$$\text{power} = Q \frac{1}{s} \times \frac{1 \text{ m}^3}{10^3 \text{ l}} \times \text{FTP Pa} \times \frac{1 \text{ N}}{\text{Pa m}^2} \times \frac{1}{\text{PF}} \times \frac{100}{\text{imp \%}} \times \frac{100}{\text{motor \%}} \times R$$

$$= \frac{Q \times \text{FTP} \times 100 \times 100}{10^3 \times \text{PF} \times \text{imp \%} \times \text{motor \%}} \times \frac{\text{N m}}{\text{s}} \times \frac{1 \text{ W s}}{1 \text{ N m}} \times \frac{1 \text{ kW}}{10^3 \text{ W}} \times \frac{1 \text{ kV A}}{1 \text{ kW}} \times R$$

$$= \frac{R \times Q \times \text{FTP}}{100 \times \text{PF} \times \text{imp \%} \times \text{motor \%}} \text{ kV A}$$

The motor input power does not drop to zero when the fan and motor are running against a closed-duct damper. Although there is zero air flow through the fan, the motor is providing the energy to overcome the air friction of the moving components and the overhead losses in the drive system and the motor itself. Fan power load is found from tests that cover the range of air flows that the fan is to be used for. At zero air flow, the fan-impeller efficiency will be low, 60% or less, the drive system, motor efficiency and power factor will remain the same, or similar to those within the normal operating range. Motor power is consumed in generating the static pressure within the fan casing or plenum chamber, and in overcoming the aerodynamic forces on the rotating components. Duct systems are not always airtight and duct dampers do not seal completely. There may be a residual air flow of 5% or so, of the design flow when all the dampers are closed. The leakage rate is tested during commissioning (Chapter 8). The fan power at the minimum air flow rate will vary from 20% to 70% of the peak power, depending upon the type of fan. The overall motor drive power ratio, R, allows for the overhead use of power at zero air flow by the fan.

EXAMPLE 9.3

A backward-curved centrifugal fan delivers air at the rate of 5.6 m³/s when the ductwork system resistance is 350 Pa. The 415 V three-phase driving motor has a power factor of 0.7. The overall efficiency of the belt drive is 95%. The fan impeller has an efficiency of 85% at the design flow. The fan power at zero air flow, when it is only generating static pressure within its casing, is 40% of the design flow value. Calculate the motor power for the design and closed damper conditions.

At the design air flow,

$$\text{supply air flow, } Q = 5600 \text{ l/s}$$

$$\text{fan total pressure, FTP} = 350 \text{ Pa}$$

$$\text{power factor, PF} = 0.7$$

$$\text{fan impeller efficiency, imp} = 85\%$$

$$\text{motor overall efficiency, motor} = 95\%$$

$$\text{motor power at zero air flow} = 40\%, 0.4$$

$$\text{overall motor power ratio, } R = 1.0 + 0.4$$

$$= 1.4$$

$$\text{Power} = R \times \frac{Q \times \text{FTP}}{100 \times \text{PF} \times \text{imp }\% \times \text{motor }\%} \text{ kV A}$$

$$= 1.4 \times \frac{5600 \times 350}{100 \times 0.7 \times 85 \times 95} \text{ kV A}$$

$$= 4.855 \text{ kV A}$$

At the closed-damper air flow,

$$\text{Power} = 0.4 \times 4.855 \text{ kV A}$$

$$= 1.942 \text{ kV A}$$

$$\text{Phase current} = \frac{\text{volt ampere}}{\sqrt{3} \times \text{volt}} \text{ ampere}$$

$$= \frac{1.942 \times 10^3}{\sqrt{3} \times 415} \text{ ampere}$$

$$= 2.7 \text{ amperes}$$

FAN PERFORMANCE CONTROL

The methods of controlling the operation and performance of fans are:

1. local isolator – a switch that is sited adjacent to the fan for disconnecting the electrical supply during servicing work;
2. speed selector switch for multi-speed motors; the electric driving motor has pairs of poles that are wired to provide two to four possible fan speeds of 48, 24, 16 and 12 Hz; this is a low- to moderate-cost method of fan speed control;
3. star/delta starter switch – three-phase fan motors that can be manually started in star connection, which operates the motor at up to 70% of its full speed; while the motor is connected in star formation, the power consumption is half of its maximum; the motor can then be manually switched over to run in delta connected mode; this is a low-cost method of speed and power-input control;
4. star/delta start controller – the same operation and benefits as the star/delta switch; the switching is achieved with an automatic controller;
5. auto-transformer control of the motor supply voltage – the motor is started at a low voltage through the transformer and then stepped up to the full 415 phase voltage in steps as the load on the fan increases; a speed-ratio reduction of up to 3:1 can be provided at moderate cost;
6. triac semiconductor motor-speed controller – a solid-state switch that alters the shape of the incoming alternating current sine wave to reduce the fan speed down to 10% of its full speed; power savings of up to 70% are achievable; the triac generates radio interference frequencies within the electrical system;
7. frequency inverter – the highest capital-cost method of fan-speed control; the fan-driving motor is started at a low frequency and can be run up to full speed over any desired time period; the fan-motor speed can be controlled at any time during the operation of the equipment; it is the method used to provide variable air flows in response to temperature, economy, pressure or atmospheric contaminant sensors; the frequency inverter provides an output of 0–50 Hz as required; the inverter generates noise and it has electrical losses of 4% or more of the input power; these losses generate heat output into the plant room; each fan start takes place at the pre-set minimum frequency.

Standard electric motors are designed to operate in ambient air temperatures of up to 50°C. Fan-drive motors are protected with a bimetallic thermal cut-out or a thermistor temperature sensor which is embedded in the motor windings. This thermal contact opens the motor contactor when an excessive temperature occurs in the windings. Such overheat protection can be reset either manually or through an auto-reset contactor once the windings have cooled. A manual reset is preferable so that the cause of the overheating can be investigated. A computer-based monitoring system will log such faults as these and generate an alarm signal.

Each fan-drive motor is protected against excessive current by an overload circuit-breaker or a high rupturing capacity fuse. In three-phase motors, unless all three fuses overheat and rupture simultaneously, the remaining windings can burn out during single phasing. A miniature circuit-breaker, MCB, uses the heating effect of an excess current to open a bimetallic strip and switch the current off. It can operate within a few seconds of the fault occurring.

Motors can be tropic-proofed by coating the windings with anti-fungal treatment. An explosion-proof motor is used when exhausting explosive mixtures of gases or dusts. The ingress-protection standard IP54 is for electric motors in non-hazardous areas. IP54 provides complete protection against contact with live or moving parts inside the enclosure and water splashed against the machine will have no harmful effect. IP55 provides the additional protection against water projected by a nozzle from any direction. IP56 provides protection from sea water. IP65 provides protection against the ingress of dust and is for ignition-proof electric-motor installations.

The performance of the ventilation or air-conditioning system is controlled by either varying the temperature of the air at the heating and cooling coils, or by varying the volume flow rate of the air. Atmospheric contaminants, such as carbon monoxide or smoke density in a vehicle tunnel, can be the parameter that is used to generate the fan system control signal. Variable-volume air-conditioning systems use duct air static pressure to control the supply and recirculation fan speeds. Modulating air volume flow control dampers, VCD, are used to vary the proportion of outdoor air that is admitted into the occupied building from, say, 10% to 100% of the supply-air quantity, to provide an economy cycle of free cooling with outdoor air. Modulating air volume flow control dampers are located in the supply-air ducts to variable air volume air-conditioned rooms in response to comfort requirements.

Reducing the downstream air flows has to be accompanied by the control of the fan output to avoid the generation of air noise at the dampers and unnecessary fan operating cost. Modulating dampers act to throttle the flow of air through the duct system. They increase the frictional resistance of the duct system and move the fan-system operating point along the fan performance curve to the controlled flow. The fan-system operating point will move away from the design efficiency and almost certainly result in a lower fan efficiency. Increased noise generation at the fan is likely.

The methods of controlling the flow of air through a ventilation system are:

1. manually adjusted volume-control dampers;
2. automatically controlled volume-control dampers;
3. multi-step fan speeds;
4. multiple fan installation;
5. variable-frequency control of fan speed;
6. variable-pitch angle axial flow fans.

Many comfort air-conditioning systems and most ventilation systems operate with a constant fan speed throughout the year. This is the result of the designer's efforts to minimize the initial capital cost. The same applies to heating and cooling water-circulation systems, with a constant pump speed through the year. The excess flow of water bypasses the terminal heating or cooling unit at a three-way diverting flow-control valve and is harmlessly recycled. Exhaust ventilation systems from toilets and kitchens are operated from on–off switches at the design air flow irrespective of the imposed ventilation requirement. Running a fan or pump at its maximum design rotational speed and throttling the fluid flow in response to the comfort requirements is the most expensive method of operating the plant. Rotody-namic machines consume less electrical power when they are run at reduced speed. Starting and operating fans and pumps at less than their maximum speed reduces wear on the shaft bearings and the belt drive, reduces noise and vibration, reduces the input electrical current, motor and cable temperatures and generally prolongs the useful service period of the equipment. The largest air-pressure drop in a ducted air-conditioning system, 35 to 150 Pa, often occurs through the air filter. Minimizing the system air flow greatly reduces the filter resistance, with a consequent saving in fan power and by extending the filter media use. Example 9.4 analyses the annual cost of the fans in an air-conditioning system for the methods of performance control.

EXAMPLE 9.4

A single-duct air-conditioning system has a design air flow of $8\,m^3/s$. The supply-air duct system frictional resistance is 300 Pa. The supply-air filter has a pressure drop of 50 Pa when clean and 175 Pa when it is ready to be changed. Use the average filter-pressure drop for the analysis. The air-conditioning system is operated for 260 hours per month. The return-air ducted system has a design air flow of $8\,m^3/s$ at a pressure drop of 150 Pa. The fan handles air at an average density of $1.2\,kg/m^3$ through the year. The fan-impeller efficiency is 90%, motor efficiency is 60% and the motor-power factor is 0.9. Assume that the overall efficiency remains constant at any air flow. The motor power used at zero supply-air flow is 1.2 kW. The fan-discharge duct is $1000\,mm \times 785\,mm$. Figure 9.14 shows the supply-air fan performance curves. The building is in southern England, where the load profile on the air-conditioning system is indicated in Table 9.2. The supply-air requirement for each month is the average for the month to match thermal loads. A higher temperature differential between the supply- and room-air is maintained in winter and this reduces the supply-air quantity. Free cooling is used in the mild weather. Electricity costs 9 p/kV A h.

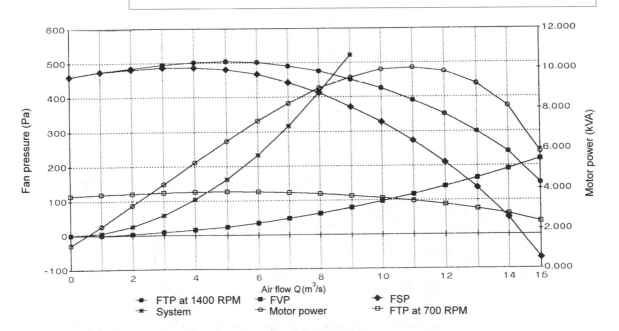

Fig. 9.14 Fan example 9.4, centrifugal fan at 1400 and 700 RPM.

Table 9.2 Monthly supply-air flow required

Month	Heating/cooling load (%)	Supply air required Q (m^3/s)
January	30	2.4
February	40	3.2
March	55	4.4
April	75	6.0
May	95	7.6
June	100	8.0
July	96	7.7
August	90	7.2
September	75	6.0
October	54	4.3
November	36	2.9
December	26	2.1

Compare the annual costs for the supply fan of the following:

1. no fan-performance control;
2. supply-duct air-volume control-damper;
3. two-speed fan motor producing 1400 and 700 RPM;
4. variable-frequency fan-speed control.

Supply-air duct-system design pressure drop $\Delta p = 300 + 175$ Pa

$$= 475\ \text{Pa}$$

Average system $\Delta p = 300 + 0.5 \times (50 + 175)$ Pa

$$= 413\ \text{Pa}$$

Design air flow $Q = 8\ m^3$/s

System characteristic $K = \dfrac{\Delta p_1}{Q_1^2} = \dfrac{413}{8^2} = \dfrac{\Delta p_2}{Q_2^2}$

So, $\Delta p_2 = 413 \times \dfrac{Q_2^2}{8^2}$ Pa

At an air flow of 4 m^3/s, the supply-duct system average pressure drop is

$$\Delta p_2 = 413 \times \frac{Q_2^2}{8^2}\ \text{Pa}$$

$$= 413 \times \frac{4^2}{8^2}\ \text{Pa}$$

$$= 103\ \text{Pa}$$

The data for the system resistance curve are shown in Table 9.3 and in Figs 9.14 and 9.15.

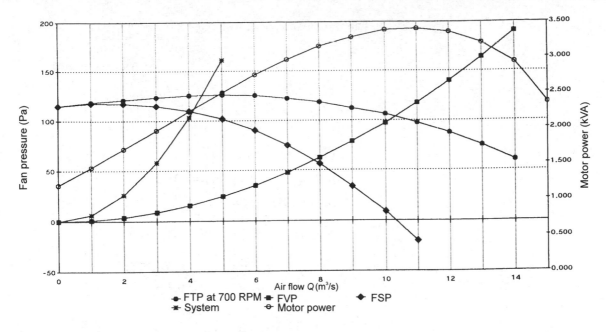

Fig. 9.15   Fan example 9.4, centrifugal fan at 700 RPM.

Table 9.3   Duct system resistance data for Example 9.4

Air flow Q (m³/s)	System resistance Δp (Pa)
0	0
1	6
2	26
3	58
4	103
5	161
6	232
7	316
8	413
9	523
10	645

The system operating point, for the average filter-pressure drop, is at a flow of 8.5 m³/s and a fan total pressure of 465 Pa.

$$FTP = 465 \, Pa$$

$$Q = 8.5 \, m^3/s$$

$$= 8500 \, l/s$$

Fan-discharge air velocity $v = \dfrac{Q \, m^3/s}{A \, m^2}$

$$= \frac{8.5 \, m^3/s}{1 \, m \times 0.785 \, m}$$

$$= 10.8 \, m/s$$

$$\text{Fan velocity pressure, FVP} = 0.5 \times \rho \, \frac{Q^2}{\text{area}^2 \times 10^6} \, \text{Pa}$$

$$= 0.5 \times 1.2 \, \frac{8500^2}{(1.0 \times 0.785)^2 \times 10^6} \, \text{Pa}$$

$$= \frac{0.9737}{10^6} \times 8500^2 \, \text{Pa}$$

$$= 70 \, \text{Pa}$$

Power factor, PF = 0.9

Impeller efficiency, imp = 90%

Motor efficiency = 60%

$$\text{Fan static pressure, FSP} = \text{FTP} - \text{FVP}$$

$$= 465 - 70 \, \text{Pa}$$

$$= 395 \, \text{Pa}$$

$$\text{Motor input power} = \frac{Q \times \text{FTP}}{100 \times \text{PF} \times \text{imp} \, \% \times \text{motor} \, \%} + 1.2 \, \text{kVA}$$

$$= \frac{8500 \times 465}{100 \times 0.9 \times 90\% \times 60\%} + 1.2 \, \text{kVA}$$

$$= \frac{8500 \times 465}{486\,000} + 1.2 \, \text{kVA}$$

$$= 9.333 \, \text{kVA}$$

The data and formulae for this example have been entered on to the spreadsheet file VFC.WKS. The complete listing of the cell contents is shown in Fig. 9.16 at the end of this example. The user can enter the file into a spreadsheet program and use it to solve this example, the relevant questions and for similar analysis in industry. Format the cell width to ten (10) characters for the whole worksheet. Copy the repeated lines, rather than typing each new formula and check the cell contents against the published list. Using this file allows any of the data to be changed easily and the new results discovered.

1. There is no control of the supply-fan output and the fan is operated at a constant speed of 1400 RPM. This is the constant-volume, variable-temperature air-conditioning system that is frequently used. The system operating point, for the average filter-pressure drop, is at a flow of 8.5 m³/s and a fan total pressure of 465 Pa. The motor power input averages 9.333 kVA. The monthly fan operating energy use is

$$\text{energy} = 9.333 \, \text{kVA} \times 260 \, \frac{\text{hour}}{\text{month}}$$

$$= 2426.6 \, \text{kVA h/month}$$

The annual electrical energy consumption is

$$\text{annual energy} = 2426.6\frac{\text{kVAh}}{\text{month}} \times 12\frac{\text{month}}{\text{year}}$$

$$= 29\,119\,\text{kVAh}$$

The annual electrical energy cost is

$$\text{annual energy cost} = 29\,119\frac{\text{kVAh}}{\text{year}} \times \frac{9\,\text{p}}{\text{kVAh}} \times \frac{£1}{100\,\text{p}}$$

$$= £2620.71$$

2. This is a variable air-volume air-conditioning system. Modulating dampers are located in the terminal units. A supply-duct damper controls the system supply-air flow through the year. The fan runs at a constant 1400 RPM and the dampers absorb the excess fan static pressure to provide the building with the required air flow. The fan total pressures are read from Fig. 9.14 and are listed in Table 9.4. Fan velocity pressure, fan static pressure, motor power and the monthly electrical energy consumption are calculated as previously and are also shown in Table 9.4.

Table 9.4 Monthly fan data

Month	Q (l/s)	FTP (Pa)	FVP (Pa)	FSP (Pa)	kVA	kVAh
January	2400	490	5	485	3.62	941.1
February	3200	500	10	490	4.49	1168.0
March	4400	510	18	492	5.82	1512.5
April	6000	510	34	476	7.5	1949.0
May	7600	485	54	431	8.78	2283.9
June	8000	475	60	415	9.0	2344.9
July	7700	485	56	429	8.88	2309.9
August	7200	480	49	431	8.31	2160.9
September	6000	510	34	476	7.5	1949.0
October	4300	510	17	493	5.71	1485.2
November	2900	495	8	487	4.15	1080.0
December	2100	485	4	481	3.3	856.9

Total: 20 041.3

$$\text{Annual electrical energy} = 20\,041.3\,\text{kVAh}$$

Annual electrical energy cost is

$$\text{annual energy cost} = 20\,041.3\frac{\text{kVA}}{\text{year}} \times \frac{9\,\text{p}}{\text{kVA}} \times \frac{£1}{100\,\text{p}}$$

$$= £1803.72$$

$$\text{Annual saving} = £2620.71 - £1803.72$$

$$= £816.99$$

3. A two-speed fan motor is used to lower the electricity consumption when the lower flow rates are needed. The fan total pressure that will be produced by the supply fan at the lower speed is

$$\Delta p_2 = \Delta p_1 \times \frac{N_2^2}{N_1^2}$$

In the case where $\Delta p = \text{FTP Pa}$

$$\text{FTP}_2 = \text{FTP}_1 \times \frac{N_2^2}{N_1^2}$$

From the fan characteristic curve, at an air flow rate of 4 m³/s, the FTP at 1400 RPM is 503 Pa. When the fan speed is reduced to 700 RPM, the FTP will be

$$\text{FTP}_2 = \text{FTP}_1 \times \frac{N_2^2}{N_1^2} \text{ Pa}$$

$$= 503 \times \frac{700^2}{1400^2} \text{ Pa}$$

$$= 503 \times 0.25 \text{ Pa}$$

$$= 126 \text{ Pa}$$

$$\text{FVP}_2 = 0.5 \times \rho \frac{Q^2}{\text{area}^2 \times 10^6} \text{ Pa}$$

$$= 0.5 \times 1.2 \frac{4000^2}{(1.0 \times 0.785)^2 \times 10^6} \text{ Pa}$$

$$= \frac{0.9737}{10^6} \times 4000^2 \text{ Pa}$$

$$= 16 \text{ Pa}$$

$$\text{FSP}_2 = \text{FTP}_2 - \text{FVP}_2 \text{ Pa}$$

$$= 126 - 16 \text{ Pa}$$

$$= 110 \text{ Pa}$$

$$P_2 = \frac{Q \times \text{FTP}_2}{100 \times \text{PF} \times \text{imp \%} \times \text{motor \%}} + 1.2 \text{ kVA}$$

$$= \frac{4000 \times 126}{100 \times 0.9 \times 90 \text{ \%} \times 60 \text{ \%}} + 1.2 \text{ kVA}$$

$$= \frac{4000 \times 126}{486\,000} + 1.2 \text{ kVA}$$

$$= 2.237 \text{ kVA}$$

The calculated data for a fan speed of 700 RPM is shown in Table 9.5. Suffix 1 is for 1400 RPM and suffix 2 is for 700 RPM fan data. Intervals of 1 m³/s have been selected. FTP_1 has been read from Fig. 9.14. FTP_2, FVP_2 and P_2 have been calculated as shown above.

The fan and system curves for the two fan speeds of 700 RPM and 1400 RPM are shown in Table 9.6 and are plotted on Fig. 9.15. The fan total pressure curve for the 700 RPM is also shown in Fig. 9.14. The lower fan speed is used up to the intersection of the duct system curve and the 700 RPM fan curve; this occurs at a flow of $4.5\,\mathrm{m^3/s}$. When the air conditioning system requires an air-flow that exceeds $4.5\,\mathrm{m^3/s}$, the higher fan speed must be selected. FTP_2 is read from the characteristic curves. Motor power P_2 and the monthly energy consumed are calculated as previously.

Table 9.5 Data for supply fan at 700 RPM

Q (l/s)	FTP_1 (Pa)	FTP_2 (Pa)	FVP_2 (Pa)	FSP_2 (Pa)	P_2 (kVA)
0	460	115	0	115	1.2
1000	475	119	1	118	1.445
2000	485	121	4	117	1.7
3000	495	124	9	115	1.964
4000	502	126	16	110	2.233
5000	505	126	24	102	2.5
6000	502	126	35	90	2.75
7000	490	123	48	75	2.96
8000	475	119	62	56	3.16
9000	450	113	79	34	3.28
10000	425	106	97	9	3.39
11000	390	98	118	−20	3.41
12000	350	88	140	—	3.36
13000	300	75	164	—	3.21
14000	240	60	191	—	2.93
15000	150	38	219	—	2.36

Table 9.6 Monthly fan data for two fan speeds

Month	Q (l/s)	*Fan speed*	FTP_2 (Pa)	P_2 (kVA)	kVA h
January	2400	700	123	1.807	470.0
February	3200	700	125	2.023	526.0
March	4400	700	126	2.341	608.6
April	6000	1400	502	7.4	1923.4
May	7600	1400	485	8.784	2284.0
June	8000	1400	475	9.02	2344.9
July	7700	1400	485	8.884	2309.9
August	7200	1400	480	8.31	2160.9
September	6000	1400	502	7.4	1923.4
October	4300	700	126	2.315	601.9
November	2900	700	124	1.94	504.4
December	2100	700	122	1.727	449.0

Total: 16 106.4

Annual electrical energy = 16 106.4 kVA h

Annual electrical energy cost is

$$\text{Annual energy cost} = 16\,106.4\,\frac{\text{kVA}}{\text{year}} \times \frac{9\,\text{p}}{\text{kVA}} \times \frac{£1}{100\,\text{p}}$$

$$= £1449.58$$

$$\text{Annual saving} = £2620.71 - £1449.58$$

$$= £1171.13$$

4. The variable-frequency controller generates the correct fan speed to supply the air flow that is required by the system. The fan total pressure is the same as the system air pressure drop as dampers are not used to absorb excessive fan pressure as when fixed speeds are used. FTP_2 can be read from Fig. 9.14, or, more accurately, calculated from:

$$FTP_2 = 413 \times \frac{Q_2^2}{8000^2} \text{ Pa}$$

The fan total pressure at the system air flow of 8000 l/s is 413 Pa. The fan data calculated, as previously, are shown in Table 9.7.

Table 9.7   Monthly fan data for variable fan speed

Month	Q (l/s)	FTP_2 (Pa)	P_2 (kVA)	kVAh
January	2400	37	1.38	359.5
February	3200	66	1.64	425.0
March	4400	125	2.33	606.2
April	6000	232	4.06	1056.7
May	7600	373	7.03	1828.6
June	8000	413	8.0	2079.6
July	7700	383	7.27	1889.7
August	7200	335	6.16	1602.4
September	6000	232	4.06	1057.7
October	4300	119	2.25	585.7
November	2900	54	1.52	395.8
December	2100	29	1.33	344.6
			Total:	12 231.5

Annual electrical energy = 12 231.5 kVA h

Annual electrical energy cost is,

$$\text{annual energy cost} = 12\ 231.5 \frac{\text{kVA}}{\text{year}} \times \frac{9\,\text{p}}{\text{kVA}} \times \frac{£1}{100\,\text{p}}$$

$$= £1100.84$$

Annual saving = £2620.71 − £1100.84

$$= £1519.87$$

The annual cost savings for the different methods are

damper volume control	£ 816.99
two-speed fan motor	£1171.13
variable-frequency fan-speed control	£1519.87

```
A1:    (G)    'Analysis o
B1:    (G)    'f Variable
C1:    (G)    ' Speed Fan
D1:    (G)    ' Drive
A2:    (G)    '**********
B2:    (G)    '**********
C2:    (G)    '**********
D2:    (G)    '******
A3:    (G)    'File   :
B3:    (G)    'VFC.wks
A4:    (G)    'Date   :
B4:    (G)    '8/5/96
A6:    (G)    'Fan type
B6:    (G)    '     :
C6:    (G)    'centrifuga
D6:    (G)    'l
A7:    (G)    'Design air
B7:    (G)    ' flow  :
C7:    (G)    8000
D7:    (G)    'litre/s
A8:    (G)    'Duct syste
B8:    (G)    'm     :
C8:    (G)    413
D8:    (G)    'Pa
A9:    (G)    'Hours per
B9:    (G)    'month  :
C9:    (G)    260
D9:    (G)    'h
E9:    (G)    'system ope
F9:    (G)    'ration
A10:   (G)    'Air densit
B10:   (G)    'y     :
C10:   (G)    1.2
D10:   (G)    'kg/m3
A11:   (G)    'Motor kVA,
B11:   (G)    ' Q=0  :
C11:   (G)    1.2
D11:   (G)    'kVA
A12:   (G)    'Impeller e
B12:   (G)    'ff.   :
C12:   (G)    90
D12:   (P0)   '%
A13:   (G)    'Motor eff.
B13:   (G)    '     :
C13:   (G)    60
D13:   (P0)   '%
A14:   (G)    'Power fact
B14:   (G)    'or    :
C14:   (G)    0.9
A15:   (G)    'Fan outlet
B15:   (G)    ' width :
C15:   (G)    1000
D15:   (G)    'mm
A16:   (G)    'Fan outlet
B16:   (G)    ' depth :
C16:   (G)    785
```

Fig. 9.16   (pp. 291–298).

D16:	(G)	'mm
A17:	(G)	'Electricit
B17:	(G)	'y cost :
C17:	(G)	9
D17:	(G)	'p/kVAh
A20:	(G)	'Fan speed
B20:	(G)	' N :
C20:	(G)	1400
D20:	(G)	'RPM
E20:	(F0)	+C20/60
F20:	(G)	'Hz
A21:	(G)	'Operating
B21:	(G)	'point Q:
C21:	(G)	8500
D21:	(G)	'litre/s
B22:	(G)	' FTP :
C22:	(G)	465
D22:	(G)	'Pa
A25:	(G)	'No volume
B25:	(G)	'flow contr
C25:	(G)	'ol
A26:	(G)	'**********
B26:	(G)	'**********
C26:	(G)	'**
A27:	(G)	'Fan speed
B27:	(G)	' N :
C27:	(G)	1400
D27:	(G)	'RPM
E27:	(F0)	+C27/60
F27:	(G)	'Hz
A28:	(D1)	'Fan Supply
B28:	(G)	' air Q =
C28:	(G)	+C21
D28:	(F3)	'litre/s
B29:	(G)	' FTP =
C29:	(G)	465
D29:	(G)	'Pa
A30:	(D1)	'Motor powe
B30:	(G)	'r =
C30:	(F3)	+$C$11+(C28*C29)/(100*$C$14*$C$12*$C$13)
D30:	(F3)	'kVA
A31:	(D1)	'Energy use
B31:	(G)	' =
C31:	(F1)	+C30*$C$9
D31:	(F3)	'kVAh per m
E31:	(F1)	'onth
A32:	(D1)	'Energy use
B32:	(G)	' =
C32:	(F1)	+C31*12
D32:	(G)	'kVAh per y
E32:	(F1)	'ear
D34:	(G)	'Total =
E34:	(F1)	+$C$32
F34:	(G)	'kVAh per a
G34:	(G)	'nnum
D35:	(G)	'Cost =
E35:	(C2)	+$E$34*$C$17/100

```
F35:   (G)   'per annum
A37:   (G)   'Modulating
B37:   (G)   ' Damper Co
C37:   (G)   'ntrol
A38:   (G)   '**********
B38:   (G)   '**********
C38:   (G)   '****
A39:   (G)   'Month
B39:   (G)   ' Q litre/s
C39:   (G)   '  FTP Pa
D39:   (G)   ' Motor kVA
E39:   (G)   ' kVAh
A40:   (G)   '—————
B40:   (G)   '—————
C40:   (G)   '  ——
D40:   (G)   '—————
E40:   (G)   ' ——
A41:   (D1)  35065
B41:   (G)   2400
C41:   (G)   490
D41:   (F3)  +$C$11+(B41*C41)/(100*$C$14*$C$12*$C$13)
E41:   (F1)  +D41*$C$9
A42:   (D1)  35096
B42:   (G)   3200
C42:   (G)   500
D42:   (F3)  +$C$11+(B42*C42)/(100*$C$14*$C$12*$C$13)
E42:   (F1)  +D42*$C$9
A43:   (D1)  35125
B43:   (G)   4400
C43:   (G)   510
D43:   (F3)  +$C$11+(B43*C43)/(100*$C$14*$C$12*$C$13)
E43:   (F1)  +D43*$C$9
A44:   (D1)  35156
B44:   (G)   6000
C44:   (G)   510
D44:   (F3)  +$C$11+(B44*C44)/(100*$C$14*$C$12*$C$13)
E44:   (F1)  +D44*$C$9
A45:   (D1)  35186
B45:   (G)   7600
C45:   (G)   485
D45:   (F3)  +$C$11+(B45*C45)/(100*$C$14*$C$12*$C$13)
E45:   (F1)  +D45*$C$9
A46:   (D1)  35217
B46:   (G)   8000
C46:   (G)   475
D46:   (F3)  +$C$11+(B46*C46)/(100*$C$14*$C$12*$C$13)
E46:   (F1)  +D46*$C$9
A47:   (D1)  35247
B47:   (G)   7700
C47:   (G)   485
D47:   (F3)  +$C$11+(B47*C47)/(100*$C$14*$C$12*$C$13)
E47:   (F1)  +D47*$C$9
A48:   (D1)  35278
B48:   (G)   7200
C48:   (G)   480
D48:   (F3)  +$C$11+(B48*C48)/(100*$C$14*$C$12*$C$13)
E48:   (F1)  +D48*$C$9
```

Fig. 9.16 *cont'd.*

```
A49:   (D1)   35309
B49:   (G)    6000
C49:   (G)    510
D49:   (F3)   +$C$11+(B49*C49)/(100*$C$14*$C$12*$C$13)
E49:   (F1)   +D49*$C$9
A50:   (D1)   35339
B50:   (G)    4300
C50:   (G)    510
D50:   (F3)   +$C$11+(B50*C50)/(100*$C$14*$C$12*$C$13)
E50:   (F1)   +D50*$C$9
A51:   (D1)   35370
B51:   (G)    2900
C51:   (G)    495
D51:   (F3)   +$C$11+(B51*C51)/(100*$C$14*$C$12*$C$13)
E51:   (F1)   +D51*$C$9
A52:   (D1)   35400
B52:   (G)    2100
C52:   (G)    485
D52:   (F3)   +$C$11+(B52*C52)/(100*$C$14*$C$12*$C$13)
E52:   (F1)   +D52*$C$9
D53:   (G)    'Total  =
E53:   (F1)   @SUM($E$41..$E$52)
F53:   (G)    'kVAh
D54:   (G)    'Cost   =
E54:   (C2)   +$E$53*$C$17/100
F54:   (G)    'per annum
A55:   (G)    'Two Speed
B55:   (G)    'Fan Motor
A56:   (G)    '**********
B56:   (G)    '*********
A57:   (G)    'Month
B57:   (G)    ' Q litre/s
C57:   (G)    '  Fan RPM
D57:   (G)    ' FTP2 Pa
E57:   (G)    ' Motor kVA
F57:   (G)    ' kVAh
A58:   (G)    '———————
B58:   (G)    '———————
C58:   (G)    ' ———————
D58:   (G)    ' ———————
E58:   (G)    '———————
F58:   (G)    ' ———————
A59:   (D1)   35065
B59:   (G)    2400
C59:   (G)    700
D59:   (G)    123
E59:   (F3)   +$C$11+(B59*D59)/(100*$C$14*$C$12*$C$13)
F59:   (F1)   +E59*$C$9
A60:   (D1)   35096
B60:   (G)    3200
C60:   (G)    700
D60:   (G)    125
E60:   (F3)   +$C$11+(B60*D60)/(100*$C$14*$C$12*$C$13)
F60:   (F1)   +E60*$C$9
A61:   (D1)   35125
B61:   (G)    4400
C61:   (G)    700
```

```
D61:  (G)   126
E61:  (F3)  +$C$11+(B61*D61)/(100*$C$14*$C$12*$C$13)
F61:  (F1)  +E61*$C$9
A62:  (D1)  35156
B62:  (G)   6000
C62:  (G)   1400
D62:  (G)   502
E62:  (F3)  +$C$11+(B62*D62)/(100*$C$14*$C$12*$C$13)
F62:  (F1)  +E62*$C$9
A63:  (D1)  35186
B63:  (G)   7600
C63:  (G)   1400
D63:  (G)   485
E63:  (F3)  +$C$11+(B63*D63)/(100*$C$14*$C$12*$C$13)
F63:  (F1)  +E63*$C$9
A64:  (D1)  35217
B64:  (G)   8000
C64:  (G)   1400
D64:  (G)   475
E64:  (F3)  +$C$11+(B64*D64)/(100*$C$14*$C$12*$C$13)
F64:  (F1)  +E64*$C$9
A65:  (D1)  35247
B65:  (G)   7700
C65:  (G)   1400
D65:  (G)   485
E65:  (F3)  +$C$11+(B65*D65)/(100*$C$14*$C$12*$C$13)
F65:  (F1)  +E65*$C$9
A66:  (D1)  35278
B66:  (G)   7200
C66:  (G)   1400
D66:  (G)   480
E66:  (F3)  +$C$11+(B66*D66)/(100*$C$14*$C$12*$C$13)
F66:  (F1)  +E66*$C$9
A67:  (D1)  35309
B67:  (G)   6000
C67:  (G)   1400
D67:  (G)   502
E67:  (F3)  +$C$11+(B67*D67)/(100*$C$14*$C$12*$C$13)
F67:  (F1)  +E67*$C$9
A68:  (D1)  35339
B68:  (G)   4300
C68:  (G)   700
D68:  (G)   126
E68:  (F3)  +$C$11+(B68*D68)/(100*$C$14*$C$12*$C$13)
F68:  (F1)  +E68*$C$9
A69:  (D1)  35370
B69:  (G)   2900
C69:  (G)   700
D69:  (G)   124
E69:  (F3)  +$C$11+(B69*D69)/(100*$C$14*$C$12*$C$13)
F69:  (F1)  +E69*$C$9
A70:  (D1)  35400
B70:  (G)   2100
C70:  (G)   700
D70:  (G)   122
E70:  (F3)  +$C$11+(B70*D70)/(100*$C$14*$C$12*$C$13)
F70:  (F1)  +E70*$C$9
```

Fig. 9.16   *cont'd.*

```
E71:   (G)    'Total  =
F71:   (F1)   @SUM($F$59..$F$70)
G71:   (G)    'kVAh
E72:   (G)    'Cost   =
F72:   (C2)   +$F$71*$C$17/100
G72:   (G)    'per annum
A73:   (G)    'Variable F
B73:   (G)    'an Speed
A74:   (G)    '**********
B74:   (G)    '*******
A75:   (G)    'Month
B75:   (G)    ' Q litre/s
C75:   (G)    'Fan N2 RPM
D75:   (G)    ' FTP2 Pa
E75:   (G)    ' Motor kVA
F75:   (G)    ' kVAh
A76:   (G)    '————————
B76:   (G)    '————————
C76:   (G)    '————————
D76:   (G)    '————
E76:   (G)    '————————
F76:   (G)    '————
A77:   (D1)   35065
B77:   (G)    2400
C77:   (F0)   +$C$20*B77/$C$21
D77:   (F0)   +$C$22*(B77/$C$21)^2
E77:   (F3)   +$C$11+(B77*D77)/(100*$C$14*$C$12*$C$13)
F77:   (F1)   +E77*$C$9
A78:   (D1)   35096
B78:   (G)    3200
C78:   (F0)   +$C$20*B78/$C$21
D78:   (F0)   +$C$22*(B78/$C$21)^2
E78:   (F3)   +$C$11+(B78*D78)/(100*$C$14*$C$12*$C$13)
F78:   (F1)   +E78*$C$9
A79:   (D1)   35125
B79:   (G)    4400
C79:   (F0)   +$C$20*B79/$C$21
D79:   (F0)   +$C$22*(B79/$C$21)^2
E79:   (F3)   +$C$11+(B79*D79)/(100*$C$14*$C$12*$C$13)
F79:   (F1)   +E79*$C$9
A80:   (D1)   35156
B80:   (G)    6000
C80:   (F0)   +$C$20*B80/$C$21
D80:   (F0)   +$C$22*(B80/$C$21)^2
E80:   (F3)   +$C$11+(B80*D80)/(100*$C$14*$C$12*$C$13)
F80:   (F1)   +E80*$C$9
A81:   (D1)   35186
B81:   (G)    7600
C81:   (F0)   +$C$20*B81/$C$21
D81:   (F0)   +$C$22*(B81/$C$21)^2
E81:   (F3)   +$C$11+(B81*D81)/(100*$C$14*$C$12*$C$13)
F81:   (F1)   +E81*$C$9
A82:   (D1)   35217
B82:   (G)    8000
C82:   (F0)   +$C$20*B82/$C$21
D82:   (F0)   +$C$22*(B82/$C$21)^2
E82:   (F3)   +$C$11+(B82*D82)/(100*$C$14*$C$12*$C$13)
```

```
F82:  (F1)  +E82*$C$9
A83:  (D1)  35247
B83:  (G)   7700
C83:  (F0)  +$C$20*B83/$C$21
D83:  (F0)  +$C$22*(B83/$C$21)^2
E83:  (F3)  +$C$11+(B83*D83)/(100*$C$14*$C$12*$C$13)
F83:  (F1)  +E83*$C$9
A84:  (D1)  35278
B84:  (G)   7200
C84:  (F0)  +$C$20*B84/$C$21
D84:  (F0)  +$C$22*(B84/$C$21)^2
E84:  (F3)  +$C$11+(B84*D84)/(100*$C$14*$C$12*$C$13)
F84:  (F1)  +E84*$C$9
A85:  (D1)  35309
B85:  (G)   6000
C85:  (F0)  +$C$20*B85/$C$21
D85:  (F0)  +$C$22*(B85/$C$21)^2
E85:  (F3)  +$C$11+(B85*D85)/(100*$C$14*$C$12*$C$13)
F85:  (F1)  +E85*$C$9
A86:  (D1)  35339
B86:  (G)   4300
C86:  (F0)  +$C$20*B86/$C$21
D86:  (F0)  +$C$22*(B86/$C$21)^2
E86:  (F3)  +$C$11+(B86*D86)/(100*$C$14*$C$12*$C$13)
F86:  (F1)  +E86*$C$9
A87:  (D1)  35370
B87:  (G)   2900
C87:  (F0)  +$C$20*B87/$C$21
D87:  (F0)  +$C$22*(B87/$C$21)^2
E87:  (F3)  +$C$11+(B87*D87)/(100*$C$14*$C$12*$C$13)
F87:  (F1)  +E87*$C$9
A88:  (D1)  35400
B88:  (G)   2100
C88:  (F0)  +$C$20*B88/$C$21
D88:  (F0)  +$C$22*(B88/$C$21)^2
E88:  (F3)  +$C$11+(B88*D88)/(100*$C$14*$C$12*$C$13)
F88:  (F1)  +E88*$C$9
E89:  (G)   'Total  =
F89:  (F1)  @SUM($F$77..$F$88)
G89:  (G)   'kVAh
E90:  (G)   'Cost   =
F90:  (C2)  +$F$89*$C$17/100
G90:  (G)   'per annum
A91:  (G)   'Cost Summa
B91:  (G)   'ry
A92:  (G)   '**********
B92:  (G)   '**
A94:  (G)   'No Fan Per
B94:  (G)   'formance C
C94:  (G)   'ontrol
D94:  (G)   'Total  =
E94:  (F1)  +$C$32
F94:  (G)   'kVAh per a
G94:  (G)   'nnum
D95:  (G)   'Cost   =
E95:  (C2)  +$E$34*$C$17/100
F95:  (G)   'per annum
```

Fig. 9.16 *cont'd.*

A97:   (G)   'Modulating
B97:   (G)   ' Damper Co
C97:   (G)   'ntrol
D97:   (G)   'Total =
E97:   (F1)   @SUM($E$41..$E$52)
F97:   (G)   'kVAh
D98:   (G)   'Cost =
E98:   (C2)   +$E$53*$C$17/100
F98:   (G)   'per annum
A100: (G)   'Two Speed
B100: (G)   'Fan Motor
D100: (G)   'Total =
E100: (F1)   @SUM($F$59..$F$70)
F100: (G)   'kVAh
D101: (G)   'Cost =
E101: (C2)   +$F$71*$C$17/100
F101: (G)   'per annum
A103: (G)   'Variable F
B103: (G)   'an Speed
D103: (G)   'Total =
E103: (F1)   @SUM($F$77..$F$88)
F103: (G)   'kVAh
D104: (G)   'Cost =
E104: (C2)   +$F$89*$C$17/100
F104: (G)   'per annum
A106: (G)   'Savings fo
B106: (G)   'r variable
C106: (G)   ' speed con
D106: (G)   'trol over
E106: (G)   'no control
F106: (G)   ',
D107: (G)   '   =
E107: (C2)   +E95-E104
F107: (G)   'per annum
E108: (G)   '==========
F108: (G)   '==========

9.16   Spreadsheet cell contents for the energy analysis for variable-speed fan drive.

EXAMPLE 9.5

The centrifugal supply-air fan is being run at a speed of 1600 RPM to deliver the required system air flow of 1500 l/s. The fan total pressure rise at this air flow is 350 Pa. The inlet air to the fan does not come into contact with the heat emitted from the drive motor. The overall fan efficiency is 55% and this includes the losses in the impeller and the fan casing. Air enters the fan at 20°C d.b. Estimate the temperature of the supply air as it leaves the fan and comment on the result.

The motor and drive belt remain outside the air stream, so the only source of heating to the supply air is the turbulence and friction that are generated within the fan casing and as air passes through the impeller. The fan shaft bearings should have a high efficiency and not be generating noticeable heat output.

$$Q = 1500 \, \text{l/s}$$

$$FTP = 350 \, \text{Pa}$$

$$\text{Impeller efficiency} = 55\%$$

The impeller is providing useful work, so the loss of power from the impeller and the fan casing is

$$\text{Overall loss} = 100 - \text{efficiency}\%$$

$$= 100 - 55\,\%$$

$$= 45\,\%$$

The overall loss applies to the input power to the fan shaft.

$$\text{Fan shaft input power loss} = Q\ \text{m}^3/\text{s} \times \text{FTP Pa} \times \frac{100}{\text{loss}\%}$$

$$= 1.5\ \text{m}^3/\text{s} \times 350\ \text{Pa} \times \frac{100}{45\%}$$

$$= 1166\ \text{W}$$

This power loss appears as heat energy that is carried away by the air stream through the fan and into the supply-air duct. Heat loss from the casing of the fan into the plant room may be significant; assuming that it is negligible, the heat balance to be used is

$$\text{Heat output by fan} = \text{heat absorbed by air stream}$$

$$\text{Take, air density} = 1.2\ \text{kg/m}^3$$

$$\text{SHC of air} = 1.012\ \text{kJ/kg K}$$

$$\text{Air inlet to fan} = t_1\ ^\circ\text{C d.b.}$$

$$\text{Air outlet from fan} = t_2\ ^\circ\text{C d.b.}$$

$$1166\ \text{W} = 1.5\,\frac{\text{m}^3}{\text{s}} \times 1.2\,\frac{\text{kg}}{\text{m}^3} \times 1.012\,\frac{\text{kJ}}{\text{kg K}} \times (t_2 - t_1)\ \text{K} \times \frac{1\ \text{kW s}}{1\ \text{kJ}} \times \frac{10^3\ \text{W}}{1\ \text{kW}}$$

$$t_2 - t_1 = \frac{1166}{1.5 \times 1.2 \times 1.012 \times 10^3}\ \text{K}$$

$$= 0.64\ \text{K}$$

$$t_2 = (t_1 + 0.64)\ ^\circ\text{C d.b.}$$

$$= (20 + 0.64)\ ^\circ\text{C d.b.}$$

$$= 20.64\ ^\circ\text{C d.b.}$$

The loss of energy between the shaft input power and the useful power absorbed by the air produces an increase in the temperature of the air flowing through the fan. This will occur in most cases and also in fluid pumps. Should the air flow be reduced to 10% of the current value, the air temperature increase would rise tenfold, to 6.4 K. The designer needs to take account of the heat-release rate into the fluid flowing. During heating processes, the fan and pump fluid temperature increase is beneficial to the system. For cooling applications, a temperature increase at the fan or pump is undesirable.

COMMISSIONING AND MAINTENANCE

The commissioning of a new installation and its routine maintenance include:

1. ensuring the correct direction and free rotation of the fan impeller;
2. checking that electrical wiring is securely connected;
3. treating corrosion of metalwork and the wiring system;
4. ensuring that the air ductwork and fan intake are clear of debris and are clean;
5. ensuring the correct installation of the local isolator, fuses and circuit-breakers;
6. ensuring the correct operation of switches, fan interlock contactors, fuse and circuit-breakers; testing safety devices;
7. checking that the running motor current does not exceed the name-plate rating;
8. checking that all bolted and riveted fastenings are secure;
9. checking that anti-vibration mountings are in place and are operational;
10. checking that protective guards are in place for all rotating components and drive belts;
11. checking that drive belts are correctly tensioned and pulleys are aligned;
12. checking that flexible duct connections are functioning properly;
13. where the fan shaft bearings are fitted with grease nipples, use the recommended lubricant at a maximum interval of 12 months or before the working hours limit has been reached; most modern fans have ball-bearings that are 'sealed for life'; the life expectancy of sealed bearings is 40 000 hours; standard ventilation fans are usually lubricated with lithium-based grease that is suitable for continuous operation at up to 130°C; fans for spilling smoke require a silicon-based grease that is suitable for continuous operation at up to 200°C; clean the grease nipples before use; purge the old grease through the relief port with the incoming grease pressure; only use compatible types of grease in a bearing;
14. isolate the fan drive motor from the electricity supply before any work is conducted on the rotating equipment.

Questions

1. State the functions of fans within building services.
2. Explain how fans create movement in air.
3. State the limiting factors that are included in the design of fan and duct systems.
4. List the components of fan and drive systems. Comment upon how they are manufactured and their relative cost to the user.
5. List the types of fan and their principal applications.
6. Discuss the advantages and disadvantages of installing a large duty centrifugal fan within a builder's work plenum when compared with a metal plenum.
7. List the safety features that are to be considered when designing a fan installation.
8. Explain how the maintenance access to fans and their associated items of plant, such as air filters, drive system and heat-exchange devices, is provided during the design of the overall system and building. State the importance of such access, the likely frequency of maintenance work and the safety precautions that are to be provided.
9. Discuss the problems that are associated with manual entry into a positively or negatively pressurized plenum or ventilated space.
10. Sketch the characteristic curves for a backward-curved centrifugal fan and ductwork system to show the relationship between fan total, static and velocity pressures, motor power and the ductwork-system resistance. Do not look at any published graphics while working on this question.
11. List the 'overhead' power requirements of a fan and electric motor drive plant. State how these minimum power requirements affect the shape of the fan power characteristic curve.
12. State the problems that are associated with running a fan against closed duct dampers.

13. What problems may be created within the ventilated rooms when starting a centrifugal supply or extract fan, or a combination of both.

14. State the designer's objectives when selecting the operating point on a fan characteristic graph.

15. Explain how fan manufacturers test fans and how they generate performance data for a range of fan sizes and rotational speeds.

16. State the electric motor speeds that are used and explain how these speeds are achieved.

17. Explain, with the aid of practical examples, how the rotational speed of a fan impeller is maintained at a different speed from that of the driving motor.

18. Explain what happens to the fan total pressure rise when the fan static pressure drops to zero.

19. Why can the duct-system characteristic curve be calculated and drawn without reference to the duct-system resistance data?

20. State the fan laws and show how they are used to generate predictions of fan performance.

21. State the data that are obtained from standard tests on fans.

22. Explain why and when fans would be connected in series with each other. Sketch the combined characteristic of two fans that are closely connected in series. Sketch the duct system resistance curve and identify the overall fan and system operational point.

23. Discuss how the equipment manufacturer and the building services designer obtain the maximum energy efficiency from a fan and drive system.

24. Explain why the fan motor power consumption does not diminish to zero when the duct-system dampers are all closed.

25. Discuss the equipment that is needed to start, control and monitor the safe operation of the speed of electric drive motors on air-conditioning system fans. State the economic and technical factors that are included in the decision as to which method is applied to an application.

26. Explain the characteristics of the different methods of controlling the flow of air through a ductwork system in relation to the energy cost of using the whole installation.

27. List the advantages that are gained by operating an air-conditioning fan at the lowest possible rotational speed while meeting the required duty.

28. Design the layout for a commissioning task sheet for the fan and duct systems within an air-conditioned building. The completed instructions will be implemented by a contract company as the result of competitive tendering for the work. The document is to state the plant that is to be commissioned and precisely what work is to be conducted.

29. Design the layout for a maintenance manual for the air-conditioned building in question 28. The completed instructions will be implemented by a contract company as the result of competitive tendering for the work. The document is to state the plant that is to be maintained, how it is to be started, stopped, shut down in an emergency and the timing for maintenance operations. State precisely what work is to be undertaken at the appropriate time intervals. The maintenance instructions must be easily understandable by the client's representative, who is not necessarily a building services engineer.

30. A forward-curved centrifugal fan maintains a static air pressure of 75 Pa above that of the atmosphere, within the fan acoustic plenum. An access door of 750 mm width and 2200 mm height allows entry for regular maintenance. The door is hinged along one vertical side and has a handle in the opposite side. Calculate the manual force that is necessary to open the door.

31. The supply-air outlet duct from a packaged roof-mounted air-conditioning unit is 500 mm wide and 700 mm high. The supply air flow to the duct system is to be 2700 l/s of air at 33°C d.b. during winter use. Calculate the velocity pressure of the supply air as it leaves the packaged unit.

32. The Daylesford Impeller Company manufacture a range of centrifugal fans for air-conditioning systems. A prototype fan was tested at an air-flow rate of passed 1800 l/s with an impeller of 420 mm diameter producing a fan static pressure of 150 Pa when the impeller was running at 1200 revolutions per minute. The measured electrical input power to the motor of the fan tested was 800 W. Predict the performance of a 700 mm diameter fan impeller that will be run at 16 Hz and is to be geometrically similar to the prototype.

33. An axial-flow fan delivers air at the rate of 12 m^3/s when the ductwork system resistance is 120 Pa. The 415 V three-phase driving motor has a power factor of 0.7. The overall efficiency of the belt drive is 96%. The fan impeller has an efficiency of 60% at the design flow. The fan power at zero air flow, when it is only generating static pressure within its casing, is 20% of the design flow value. Calculate the motor power for the design and closed damper conditions.

34. An axial-flow fan runs at 940 RPM and has a blade angle of 28°. It passes air at a density of 1.21 kg/m^3 into a duct system that has a resistance of 100 Pa when the air flow is 5.0 m^3/s. The outlet-air velocity from the fan is 11.0 m/s. The 415 V three-phase drive motor has a power factor of 0.92. The motor and drive has an overall efficiency of 65% and the impeller has an efficiency of 80%. The minimum power consumption at zero air flow

would be 0.25 kV A. The fan performance data are shown in Table 9.8. Calculate and plot the fan characteristic curves for fan velocity and static pressures, the duct-system resistance and the motor input power. Use the spreadsheet file that is shown in Fig. 9.9. Find the fan and system operating conditions and the current that will be taken by the motor.

Table 9.8   Fan data for Question 34

Q (m³/s)	FTP (Pa)
0	160
1	155
2	150
3	140
4	125
5	90
6	50
7	0

35. A mixed-flow exhaust fan runs at 1450 RPM and handles air at a density of 1.2 kg/m³. The fan-outlet diameter is 500 mm. The duct system has a resistance of 600 Pa when the air flow is 1100 l/s. The outlet-air velocity from the fan is 7.6 m/s. The 415 V three-phase drive motor has a power factor of 0.7, the motor and drive has an overall efficiency of 70% and the impeller has an efficiency of 82%. The minimum power consumption at zero air flow would be 0.4 kV A. The fan-performance data are shown in Table 9.9. Calculate and plot the fan characteristic curves for fan velocity and static pressures, the duct system resistance and the motor input power. Use the spreadsheet file that is shown in Fig. 9.9. Find the fan and system operating conditions and the current that will be taken by the motor.

Table 9.9   Fan data for question 35.

Q (l/s)	FTP (Pa)
0	750
250	800
500	850
750	850
1000	780
1250	630
1500	450
1750	200

36. A belt driven centrifugal air-conditioning supply fan runs at 700 RPM and handles air at a density of 1.18 kg/m³. The fan-outlet diameter is 750 mm. The duct system has a resistance of 500 Pa when the air

flow is 4000 l/s. The outlet-air velocity from the fan is 7.6 m/s. The 415 Volt three-phase drive motor has a power factor of 0.92. The motor and drive has an overall efficiency of 75% and the impeller has an efficiency of 71%. The minimum power consumption at zero air flow would be 1.0 kV A. The fan-performance data are shown in Table 9.10. Calculate and plot the fan characteristic curves for fan velocity and static pressures, the duct system resistance and the motor input power. Use the spreadsheet file that is shown in Fig. 9.9. Find the fan and system operating conditions and the current that will be taken by the motor.

Table 9.10   Fan data for question 36

Q (l/s)	FTP (Pa)
0	750
500	775
1000	780
1550	775
2000	765
2500	750
3000	700
3500	650
4000	576
4500	500
5000	360
5500	80

37. An air-conditioning system has a design air flow of 3 m³/s. The supply-air duct-system frictional resistance is 450 Pa. The supply-air filter has a pressure drop of 40 Pa when clean and 180 Pa when it is ready to be changed. Use the average filter pressure drop for the analysis. The air-conditioning system is operated for 200 hours per month. The fan handles air at an average density of 1.2 kg/m³ through the year. The fan impeller efficiency is 80%, motor efficiency is 66% and the motor power factor is 0.92. Assume that the overall efficiency remains constant at any air flow. The motor power used at zero supply-air flow is 0.8 kW. The fan discharge duct is 700 mm × 600 mm. Table 9.11 shows the supply-air fan performance data. The building is in southern England, where the load profile on the air-conditioning system is indicated in Table 9.12. The supply-air requirement for each month is the average for the month to match the thermal loads. A higher temperature differential between the supply and room air is maintained in winter and this reduces the supply-air quantity. Free cooling is used in the mild weather. Electricity costs 9 p/kV A h. The fan is operated at a constant speed of 960 RPM.

Table 9.11 Supply-air fan-performance data

Air flow Q (l/s)	FTP (Pa)
0	800
500	810
1000	820
1500	820
2000	800
2500	750
3000	650
3500	500
4000	300
4500	50

Table 9.12 Monthly supply-air flow required

Month	Heating/ cooling load (%)	Supply air required Q (m^3/s)
January	35	1.05
February	42	1.26
March	56	1.68
April	76	2.28
May	97	2.91
June	100	3.0
July	97	2.91
August	90	2.7
September	77	2.31
October	57	1.71
November	40	1.2
December	31	0.93

Compare the annual costs of the following:

(a) no fan performance control;
(b) supply-duct air-volume-control damper;
(c) two-speed fan motor producing 960 and 550 RPM;
(d) variable-frequency fan-speed control.

The data can be calculated manually and with the aid of the spreadsheet file that is listed on figure 9.16.

Answers

CHAPTER 1

11.

Side	Glass (m^2)	Peak (W/m^2)	Peak (kW)	Date	Time (h)
S	80	510	40.8	Sept	1300
E	80	477	38.16	June	0900
W	80	477	38.16	June	1700
N	80	161	12.88	July	1300

Plant refrigeration load peak is not found from the sum of the peaks of each side.

CHAPTER 2

1. 46%, 0.846 m³/kg, 10.1°C, 41.7 kJ/kg, 0.0077 kg/kg.
2. Entry: 0.0038 kg/kg, 9.45 kJ/kg, 0.778 m³/kg, 0°C dew point. Exit: 30°C d.b., 14.5°C w.b., 14% saturation, 0.864 m³/kg, 40 kJ/kg, 0°C dew point.
3. Entry: 12°C d.b., 4.1°C w.b., 20%, 0.00175 kg/kg, 16.5 kJ/kg, 0.81 m³/kg, –9°C dew point. Heated to 37°C d.b.: 15.3°C w.b., 4%, 0.00175 kg/kg, 41.5 kJ/kg, 0.881 m³/kg, –9°C dew point. Humidified: 15.8°C d.b., 14.8°C w.b., 90%, 0.0102 kg/kg, 41.5 kJ/kg, 0.832 m³/kg, 14.4°C dew point.
4. Outdoor: 43%, 0.00118 kg/kg, 60.3 kJ/kg, 0.875 m³/kg, 16.3°C dew point. Cooled: 11.1°C w.b., 0.0079 kg/kg, 32 kJ/kg, 10.5°C dew point.
5. Outdoor: 58%, 0.0141 kg/kg, 64 kJ/kg, 0.872 m³/kg, 19.2°C dew point. Cooled: 5°C d.b., 5°C w.b., 100%, 0.0054 kg/kg, 18.6 kJ/kg, 0.794 m³/kg, 5°C dew point. Reductions of 45.4 kJ/kg and 0.0087 kg/kg.
6. 26°C d.b., 18.6°C w.b., 48%, 0.0103 kg/kg, 52.4 kJ/kg, 0.861 m³/kg, 14.4°C dew point.
7. Mixed: 17.4°C d.b., 13°C w.b., 61%, 0.0076 kg/kg, 36.5 kJ/kg, 0.833 m³/kg, 9.8°C dew point. Heated: 35°C d.b., 19.5°C w.b., 21%, 0.0076 kg/kg, 54.8 kJ/kg, 0.883 m³/kg, 9.8°C dew point. Humidified: 24°C d.b., 19.3°C w.b., 64%, 0.0121 kg/kg, 54.8 kJ/kg, 0.858 m³/kg, 16.9°C dew point. Duty: 54.922 kW.
8. 113.779 kW.
9. Inlet: 52%, 0.0134 kg/kg, 63.3 kJ/kg, 0.874 m³/kg, 18.5°C dew point. Cooled: 12.1°C w.b., 0.0084 kg/kg, 34.4 kJ/kg, 0.821 m³/kg, 11.4°C dew point. Duty: 132.265 kW.
10. 163.5 kW.
11. Mixed: 23.7°C d.b., 0.010 kg/kg, 55%, 17.5°C w.b., 49.5 kJ/kg. Cooled: 7.8°C d.b., 0.0063 kg/kg, 95%, 7.4°C w.b., 23.7 kJ/kg. Duty: 145 kW, 41.232 tonne refrigeration, condensate 74.867 kg/h.
12. –5.9°C w.b., 0.7615 m³/kg, 0.00198 kg/kg, 10%, 76.428 kW.
13. 14.2°C w.b., 91% saturation, 48.055 kW.
14. 19.6°C d.b., 0.0075 kg/kg, 38.7 kJ/kg, 0.839 m³/kg, 13.9°C w.b.
15. No. 21.2°C w.b., 0.877 m³/kg, 6.681 kW.
16. p_s 2.9808 kPa, g_s 0.01885 kg/kg, for t_{sl} = 16°C p_s 1.8159 kPa, p_v 1.276 kPa, g 0.00793 kg/kg, PS 42%, h 44.3 kJ/kg, p_v 1276 Pa, v 0.8523 m³/kg dry air, density 1.173 kg/m³, t_{dp} 11.03°C. Accuracy is acceptable for most purposes.
17. p_s 0.6564 kPa, g_s 0.00406 kg/kg, for t_{sl} = 0.5°C p_s = 0.633 kPa, p_v 0.6 kPa, g 0.0037 kg/kg, PS 91%, h 10.26 kJ/kg, v 0.781 m³/kg dry air, density 1.28 kg/m³, t_{dp} – 0.01°C.
18. p_s 4.751 kPa, g_s 0.0306 kg/kg, for t_{sl} = 22°C p_s = 2.641 kPa, p_v 1.966 kPa, g 0.0123 kg/kg, PS 40%, h 63.65 kJ/kg, v 0.881 m³/kg dry air, density 1.135 kg/m³, t_{dp} 17.3°C.

CHAPTER 3

5. 44.8 m.
6. Total height = 46.15 m + 15.593 m = 61.744 m 3.25 m each.
7. Vertical height = 20.139 m sloping face length = 20.59 m area = 617.7 m².
8. 52°, 69.3°.
9. D 69°, F 87°, 44.3 W/m².
10. D 0°, F 33°, 686.9 W/m².
11. From CIBSE A2.27, 1986: the maximum may occur on 22 Sept at 1200 h, I_{THd} = 625 W/m², I_{TVd}

$= 710 \, \text{W/m}^2$, I_{dHd} 190 W/m^2, I_{TSd} 878.4 W/m^2. Check other dates and times.

12. $I_{DVd} = 303$ W/m^2, $Q_D = 1569.5$ W, $Q_d = 1738.8$ W, total 3308.3 W.

13. $I_{DSd} = 632$ W/m^2, $Q_D = 872$ W, $Q_d = 256$ W, total = 1128 W.

14. $t_g = 39.8°C$, Q 324 W/m^2, total T 0.52.

15. $t_g = 4.8°C$, $Q = -59$ W/m^2, total $T = -0.34$, heat flow outwards from room.

16. $F_u = 0.54$, $F_v = 0.92$, $F_y = 0.96$, $Q_u = 264$ W, $Q\tilde{u} = 2402$ W, net $Q_u = 2666$ W.

17. $F_u = 0.98$, $F_v = 0.5$, $F_y = 0.88$, $Q_u = 466$ W, $Q\tilde{u} = -12$ W, net $Q_u = 454$ W.

18. Solar gain through the glazing 4304 W, 24 hour mean conduction through the structure −151 W, swing in the conduction gain −329 W, ventilation air infiltration −820 W, occupancy gain 180 W, electrical equipment emission 500 W, total heat gain 3684 W.

19. $F_u = 0.94$, $F_v = 0.93$, $F_y = 0.89$, $F_2 = 1.17$, solar gain through the glazing 14 208 W, 24 hour mean conduction through the structure −1554 W, net swing in the conduction gain 2324 W, outdoor ventilation air infiltration −912 W, indoor air infiltration 1216 W, occupancy gain 3600 W, electrical equipment emission 5050 W, mean conduction gain from below 6545 W, total net gain 30 477 W, 30.477 kW, 14.5 W/m^3 of the office volume.

20. 485 W.

22. Total mean gain Q 1586 W, mean $t_{ei} = 24.7°C$, total swing in gains $Q_{\overline{t}}$ 4095 W, $t_{\tilde{ei}}$ 3.2°C, peak t_{ei} 27.9°C.

23. Mean internal gain 48 kW, mean t_{ci} 21.2°C, total swing in gains $Q_{\overline{t}}$ 399.375 kW, $t_{\tilde{ei}} = 3.3°C$, peak environmental temperature t_{ei} 24.5°C.

25. 73 mm.

26. 5.473 m^2.

28. Net gain of 531 W.

32. Approximately 10.6°C.

CHAPTER 4

2. 1.014 kg/s.

3. 15 mm.

4. 15 mm: 0.107 kg/s; 22 mm: 0.311 kg/s; 28 mm: 0.624 kg/s; 35 mm: 1.12 kg/s; 42 mm: 1.881 kg/s; 54 mm: 3.801 kg/s.

5. 1.264 kg/s.

6. 0.49 l/s.

7. 22 mm: 1.13 l/s ; 28 mm: 2.3 l/s ; 35 mm: 4.19 l/s ; 42 mm: 7.11 l/s ; 54 mm: 14.56 l/s.

8. 0.507 m^3/s.

9. Q 0.782 m^3/s, v 4.92 m/s.

10.

Q (m^3/s)	d (mm)
0.12	200
0.3	250
0.6	350
1	450
2	600

11 For LTHW and 3.5 kW, M 0.07 kg/s; for LTHW and 420 kW, M 8.345 kg/s. For HTHW and 3.5 kW, M 0.027 kg/s; for HTHW and 420 kW, M 3.296 kg/s.

12 M 0.497 kg/s, 28 mm pipe, Δp 330 Pa/m, v 0.941 m/s.

13. M 4.316 kg/s, 54 mm, Δp 630 Pa/m, v 2.11 m/s.

14. For Table X copper pipes, LTHW: M 1.987 kg/s, d 54 mm, Δp 150 Pa/m, v 0.96 m/s. For galvanized sheet metal duct: M 8.235 kg/s, Q 7.331 m^3/s, d 1.1 m Δp 0.475 Pa/m, v 7.72 m/s. Distribute the air-handling plant around the building to minimize air-duct runs and supply each plant with LTHW pipework.

15. For Table X copper pipe, LTHW, M 1.49 kg/s, d 35 mm, Δp 840 Pa/m, v 1.83 m/s, pump head H 157.5 kPa, pump power consumption 370.3 W. For galvanized steel air duct: Q 5.498 m^3/s, d 0.9 m Δp 0.75 Pa/m, v 8.64 m/s, fan pressure rise H 196.875 Pa, fan power consumption 1.665 kW.

17. Stored energy 35.2 kWh, heater input power 4.4 kW.

18. Storage capacity 864 MJ, heater power 30 kW, tank needs to be 2.907 m × 2.907 m × 1 m high.

19. Cooling load 115 kW, compressor runs for 30 minutes per hour for 4 runs of 7.5 minutes each, chilled water storage tank size is 4.44 m × 4.44 m × 1 m.

CHAPTER 5

1. 1.999 m^3/s.

2. Q 1.2 m^3/s, SH 7.287 kW.

3. Q 2.25 m^3/s, t_s 15.1°C.

4. Q 3.891 m^3/s.

5. Q 2 m^3/s, SH 16.49 kW.

6. Q 1.68 m^3/s, t_s 33°C.

7. Summer: Q 5.315 m^3/s, N 12.8 air changes/h, v 0.822 m^3/kg, M 6.466 kg/s. Winter: t_s 28.3°C d.b., v 0.863 m^3/kg, Q 5.58 m^3/s, recalculated t_s 28°C d.b

8. Q 1.454 m^3/s, g_s 0.008825 kg/kg.

9. S/T ratio 0.96, Q 6.722 m^3/s, g_s 0.007709 kg/kg.

10. All air flows l/s : fresh air inlet 960, total supply 12 821, office supply 10 630, office extract duct 9942, corridor supply and extract 1962, toilet

supply duct 229, transfer grille into corridor and toilet 688, toilet separate exhaust 917, plant exhaust 43, plant recirculation duct 11 861.

11. 9.846 m³/s, 0.0073 kg/kg.
12. Q 0.4 m³/s, t_s 7°C d.b., 0.007 kg/kg.
13. SH gain 13.75 kW, Q 2.83 m³/s, 0.00725 kg/kg.
14   Summer: SH gain 12.974 kW, 14.4 W/m³, Q 1.331 m³/s, N 5.3 air changes/h, g_r 0.008905 kg/kg, g_s 0.008590 kg/kg, fresh air proportion 22.5%, t_m 24.4°C d.b., t_m 17.8°C w.b., h_m 50 kJ/kg, 0.856 m³/kg, supply 15°C d.b., 13°C w.b., cooling coil 20.665 kW. Winter: SH loss 9.57 kW, 10.6 W/m³, Q 1.331 m³/s, mixed air t_m 14.5°C d.b., g_m 0.006 kg/kg, room 45% saturation produced, h_m 29.75 kJ/kg, 0.822 m³/kg, supply 25.1°C d.b., 14.7°C w.b., h_s 40.55 kJ/kg, heating coil 17.488 kW.
15. Summer: SH gain 2.928 kW, 19.5 W/m³, N 9.6 air changes/h, total Q 402 l/s (where 121 l/s enter from the hot duct at 23°C and 281 l/s from the cold duct at 13°C), g_s 0.008282 kg/kg, fresh air 6%. Winter: t_s 23°C d.b., t_m 19°C d.b., total supply Q 402 l/s (comprising 268 l/s at 25° from the hot duct plus 134 l/s at 19°C from the cold duct.

CHAPTER 6

5. 1.183 kg/m³.
6. 46.6°C d.b.
7. 245 Pa, 5 kPa, 1226 Pa, 2942 Pa, 25 kPa.
8. 3.82 m/s.
9. 1 m³/s, 2.26 m³/s, 6.28 m³/s, 25.13 m³/s.
10. 1.145 kg/m³, 28 Pa, −413 Pa.
11. 1.181 kg/m³, 10.2 m/s, 361 Pa.

12.

Node	P_t (Pa)	P_v (Pa)	P_s (Pa)
1	600	22	578
2	595	22	574
3	591	62	529
4	573	62	511

13.

Node	P_t (Pa)	P_v (Pa)	P_s (Pa)
1	250	29	221
2	240	29	211
3	237	146	91
4	42	146	−104

14.

Node	P_t (Pa)	P_v (Pa)	P_s (Pa)
1	400	221	179
2	320	221	99
3	144	38	106
4	126	38	89

Static regain $P_{s3} - P_{s2} = 6$ Pa

15.

Node	P_t (Pa)	P_v (Pa)	P_s (Pa)
1	200	36	164
2	165	36	129
3	145	2	143
4	145	2	143

Static regain $P_{s3} - P_{s2} = 13$ Pa

16.

Node	P_t (Pa)	P_v (Pa)	P_s (Pa)
1	400	21	379
2	391	21	370
3	389	30	360
4	379	30	349
5	375	23	352
6	365	23	342

Static regain $P_{s3} - P_{s2} = -10$ Pa

17.

Node	P_t (Pa)	P_v (Pa)	P_s (Pa)
1	200	36	164
2	183	36	147
3	183	16	167
4	175	16	159
5	170	38	132
6	166	38	128

Static regain $P_{s3} - P_{s2} = 20$ Pa

18. $v_1 = 15.3$ m/s, $p_{v1} = 136$ Pa, $v_2 = 12.5$ m/s, $p_{v2} = 91$ Pa, $p_{s1} = -883$ Pa, $p_{t1} = -747$ Pa, $p_{s2} = 1162$ Pa, $p_{t2} = 1253$ Pa, FTP = 2000 Pa, FVP = 91 Pa, FSP = 1909 Pa.
19. There is more than one correct solution to this question. Duct sizes can be: section 1–2, 1200 mm × 1500 mm; section 3–4, 600 mm × 600 mm; section 5–6, 1200 mm × 700 mm; section 6–7, 800 mm × 400 mm; section 6–8, 1200 mm × 500 mm.

Sect	length l (m)	$\Delta p/l$ (Pa/m)	v (m/s)	p_v (Pa)	k	Fit (Pa)	Duct (Pa)	Total (Pa)	Index
1–2	0	0.05	2.3	3	3	10	0	10	*
2–3	3	0.07	2.3	3	0.07	105	0	105	*
3–4	0	2.16	11.7	82	0.04	3	0	3	*
4–5	0	0.24	5.1	15	0.36	29	0	29	*
5–6	16	0.24	5.1	15	0.35	5	4	9	*
6–7	0	0.42	5	15	1.25	49	0	49	*
6–8	15	0.31	4.5	12	0.09	31	5	36	

Total Δp for index route = 206 Pa.

20. There is more than one correct solution to this question. Duct sizes can be: section 1–2 2000 mm × 1600 mm; section 2–3 1350 mm × 1600 mm; section 3–4 650 mm × 650 mm; section 5–6 1000

mm × 800 mm; section 6–7 800 mm × 600 mm; section 6–8 800 mm × 450 mm.

Sect	length l (m)	$\Delta p/l$ (Pa/m)	v (m/s)	p_v (Pa)	k	Fit (Pa)	Duct (Pa)	Total (Pa)	Index
1–2	0	0.02	2	2	2.1	5	0	5	*
2–3	9	0.06	2.9	5	0.07	225	1	226	*
3–4	0	2.9	14.9	133	0.45	60	0	60	*
4–5	0	0.6	7.9	37	0.04	5	0	5	*
5–6	48	0.6	7.9	37	0.35	13	29	42	*
6–7	0	0.81	7.7	35	1.25	114	0	78	
6–8	45	1	7.4	33	0.09	73	45	118	*

Total Δp for index route = 456 Pa.

CHAPTER 7

4. 6.25 V, 1250 Ω.
7. 500 Ω.

CHAPTER 8

1. Allowable leakage 169 l/s , test leakage 161 l/s , duct passes test.

2. Assume duct air p_s is 300 Pa to estimate ρ 1.234 kg/s, Δp 2256 Pa, test orifice flow Q 100 l/s; for a duct air leakage of 100 l/s; ps is 290 Pa; watch for $X^{1/Y}$ function, h is 29.6 mm H_2O.
3. Allowable leakage 151 l/s , leakage flow on test 185 l/s, duct fails test.

CHAPTER 9

30. F_2 6.3 kg.
31. FVP 34 Pa.
32. N_1 20 Hz, N_2 16 Hz, Q_2 6667 l/s, Δp_2 267 Pa, p_2 5267.5 W.
33. At the design air flow, power 4.286 kVA at the closed damper air flow, power 0.857 kVA, phase current 1.19 amperes.
34. Q 4.9 m^3/s, FTP 90 Pa, 1.2 kVA, 1.7 amp.
35. Q 1200 l/s, FTP 680 Pa, 2.3 kVA, 3.2 amp.
36. Q 4200 l/s, FTP 550 Pa, 5.6 kVA, 7.8 amp.
37. Annual energy costs: (a) £979.87, (b) £822.35, (c) £620.30, (d) £426.53. Annual cost savings: (a) £157.52, (b) £359.57, (c) £553.34.

References

Awbi, H. B. (1991) *Ventilation of Buildings.* E & F N Spon, London.

BSI (1971) BS 2871: Specification for copper and copper alloys. Tubes. Part 1, Copper tubes for water, gas and sanitation. British Standards Institution.

BSI (1986) BS 5588: Fire precautions in the design and construction of buildings. Part 4, Code of practice for smoke control in protected escape routes using pressurization. British Standards Institution.

Chadderton, D. V. (1995) *Building Services Engineering*, 2nd edn. E & F N Spon, London.

Chadderton, D. V. (1997) *Building Services Engineering Spreadsheets.* E & FN Spon, London.

CIBSE (1971) *Commissioning Code A, Air Distribution.* Chartered Institution of Building Services Engineers.

CIBSE (1973) *Commissioning code for automatic control.* Chartered Institution of Building Services Engineers.

CIBSE (1975) *Commissioning code for boiler plant.* Chartered Institution of Building Services Engineers.

CIBSE (1986a) CIBSE Guide, Vol. A, *Design Data.* Chartered Institution of Building Services Engineers.

CIBSE (1986b) CIBSE Guide, Vol. B, *Installation and Equipment Data.* Chartered Institution of Building Services Engineers.

CIBSE (1986c) CIBSE Guide, Vol. C, *Reference Data.* Chartered Institution of Building Services Engineers.

CIBSE (1989) *Commissioning code for water distribution systems.* Chartered Institution of Building Services Engineers.

CIBSE (1991) *Commissioning code for refrigeration systems.* Chartered Institution of Building Services Engineers.

CIBSE (1991) *Building Services*, The CIBSE Journal, April.

CIBSE (1992) *OPUS '92.* Chartered Institution of Building Services Engineers, p. 11.

Encyclopedia Britannica (1980)

Health and Safety Executive (1990) Occupational Exposure Limits. Guide note EH40/90, HMSO, London.

HMSO (1991) *The control of* Legionella *in health care buildings: a code of practice*. HMSO, London.

HVCA (1982) *DW 142*. Heating and Ventilating Contractors Association.

Jones, W. P. (1985) *Air Conditioning Engineering*, 3rd edn. Edward Arnold, London, p.288.

Moss, B. (1991) A New Approach to Airports. *The CIBSE Journal*, **13** (10), October, p.22.

Rogers, G. F. C. and Mayhew Y. R. (1987) *Thermodynamic and Transport Properties of Fluids*, 3rd edn (SI units). Basil Blackwell.

Smith, T. (1991) Presidential address. *The CIBSE Journal*, **13** (6), June

Index